**PERFECTLY REASONABLE DEVIATIONS** *from the* **BEATEN TRACK**

*Edited and with Additional Commentary by*

**Michelle Feynman**

*Foreword by*

**Timothy Ferris**

A MEMBER OF THE PERSEUS BOOKS GROUP

NEW YORK

# PERFECTLY REASONABLE DEVIATIONS

# *from the*

# BEATEN TRACK

## *The Letters of Richard P. Feynman*

Published by Basic Books
A Member of the Perseus Books Group

*Designed by Lovedog Studio*

ISBN 0-7382-0636-9

*For my parents, Richard and Gweneth*

*For my nieces, Rachel and Emma,*
*and my children, Ava and Marco,*
*so that they may better know*
*their grandfather*

# CONTENTS

# FOREWORD

Richard P. Feynman was both a first-rate and a famous physicist. The combination is by no means inevitable. For every Nobel Laureate in physics who became a public figure—for every Marie Curie, Albert Einstein, or Werner Heisenberg—others of exemplary stature did not: Paul Dirac, Wolfgang Pauli, and Subrahmanyan Chandrasekhar, for instance. It's not always clear what made the difference. Why was Feynman famous?

Extraneous circumstances can, of course, determine on whom the spotlight falls. Heisenberg's "uncertainty principle" resonated with an age of philosophical and psychological uncertainty for reasons having little to do with its importance to quantum mechanics. Marie Curie's research into radioactivity demonstrated that a woman could advance to the loftiest tiers of science. The spectacle of an English expedition confirming Einstein's general theory of relativity—by measuring the bending of starlight passing near the sun during the total solar eclipse of 1919—raised hopes of a more rational, peaceful future among nationalities exhausted by the carnage of the Great War. But such considerations do little to explain the fame of Feynman, who worked on the Manhattan Project but who otherwise seldom made headlines, and whose research, though highly regarded by his fellow physicists, could scarcely be understood, much less appreciated, by the general public.

Personality can play a role, and some of Feynman's critics accused him of carefully crafting a vivid persona in order to raise his profile. One close colleague objected to what he called Feynman's "extraordinary efforts to be different—especially from his friends and colleagues," adding that Feynman

"surrounded himself with a cloud of myth, and he spent a great deal of time and energy generating anecdotes about himself." It is true that Feynman spoke unvarnished, blue-collar Brooklynese, that he played the bongos, liked to sketch dancers in topless bars, and otherwise presented himself as something of a raffish guy. But such behavior is commonplace. As Oscar Wilde observed, "The first duty in life is to assume a pose. What the second duty is, no one has yet found out." Since every faculty has its share of poseurs, to accuse Feynman of posing hardly accounts for his having been the subject of plays, movies, and bestselling books.

Speaking as someone who knew him slightly, heard him speak on a number of occasions, and followed his research fairly closely, it seems to me that the key to understanding Feynman's appeal resides less in external circumstances or extracurricular activities than in his core conduct as a working scientist—specifically, that his freedom, integrity, and enthusiasm reflected the spirit of science in action.

Scientists cherish freedom as both a prerequisite and a perquisite of their chosen careers, and Feynman certainly reveled in his liberty. "I am completely free, and there are no levers that can be used to influence me," he wrote home in 1986, while serving on the presidential commission investigating the *Challenger* space shuttle crash. There his impromptu demonstration—dropping a piece of O-ring material into a glass of ice water to show that, contrary to expert testimony, the cold temperatures at the launch site that frigid morning could indeed have rendered the rubber sealing the shuttle's solid-rocket boosters dangerously brittle—constituted one of the great twentieth-century demonstrations of the power of scientific experiment. Feynman was emphatic about encouraging students to feel free to pursue whatever most interested them, without worrying overmuch about curriculum demands, the opinions of their elders, or the need to land a job. Writing to an old friend who wondered whether his fifteen-year-old son ought to go into engineering or science, Feynman urged that "he must have freedom to pursue his delight." He advised another student, who later became a NASA scientist, "Work hard to find something that fascinates you." To a woman who fretted that she had studied physics in a relatively haphazard way, he responded, "So much the better. . . study hard what interests you the most in the most undisciplined, irreverent and original manner possible."

As to integrity, many noteworthy scientists eventually lapse into defending their views by wielding their status as recognized authorities. (Einstein used to joke that "to punish me for my contempt for authority, Fate made

me an authority myself.") Feynman eluded that trap. Although he became a celebrated lecturer and an authority in several fields of physics, he remained an instinctive rebel who sympathized more with the students in the hall than the sage on the stage. He preferred clarity to profundity, questions to answers, the hunt to the trophy on the wall. "Don't pay attention to 'authorities,' he advised a nineteen-year-old undergraduate. "Think for yourself." When a Caltech student asked the eminent cosmologist Michael Turner what his "bias" was in favoring one or another particle as a likely candidate to comprise the dark matter in the universe, Feynman snapped, "Why do you want to know *his* bias? Form your own bias!"

The price for rejecting easy answers and reliance on authorities is that one must be willing to tolerate ambiguity and accept one's own ignorance. This didn't bother Feynman. "I can live with doubt and uncertainty," he said. "I think it's much more interesting to live not knowing than to have answers which might be wrong." He once defined science as belief in the ignorance of experts.

Although he claimed to be an "utter ignoramus about politics," Feynman was one of the few scientists of his generation (Carl Sagan toward the end of his life was another) to grasp the importance of science in promoting democracy and human rights. He told a Seattle audience in 1963 that the need of scientists to freely investigate the world while "dealing with doubt and uncertainty . . . is of very great value, [one that] extends beyond the sciences." He added,

> I feel a responsibility as a scientist who knows the great value of a satisfactory philosophy of ignorance, and the progress made possible by such a philosophy, progress which is the fruit of freedom of thought . . . to proclaim the value of this freedom and to teach that doubt is not to be feared, but that it is to be welcomed as the possibility of a new potential for human beings. If you know that you are not sure, you have a chance to improve the situation. I want to demand this freedom for future generations.

During the Cold War, when many academics were beguiled by Marxism, Feynman clear-headedly viewed the conflict as not "between socialism and capitalism, but rather between suppression of ideas and free ideas." Declining an invitation to attend a 1964 conference staged by the Soviet Joint Institute for Nuclear Research, he wrote, "I would feel uncomfortable at a scientific conference in a country whose government respects

neither freedom of opinion of science, nor the value of objectivity, nor the desire of many of its scientist citizens to visit scientists in other countries."

In addition to exercising the freedom and demonstrating the integrity one expects from a first-rate scientist, Feynman expressed an infectious enthusiasm that rubbed off on his students even if they couldn't always comprehend the science itself. The celebrated *Feynman Lectures on Physics* often proved too difficult for the Caltech freshmen at whom they were aimed (many of whom soon slunk away, their seats taken over by rapt faculty members) but recordings of them remain popular today, in part due to Feynman's obvious delight in his subjects. Most popularizers seek to humanize science by dressing it up in costumes of poetry, art, or philosophy; Feynman preferred to strip science bare, presenting it with the raw immediacy of a wild animal glimpsed in its natural habitat. He was unapologetic in standing up for what he saw as the superiority of a scientific approach to studying the world. "Experiment and observation is the sole and ultimate judge of the truth of an idea," he wrote. "It is not philosophy we are after, but the behavior of real things . . . I like science because when you think of something you can check it by experiment; 'yes' or 'no,' Nature says, and you go on from there progressively. Other wisdom has no equally certain way of separating truth from falsehood." Uninterested in building bridges between what C. P. Snow called a gap between the "two cultures" of science and the humanities, Feynman staunchly if controversially maintained that scientists alone can fully appreciate the "intense beauty" of nature.

Feynman's rough-hewn style of lecturing ("I don't speak writable English") had a lot to do with the fact that he preferred to think out loud rather than to present the polished products of prior thinking. I once saw him deliver a rather shambling talk—something to do with Bose condensates, if memory serves—during which he circled and stalked his prey at length before ultimately admitting defeat. "I've given a lecture on this subject every five years," he said, "each time thinking that if I give this one more lecture, I'll figure it out," but once again he had failed. His willingness to keep trying, in public, reminded me of Niels Bohr's dictum, "Never express yourself more clearly than you think," and of Carl Friedrich von Weizsacker's saying of Bohr, "I have seen a physicist for the first time. He suffers as he thinks."

His intensity was accompanied by a robustly self-deprecatory sense of humor. Echoing Groucho Marx's remark, "I refuse to join any club that would have me as a member," Feynman objected to an upcoming scientific

conference on the grounds that its planners had lowered their standards by inviting him: "What the hell is Feynman invited for? He is not up to the other guys and is doing nothing as far as I know. If you clean up the invitation list, to just the hard-core workers, I might begin to think about attending." He teased those naïve enough to be shocked when he made mistakes: "I goofed. And you goofed too, for believing me. We both had bad luck." He frequently called himself a "fool," as indeed he does several times in these letters.

Some thought it affected that one of the smartest and most capable men alive would habitually describe himself as lazy, inconsistent, and foolish, but it could be that Feynman was simply reflecting his appreciation of the human condition. The remarkable thing about great attainments in art or science is not that their creators are perfect but that they have as many limitations as the rest of us—a point Isaac Newton made rather nicely when he likened himself to "a boy playing on the seashore . . . diverting myself in now and then finding a smoother pebble or a prettier shell than ordinary, whilst the great ocean of truth lay all undiscovered before me."

Mindful of his own shortcomings, Feynman could be admirably indulgent of the shortcomings of others: One learns from this collection of letters that he frequently took the trouble to reply to amateur scientists and even "cranks," so long as their inquiries struck him as arising from honest intentions. Characteristically charming is the letter he wrote in response to a missive from one Bernard Hanft, enclosing a washer and thread the unprompted spinning of which, Hanft proposed, was due to a new force that he rather immodestly dubbed "The Hanft Force." Many scientists would have consigned such a correspondence to the circular file, but Feynman—reacting, perhaps, to Hanft's earnest, hands-on approach—performed the experiment himself and replied with a fulsome, sunny letter, showing that the washer's twirl had a simpler explanation and concluding with a gracious salutation: "Thank you again for calling my attention to these entertaining phenomena."

There is considerable charm in these letters, along with love, heartbreak, and occasional bolts of intellectual lightning vivid enough to remind us that their author was, after all, one of the most powerful thinkers of his time. Observing that the vast universe discovered by science tends to dwarf Biblical tales in which "God can watch human beings struggle for good and evil," Feynman remarks, "The stage is too big for the drama"—which puts the case about as well, in eight short words, as anyone ever has. In the middle of a letter to the editor of the *California Tech* in 1976, supporting a

female colleague who despite harsh treatment by the administration would stick to her guns and become Caltech's first tenured female professor, Feynman suddenly digresses to say, "In physics the truth is rarely perfectly clear, and that is certainly universally the case in human affairs. Hence, what is not surrounded by uncertainty cannot be the truth."

There are groaning shelves full of respected philosophical tomes that contain not one sentence as incisive and original as that offhand aphorism: "What is not surrounded by uncertainty cannot be the truth." Moreover it sums up the main components of Feynman's world view—the pursuit of truth, amid a welter of uncertainty, by a researcher who will neither abandon the quest nor lose sight of his doubts. It was that spirit, the spirit of science, to which the world responded by cheering Feynman on, and he will be remembered as long as science thrives.

———Timothy Ferris

# GENIUS TO YOU, FATHER TO ME: MY LIFE WITH RICHARD P. FEYNMAN

---

When I was very young, I thought my father knew everything. Indeed, *Omni* magazine once declared him "The Smartest Man in the World." Upon hearing this, his mother, who was immensely proud of her son—and who had a great sense of humor—threw up her hands and exclaimed, "If Richard is the smartest man in the world, God help the world!" My father was the first one to laugh.

As I grew older, I began to see only what my father *didn't* know, and soon thought *I* knew everything. He would ask me questions whose answers I found to be painfully obvious, such as, "Hey, Michelle, where do we keep the spoons around here?" I discovered the real truth in my late teens: my father was a wise man with a tremendous appetite for life and aptitude for teaching. He had interesting, often profound ideas about life and the world and was highly engaging to listen to. I was eager to work on this book because I wanted to revisit him. It was fun to spend time with him again. I had no idea, but should not have been surprised, that he would still be teaching me things today.

Here, more objectively, are the basic facts of his life. Richard Phillips Feynman was born in New York City in 1918 and grew up in Far Rockaway, Queens. He attended the Massachusetts Institute of Technology as an undergraduate, and he received his Ph.D. from Princeton University. In 1942 he married his high school sweetheart, Arline Greenbaum, despite the fact that she was ill with tuberculosis. That same year Richard was asked to join the Manhattan Project; he accepted and went on to become a group leader at Los Alamos, New Mexico. Arline died in 1945. After the war, he

became a professor of theoretical physics at Cornell University in Ithaca, New York. In 1950 he joined the faculty of the California Institute of Technology and spent the remainder of his career there. He had a brief marriage in the early fifties that did not work out. He married my mother, Gweneth Howarth, in 1960. My brother, Carl, was born in 1962, and I was adopted in 1968.

Though he remained forever ambivalent about it, his most public achievement came in 1965, when he won the Nobel Prize in Physics, sharing it with Julian Schwinger and Shin'ichiro Tomonaga for their independent work in quantum electrodynamics. In 1986, he was again in the public eye, this time working on the commission investigating the explosion of the Challenger space shuttle. He died in 1988 after a long battle with abdominal cancer. It is no surprise to any of us who loved him that his memorial service at Caltech was attended by thousands. Anticipating large crowds on that day, Caltech planned to hold the memorial service twice. Even with such forethought, the auditorium quickly reached capacity both times.

He has been the subject of countless interviews, books, articles, a few stage plays, several documentaries, and one motion picture. He was sought after not only for his scientific achievements but also for his outsized curiosity, his irrepressible love of puzzles, and his embrace of life at large. He was an adventurer who made a hobby of cracking safes while working on the atomic bomb, who played bongo drums for a San Francisco ballet, and who decided to learn to draw in his forties—and became remarkably good at it.

As a result of the admiration people have felt for my father, a great many wonderful and interesting people, treasured friendships, and rare opportunities have come my way. But as with any privilege, being his heir comes with a significant responsibility. My brother and I find ourselves faced with a large demand for all things having to do with Richard P. Feynman, and we strive to balance that demand with maintaining his legacy in a conscientious manner. I hope that this book not only does justice to the spirit in which he worked, but also reveals more of the personality behind his many accomplishments.

For all the Feynman anecdotes that have been bandied about over the years, it is the lesser-known ones that I believe are the most telling. The story of my parents' courtship, for one, says a great deal about his unusual approach. My mother was from England, and when she met my father she was living in Switzerland, intending to travel around the world working as an *au pair*. Richard promptly invited her to come to America to work as his

maid. She told him she would think about it. He met her again the next day and, fearing that a twenty-four-year-old woman would be put off by such an idea coming from a forty-year-old man, apologized for his impertinent suggestion. But she accepted. Many months later, and with one of my father's future collaborators on *The Feynman Lectures on Physics,* Matt Sands, as the necessary immigration sponsor (the government was naturally suspicious of single men importing women), my mother moved in.

Prior to her arrival, my father had written to her, saying, "I'm managing poorly without you. Come quick." Upon her arrival my mother cooked, cleaned, and once she obtained her driver's license, even drove him to Caltech while he sat in the back seat. Though they were friendly, there was as yet no romance—both were dating other people. A lightbulb had gone off in my father's head, however, the day he took her to pick up a study manual for the driver's test. They found themselves in the wrong line, and Gweneth ended up taking the test with no preparation. Impressively, she passed. He realized soon after that he was falling in love with her and that he wanted to propose. Then he faulted himself for being too impulsive. He proceeded to mark a day on the calendar a few months ahead and thought, "If I still feel the same way then, I'll ask her." The night before that day arrived, he could not stand the wait and kept her up until midnight. They were married a few months later.

The household I grew up in was similarly unusual. We played many games. On camping trips, we would go to great lengths to put ourselves in the middle of nowhere. At every fork in the road, we took the road in the worse condition, the one that looked the most interesting to us. On Sunday mornings, my father would often forego reading the newspaper in favor of a wild hour of loud, often discordant, music, drumming, and storytelling with my brother and me. When it was his turn to drive the car pool to elementary school, he would pretend to get lost, or start to drive himself to work at Caltech. "No, not that way!" the kids would scream. "Oh, all right. Is it this way?" and he would turn the wrong way again. "Nooooooo!" we would yell in utter panic, convinced we were going to be late (we always arrived just in the nick of time, of course). Of my father's many skills, this willingness to play the fool—and to let me think he could be completely outfoxed by my clever thinking—was the one that shaped my childhood more than any other.

I was simply unaware for many years that he was revered as a supreme intellect. In fact, he encouraged a certain amount of irreverence toward himself. Most of the stories he told us were ones that highlighted his ineptitude. Our dinner conversations were full of animated stories about mis-

takes he made during the day: losing his sweater, forgetting something terribly important, having complete conversations with people and not remembering their names. He would talk about his experiences away from home as well, as in the time he was so disgusted with the fancy hotel hosting a conference he was attending that he took his suitcase and slept outside in the woods. "Oh, Richard," was the invariable refrain from my mother's end of the table. Yet he always laughed at himself, and so we laughed along with him.

This is the key, in my mind, to his success as a teacher. Never condescending when he explained things, he had a knack for breaking problems down to a small, comprehensible scale. "O.K., say the earth is this apple," he would begin, holding one up to illustrate his point. The simplicity of his illustrations brought the theretofore incomprehensible down to a level more readily grasped. In the early 1960s, his love of teaching and sense of civic duty led him to the California Curriculum Commission, where he devoted countless hours to evaluating math textbooks for elementary school students. Being awarded the Oersted Medal in 1972 for his contributions to the teaching of physics gave him tremendous satisfaction. Ten years later, the Associated Students of Caltech presented him with an award for excellence in teaching, to which he responded, "I was very pleased to be honored for doing something I so thoroughly enjoy."

He was a staunch believer in public education but invariably had his frustrations with the bureaucracy and its inflexible thinking. When I was in high school, he started showing me shortcuts in my math homework that diverged from the teacher's methods. I was subsequently scolded by my Algebra II teacher for not solving the problem in the right way. My father found this ridiculous, as I had nonetheless arrived at the correct answer, and he dropped by the school to discuss the matter. The teacher did not know who he was and treated him as a bumbling idiot trying to make his life more difficult. The teacher finally accused him of knowing nothing about math. My father, biting his tongue the whole time, finally could not stand it any longer, and spoke his mind. The next day I was transferred to another class. The next year, solving problems without using the prescribed methods yielded similar results, and so I ended up studying the material at home with my father and taking the exams at school.

. . . .

These memories and more came flooding back not too long ago. I remember reading a few of my father's letters in 1990, shortly after my mother's

death. One that particularly struck me was a letter to a former secretary in which he lamented that he was not a grandfather yet, that his kids were slow. I figured out I was in high school—in eleventh grade—when he wrote that. As I laughed, it occurred to me that his letters might make for interesting reading one day.

It would take fourteen years, but in May of 2004, I started doing just that when copies of the Feynman Papers from the Caltech Archives were shipped out to me. Twelve filing cabinet drawers and thousands of sheets of paper took quite a while to go through. Most were technical in nature: notes for lectures and classes, correspondence with colleagues regarding developments in physics, attending conferences, and so forth. About a third of the documents were non-technical, and almost all of those were in the form of letters. Boxes long stored in my basement yielded even more: newspaper clippings, photographs, family snapshots, and many letters more personal in nature.

Perhaps because most of my father's books were derived from spoken materials—be it lectures or stories—which were then edited, and because he constantly complained that he could not speak "grammatically correct," I did not know what to expect from his written work. As I delved into his correspondence, however, I was completely captivated. In these letters he is articulate, insightful, considerate, humble, nurturing, funny, and charming.

That he wrote so many letters, both to scientists and to ordinary people, surprised me. According to Helen Tuck, his secretary of almost thirty years, he was in the habit of answering his own mail when she came onboard at Caltech in the mid–1960s. His desk was usually piled with stacks of both opened and unopened mail, and he would answer correspondence when he felt like it, which, according to Helen, was not all that often. She eventually persuaded him to let her sort through the mail and then they would answer the letters together, an idea he loved. She gradually learned which letters would interest him and elicit a response.

A few of his friends were surprised when I described what I had found, as my father had a reputation in the physics community for not writing letters. Why had he spent so much time corresponding with the general public and not with his fellow scientists? When I raised this question with one of my father's colleagues at Caltech, he offered the possible explanation that my father was a very kind man. Although that was true, I believe the answer lies beyond that, and it has to do with my father's great love for teaching.

In an article on education he wrote for Caltech's *Engineering and Science* he stated: "The problem is clear language. The desire is to have the idea

clearly communicated to the other person." He was speaking of mathematics textbooks at the time, but I think this statement in part explains
why he was such an effective and prodigious communicator. These letters
are testimony to his skill and desire to be plainly understood—and, of
course, to his passion and curiosity about the world. Again, his own words,
this time from a letter to a young student seeking his advice, explain it
best: "You cannot develop a personality with physics alone, the rest of life
must be worked in."

Because he was selective about which letters he answered, I feel this collection represents the best of all that came his way, as well as his most considered responses. I was even startled and touched to find an apology to that
high school algebra teacher who had given him such a hard time.

I have decided to present these letters in a roughly chronological order,
making a few exceptions in the interest of clarity and continuity. I have
kept letters together with their replies. The earliest letter, written to his
mother, dates back to 1939. Letters he wrote to his first wife, Arline, give us
a glimpse into what his life was like while working on the atomic bomb—
and paint a portrait of a sweet and tender young romance. I have grouped
together a series concerning membership in the National Academy of
Sciences, as well as a flurry of letters written on the occasion of his Nobel
Prize, as here subject takes priority. The rest of the book is organized to
allow one to fully appreciate the arc of his life.

My favorite books about my father, *Surely You're Joking Mr. Feynman!*,
*What Do You Care What Other People Think?*, and Christopher Sykes's *No
Ordinary Genius,* are based on the spoken word. In this book his letters,
written in his own hand, speak for themselves. Taken as a whole, they present his character in a way that we have not seen before. The act of writing a
letter is more deliberate than conversation and suggests a certain confidence or intimacy. But though his letters are addressed to individuals, concerning both matters of historic importance and private events of little
lasting consequence, the themes in this book are universal.

After reading through hundreds of letters, I found a compass for them in
his notes for the brief speech he delivered at the Nobel Banquet, and it is
also perhaps his most eloquent statement on the stature he was afforded in
many people's minds. In these words, he seems to have come to a temporary peace about the recognition and fanfare that so often bothered him,
and he expresses gratitude for all the goodwill shown him. In its clarity,
grace, humor, and optimism, it is perhaps the most fitting introduction to
this collection:

The work I have done has already been adequately rewarded and recognized. Imagination reaches out repeatedly trying to achieve some higher level of understanding, until suddenly I find myself momentarily alone before one new corner of nature's pattern of beauty and true majesty revealed. That was my reward.

Then, having fashioned tools to make access easier to the new level, I see these tools used by other men straining their imaginations against further mysteries beyond. There, are my votes of recognition.

Then comes the Prize, and the deluge of messages. Reports of fathers turning excitedly with newspapers in hand to wives; of daughters running up and down the apartment house ringing neighbors' door bells with news; victorious cries of "I told you so" by those having no technical knowledge—their successful prediction being based on faith alone; from friends, from relatives, from students, from former teachers, from scientific colleagues, from total strangers; formal commendations, silly jokes, parties, presents; a multitude of messages in a multitude of forms.

But, in each I saw the same two common elements. I saw in each, joy; and I saw affection (you see, whatever modesty I may have had has been completely swept away in recent days).

The Prize was a signal to permit them to express, and me to learn about, their feelings. Each joy, though transient thrill, repeated in so many places amounts to a considerable sum of human happiness. And, each note of affection released thus one upon another has permitted me to realize a depth of love for my friends and acquaintances, which I had never felt so poignantly before.

For this, I thank Alfred Nobel and the many who worked so hard to carry out his wishes in this particular way.

And so, you Swedish people, with your honors, and your trumpets, and your king—forgive me. For I understand at last—such things provide entrance to the heart. Used by a wise and peaceful people they can generate good feeling, even love, among men, even in lands far beyond your own. For that lesson, I thank you.

He showed us all how to look at the world. He showed me how to laugh. For that and so much more, I thank him.

—Michelle Feynman, 2004

# 1939-1942

Richard Feynman received his undergraduate degree from MIT in June of 1939. He then went on to Princeton University for his graduate studies (when he expressed interest in staying at MIT, Professor Slater advised him not to: "You should find out how the rest of the world is"). In these early letters, he wrote to his parents in Far Rockaway, Queens, of his first forays into teaching and the vagaries of student life: canned food, lack of funds, and irregular hours.

In addition to his devotion to physics and, later, the war effort, a young woman by the name of Arline Greenbaum was another major involvement during this time. Arline and Richard married on June 29, 1942, just two weeks after he had received his Ph.D.

Aside from the youthful voice reverberating in these early letters, several of them display themes that would echo throughout Feynman's life. They reveal a great eye for the telling detail, confidence in his decisions, and a curious ambivalence toward time. Although he was careful to note the time of his writing or the time he went to bed the night before, for example, he often noted the date as "I don't know what" or simply omitted it altogether.

Feynman was just starting his professional climb, and his letters from this period are infused with energy and enthusiasm. His first published paper dates back to this time, and incidentally, it came in the form of a letter to the *Physical Review*. Co-written with MIT Professor Manuel S. Vallarta, the paper explored stars' interference in the scattering of cosmic rays. Although it was not itself groundbreaking, it contained the thought processes that would come to characterize his work and foreshadowed his great papers of the late 1940s.

. . . .

## RICHARD P. FEYNMAN TO LUCILLE FEYNMAN, OCTOBER 11, 1939

*Richard was twenty-one years old and had just begun his studies at Princeton in New Jersey, seventy miles from his family in New York.*

Tues. Oct—1939

Dear Mom,

I like your idea about coming to visit me alone some time very much. Why don't you just jump on a train some Sunday morning and I'll meet you at the station. You don't need Pop—you'll only have to worry about whether he can get anything to eat in the restaurant, etc. Not that I don't want to see Pop—but he'll drop by alone lots on business trips—so you make a little inexpensive outing just for yourself—any Sunday you say, provided I know which one so I can meet you. In fact, if you are worried about the expense let me treat you. It will be fun.

The raincoat came O.K. It is very nice. I think they make raincoats very stupidly these days—the bottoms of your pants get all wet. Now that I have the raincoat it is hot as hell (excuse the language but hell is damn hot, so is "it" and the sun is shining brightly).

Prof. Wheeler was called away suddenly last night so I took over his course in mechanics for the day. I spent all last night preparing. It went very nicely and smoothly. It was a good experience—I guess someday I'll do a lot of that.

Things are rather uneventful. I didn't fall in the water while sculling these last two times, so maybe I won't fall in anymore—so saying he fell in.

I'll tell you all about things when I get home.

Love,

RP Feynman

## RICHARD P. FEYNMAN TO LUCILLE FEYNMAN, NOVEMBER 1939

Monday Nov. I don't know what

Dear Mom,

The best news to really write home about is, unfortunately, already known to you. Arline came to visit me. We had lousy weather but a good time.

TOP *At the Princeton Library, 1939.*

BOTTOM *Arline Greenbaum, c. 1939.*

You must come visit me, Mom, some time. You wrote me saying you wanted to, and if I know you, and I think I do, a little, you'll never quite get around to coming up—unless I keep pushing you. How about a date? Write in your next letter which week-end you're coming up *definitely*.

My academic life has its usual characteristic of being "not write home-able."

However, last week things were going fast and neat as all heck, but now I'm hitting some mathematical difficulties which I will either surmount, walk around, or go a different way—all of which consumes all my time—but I like to do very much and am very happy indeed. I have never thought so much so steadily about one problem—so if I get nowhere I really will be very disturbed—However, I have already gotten somewhere, quite far—and to Prof. Wheeler's satisfaction. However, the problem is not at completion although I'm just beginning to see how far it is to the end and how we might get there (although aforementioned mathematical difficulties loom ahead)—SOME FUN!

Even if your door is closed and locked now I'm still serious when I say it's fun.

Tell Pop I have made out a time schedule so as to efficiently distribute my time and will follow it quite closely. There are many hours when I haven't marked down just what to do but I do what I feel is most necessary then—or what I am most interested in—whether it be W.'s problem or reading Kinetic Theory of Gases, etc.

And while you're telling Pop, give him this problem in LONG DIVISION. EACH OF THE DOTS REPRESENTS SOME DIGIT (ANY DIGIT). EACH OF THE A'S REPRESENT THE SAME DIGIT (for example, a 3) NONE OF THE DOTS ARE THE SAME AS THE A (i.e., no dot can be a 3 if A is 3).

Love,

R.P. Feynman

## RICHARD P. FEYNMAN TO LUCILLE FEYNMAN, OCTOBER 1940

Oct. ? 1940

Dear Mom,

I never didn't write a letter in such a long time before. I don't know why—but I never got around to it.

Kinetic Theory of Gases, etc.

& while you're tittering pop, give him this problem in LONG DIVISION. EACH OF THE DOTS REPRESENTS SOME DIGIT (ANY DIGIT). EACH OF THE A's REPRESENT THE SAME DIGIT (for example, a 3) ~~a 4 A~~

~~#~~ NONE OF THE DOTS ARE THE SAME AS ~~X~~ THE A (ie, no dot can be a 3 if A is 3).

Love.
R. P. Feynman.

Thanks very much for answering my telegram. I will register on Tuesday for voting (Wed. for draft).

Putzie is coming up to visit me tomorrow.*

I am listening in to a course in physiology (study of life processes) in the biology department. It's a graduate course—I didn't realize how much I really picked up reading all about those things on my vacations. I don't know at all as much as the 3 other fellows in the course but I can understand and follow everything easily.

The night you left I had a fellow visit me and we finished the rice pudding and most of the grapes. I finished the grapes the next day.

A few nights later two mathematicians came to visit. We had crackers and peanut butter and jelly and pineapple juice. I had a lot of trouble opening the can because I didn't have a good can opener.

The next day I received a present of a dandy can opener from the mathematicians. A practical gift—I suppose.

Night before last a fellow came in and we had tea and crackers. I can boil water pretty well now because I bought a pot cover.

It's a lot of fun.

Well, I'd better get to work.

Love,

R.P.F.


## ARLINE GREENBAUM TO RICHARD P. FEYNMAN, JUNE 3, 1941

Richard sweetheart I love you—maybe more than I can ever tell you—and perhaps we can find a still happier plan of living—besides my happiness we must consider yours—we still have a little more to learn in this game of life and chess—and I don't want to have you sacrifice anything for me—tomorrow Dr. Treves is going to see me and tell me, according to "Woody" (Dr. Woodwood) some news—I wonder if he's going to give me that story about "glandular fever" that you said "Woody" had planned to tell me—I'll swallow it but Nan wrote and said I had the right to call in another doctor to check the diagnosis and have him look at the slides from the biopsy—she recommended a man too—I'll look into it before you come home this weekend and we could see him together—I know you must be working very hard trying to get your paper out—and do other problems on the

---

\* "Putzie" (rhymes with "footsie") was Arline's nickname.

side—I'm awfully happy tho' that you're going to publish something—it gives me a very special thrill when your work is acknowledged for its value—I want you to continue and really give the world and science all you can—if I were an artist I'd give all I could to art—but all I can do is draw a little—darling I love you—and if you receive criticisms—remember everyone loves differently—but I'll always do my best for you—your happiness is as important to me as mine is to you—the problem we're faced with confused even Aristotle. "What is the chief 'good' of man"—Darling I'll always belong to you—and always love you—no matter where or when—

I'm your
Putzie

## RICHARD P. FEYNMAN TO LUCILLE FEYNMAN, MARCH 3, 1942

*There is a scrap of paper from his mother's letter taped to the first page of his response. It reads:*

> *You wrote: I get $60*
> *pay laundry    $18 + 2*
> *Memberships    $13 + 3*
> *Mother    $10*
> *Remainder    $19*
> *Oh! Richard! How fallen art thou!!*
> *I make it $14*
> *Woe is me! Who will*
> *Balance my checkbook?*

Dear Mom,

I will balance your check book.

If you will look more carefully at my last letter you might find something such as "I get $60 I pay 1. Laundry $18 and 2 membership $13 and 3 Mother $10." The 2 and 3 are numbers which are used to number the items, and I didn't mean them to be added in.

My bill for $265 came (or did I tell you). I spent 20 minutes figuring out which of my certificates at the Postal Savings I should cash in in order to get the maximum interest. I can get a maximum of 53¢. This was interesting until I discovered that if I did it the worse way, and got the least interest possible it would be 45¢. I think 20 minutes of my time ought to be worth

*Feynman, 1942.*

more than 4¢. (In order to dispel more slurs on my calculating ability might I say that the 4¢ is right. If I didn't figure it out at all and did it at random on the average I might expect to come out with about 1/2 way between the most possible (53¢) and the least (45¢) or I should expect if I didn't figure I would get about 49¢. Since I get 53¢ by figuring I save 4¢ on the average by figuring for 20 min. It's all in the laws of chance that the 8¢ becomes 4¢).

Princeton pays me 10¢ for 20 minutes figuring on a 48 hr week. (OK, 10 5/12¢, or make it 10¢ figuring on a 50 hr week).

Now I hope your little door is slammed shut.

There isn't much else to tell about, except the letter from Pop which you know all about and which I answered today.

Love Richard

P.S. Wishing you a happy anniversary and Joan a happy birthday—in case I forget later.

## RICHARD P. FEYNMAN TO MELVILLE FEYNMAN, JUNE 5, 1942

RICHARD P. FEYNMAN
PRINCETON GRADUATE COLLEGE
PRINCETON, NEW JERSEY

Dear Pop,

As you suggested, I asked Prof. Smyth about how he thought my marriage might affect my career. He said the only thing he could think of is that possibly some might not hire me because they figured I carried too much of a burden. He said, however, that he tries as much as possible to keep out of people's private affairs and that it didn't make any difference to him at all, and he supposed it probably wouldn't make much difference to others.

I then pointed out that I would be in contact with an active case of T.B. and so he might not think it a good idea for me to teach, because it might affect the students. He said he had not thought of that at all, but, since he didn't know too much about T.B., he would call the University Doctor—Doctor York.

He told me later that he called Dr. York, and that Dr. York said that as long as the girl was in a sanatorium there was no danger to my students and myself. He said Dr. York would like to speak to me. I went over to talk to him today.

The doctor said he heard I had a little problem, and that he'd like to know if I knew a few things. First he told me that one of the most important things for a T.B. patient was freedom from worry, etc.—what he called emotional security. I told him I realized this and that it was one of the important reasons for my getting married—because she would worry far less if she were married, than the way things are now.

Then he asked if I knew that it would be very bad for an active case of T.B. to become pregnant. I said yes, again—and told him there would be no chance of it.

He said these were the main things he wanted to say and he wanted to find out if I had thought about them—and also the fact that T.B. is not always curable. He said he wanted to find out if I was responsible and I had thought about these things.

Then we discussed all kinds of things—how Arline was getting along—how long she would have it—etc. We spoke about where to put her and he warned against private places (except for a few like Saranac—which is too expensive). He asked me what the parents thought and I told him her parents saw no objection, but that mine were worried about my getting infected and indirectly carrying the germ around and infecting others—and for this and many other reasons they didn't like the idea.

He said I should understand T.B. is an infectious disease but not a contagious one (I asked him what he meant by that—the idea is the germs don't float all around in the air and you don't get it just on contact—etc. I didn't understand the difference very well—apparently it's one of degree). He said that I had less chance of getting sick in the sanatorium visiting her than when I was out in the street—because in the sanatorium they take precautions to burn the sputum, but careless people in the street just spit all around. I think he was exaggerating, but I still think you shouldn't worry that by marrying I lay myself and my friends open to a great danger.

Love from
R.P.F.

## RICHARD P. FEYNMAN TO LUCILLE FEYNMAN, DATE UNKNOWN, JUNE 1942

*The following letter is in response to one from Lucille, Richard's mother, in which she lovingly but forcefully outlined her concerns about Richard's intent to marry Arline. Arline's illness, she feared, would compromise not only his own health but his career. She was also concerned about the high cost of treatment (for oxygen, specialists, hospitalization, and so on).*

*Lucille suggested that his desire to marry stemmed from his desire to please someone he loved ("just as you used to occasionally eat spinach to please me") and recommended that they stay "engaged."*

*Richard's response came on the same embossed stationary as his previous letters, with "Ph.D." added in ink following his name.*

RICHARD P. FEYNMAN, **Ph.D.**
PRINCETON GRADUATE COLLEGE
PRINCETON, NEW JERSEY

*Graduation from Princeton, June 16, 1942.*

Dear Mom,

I really should have answered you sooner, but I have been spending the last few nights working on physics problems—and now that I'm stuck for a while I'll take time out to write to you.

I enclose your letter so you will remember which point is which that I refer to.

With regard to (1) and (2) I went to see Prof. Smyth at Pop's suggestion and the doctor here at the university. The doctor said I have less chance of getting T.B. in the sanatorium when visiting her than when I am walking around in the street. I think he was exaggerating (all this is in detail in a letter to Pop, so I won't repeat it all here). He said T.B. is infectious but not

contagious—I didn't understand the distinction he made, however. Ask Dr. Sarrow. He said in sanatoriums the patients take care of their sputum by cups or Kleenex for the purpose, but on the streets people are careless and just spit all around and when it dries the germs float into the air. He said the germs are not floating around in the air in a sanatorium. He said a lot has been found out about this in the last 25, and in particular the last 10, years. I would be no danger to my students. Prof. Smyth didn't see any objection from his point of view to hiring me if my wife is sick.

(3) If no one can make a budget for illness, how can I ever make enough to pay for it? How much is enough? Some *guesses* must be made and I *guess* I have enough. How much would you *guess* would be necessary?

(4) I wouldn't be satisfied being engaged any longer. I want the burden and responsibility of being married.

(5) It really wasn't hard at all. While I was out to lunch while waiting for somebody to come back to the courthouse in Trenton, I found myself singing—and I realized then that I really was very happy arranging things. It was, I suppose, the pleasure of arranging things for our life together—before she was sick we used to talk of the fun it would be going around ringing doorbells looking for a place to live—I guess it was similar to that idea.

I am not afraid of her parents—and if they don't trust me with their daughter let them say so now. If they get sore at my mistakes later, it's too late and it won't bother me. You are right about my lack (4) of experience—I have no answer to that.

(6) The cost here again is a guess. I want to take the chance, however, that it will be sufficient. If it isn't I'll be in difficulty as you suggest.

(7) I've already been employed at Princeton for the next year. If I must go elsewhere, I'll go where I'm needed most.

(8) I *do want* to get married. I also want to give someone I love what she wants—especially because at the same time I will be doing something I want. It is not at all like eating spinach—(also you misunderstood my motives as a small boy—I didn't want you angry at me)—I didn't like spinach.

(9) This is the problem we are discussing—I mean whether marriage is worse than engagement.

(10) I'm honestly sorry it makes you feel so bad. I bet it won't be too heavy.

Why I want go get married;

It is *not* that I want to be noble. It is *not* that I think it's the only right,

honest and decent thing to do, under the circumstances. It is *not* that I
made a promise five years ago—(under entirely different circumstances)—
and that I don't want to "back out" of the promise. That stuff is baloney. If
anytime during the five years I thought I'd rather not go thru with it—
promise or no promise I'd "back out" so fast it would make your head spin.
I'm not dopey enough to tie up my whole life in the future because of
some promise I made in the past—under different circumstances.

This decision to marry is a decision now and not one made five
years ago.

I want to marry Arline because I love her—which means I want to take
care of her. That is all there is to it. I want to take care of her.

I am anxious for the responsibilities and uncertainties of taking care of
the girl I love.

I have, however, other desires and aims in the world. One of them is to
contribute as much as to physics as I can. This is, in my mind, of even more
importance than my love for Arline.

It is therefore especially fortunate that, as I can see (guess) my getting
married will interfere very slightly, if at all with my main job in life. I am
quite sure I can do both at once. (There is even the possibility that the con-
sequent happiness of being married—and the constant encouragement and
sympathy of my wife will aid in my endeavor—but actually in the past my
love hasn't affected my physics much, and I don't really suppose it will be
too great an assistance in the future.

Since I feel I can carry on my main job, and still enjoy the luxury of taking care of someone I love—I intend to be married shortly.

Does that explain anything?

Your Son.

R.P.F. PH.D.


P.S. I should have pointed out that I know I am taking chances getting
married and may get into all kinds of pickles. I think the chances of
major disasters are sufficiently small, and the gain to me and Putzie great
enough, that the risk is well worth taking. Of course, this is just the point
we are discussing—the magnitude of the risk—so I am saying nothing
but simply asserting I think it is small. You think it is large, and therefore I
was particularly anxious to have you tell me where you thought the pitfalls were—and you have pointed out a few new ones to me. I still feel
the risk is worth taking—and the fact that we differ is due to our difference in background, experience and viewpoint. Please don't worry that,

by explaining your viewpoint, you have in any way pushed us further apart—you haven't. I only hope that my marrying directly in the face of your disapproval and your better judgment won't alienate you from me— because honestly, our judgments differ and I think you're wrong. I honestly believe we (Putzie and I) will be better off married and nobody will be hurt by it.

RPF

## RICHARD P. FEYNMAN TO DAN ROBBINS, JUNE 24, 1942

*Mr. Robbins was a fraternity brother of Feynman's. The project in question in this letter was the early phase of the race to build the atomic bomb. Feynman had received a letter a few days earlier from researchers at the University of Chicago asking whether he was interested in working on what they could only describe as "research and development work on new methods of military offense and defense." They promised that this work would be a vital factor in the successful conclusion of World War II. Feynman later recalled how Professor Robert Wilson came to his office at Princeton to encourage him to enlist in the project. He signed on shortly thereafter.*

Dear Danny,

I wrote a letter to the fraternity recently—and I tried to call you up at your home. I spoke to your mom. She tells me you are already working on a defense project at M.I.T.

I wanted to write to you especially to reitterate (how do you spell it?) to you personally the offer of a job with our group at Princeton.

The darn trouble is I can't go into any detail about what the work concerns and what we need you for. It makes it hard for me to explain or make the opportunity seem good—except in general,—and therefore (for me at least) weak, terms.

(Up to this point I've just copied stuff from another letter I just wrote to you but won't send because I told you so much about the job—indirectly that I think one could have guessed pretty closely what it was—I hope I avoid writing again and make no more mistakes).

I can only say I find the job very exciting, because the results are of such enormous consequence. You really feel you are fighting right alongside the army and you hope you can get it done in time—in time means before the other side works it out.

I'm doing most of the theoretical figuring. I try to figure out what should happen for various conditions and what the best way to build this part or the other is. I don't know if you were most interested in theoretical or experimental work or both—but we can use you and we will listen to all your ideas—(because we need more ideas) in any capacity. I urge you to come.

In making a decision, however, I can see you have a considerable problem because you are already working on a defense project. I would say the important thing is to do that work that you think will contribute the most to the war effort. Your mother said you seemed tired of the environment up there at M.I.T. and might like it out here much better. I'd like to see you and work with you too—but I don't think such personal matters carry any weight. The real thing is your usefulness. Make that as great as possible. Your problem is hard because I can't tell you what kind of work there is here and you can't tell where you'll be the most useful—here or there. If you are interested and think you might advantageously make the change but are not sure we may be able to get the opinion of someone who knows about both jobs—the one you're on and this one.

How is your personal life?

I am getting married in a few days to Arline Greenbaum (also known as Putzie). I just got my Doctor's degree. I also just got a job as a visiting associate professor at the Univ. of Wisconsin for one year—on leave without pay to do war work (it sounds silly but the idea is I've got the job if the war work stops).

Can I hear from you soon?

Brother R.P. Feynman

*Mr. Robbins was not able to accept this job offer, as he had already accepted a commission in the U.S. Navy.*

## RICHARD P. FEYNMAN TO LUCILLE FEYNMAN, DATE UNKNOWN, 1942

DR. RICHARD P. FEYNMAN
PALMER PHYSICAL LABORATORIES
PRINCETON, NEW JERSEY

Dear Mom,

I won't write you a long letter for I haven't time. Putzie wanted me to

*Mr. and Mrs. Feynman, June 29, 1942.*

write to you and her mom to tell you to excuse her for her not writing this week as she feels sick. Would you call up her mom and tell her?

You ask for a letter about myself and what I'm doing. All the letters so far were on just that subject—at least what plays I saw, etc. Other times life is very simple.

This week has been unusual. There is an especially important problem to be worked out on the project, and it's a lot of fun so I am working quite hard on it. I get up at about 10:30 AM after a good night's rest and go to work until about 12:30 or 1 AM the next morning when I go back to bed. Naturally I take off about 2 hrs for my two meals. I don't eat any breakfast, but I eat a midnight snack before I go to bed. It's been that way for 4 or 5 days. Usually I don't work nearly as hard—more like 7 1/2 or 8 hours a day.

The only difference from one day to the next is that some days when I go to lunch I take my laundry out and some days when I come from lunch I bring my laundry back.

Hey, what's this about Frances and 4 AM?* When I'd stay out that late you (or at least Pop—I never could remember which of you bawled me out for which offense) used to say you don't see how Arline's parents permit her to stay out so late, etc., etc... Isn't Frances partly in your charge when she visits from N.Y.? Perhaps I shouldn't remind you or Frances will get angry.

Mooseface is next, and I expect her to come in at dawn every day.** You can't call her down, however—you can ascribe it all to her interest in astronomy. She had better get used to staying up during the night because there are few stars visible in the daytime. Heck, wait until you become a Red Cross Night Nurse or Night Staff Assistant or what have you, then the only one who will see the green of the trees, etc. in the daytime will be Pop when he comes home from a Saturday day of rest. He'll probably go to sleep on the beach anyway.

I'd better shut up, it is 1:45 AM now.

Love to Everyone

---

* Frances Lewine, Richard's cousin.
** "Mooseface," or "Moose," was Richard's nickname for his sister, Joan.

# 1943–1945

After many months of preliminary work at a variety of institutions, the efforts of J. Robert Oppenheimer to consolidate the Manhattan Project began to take shape in early 1943. Feynman, then only twenty-four, began making plans to relocate to Los Alamos, New Mexico, along with the first wave of physicists from Princeton. They would arrive at an unfinished compound on a high mesa, and for Feynman, who had not traveled much, the setting was spectacular.

In these letters from Los Alamos, he describes the surrounding scenery with a wonderful immediacy and clarity. His letters to Arline, sent on a near daily basis, form a de facto journal of his experience there, and while there are accounts of his various antics (safe-cracking, giving the great minds of Los Alamos a demonstration on "some interesting properties of numbers," and showing the night watchman the fence full of holes), the impression is one of a sober young man hard at work. Writing from her bed in an Albuquerque sanatorium, Arline was also a faithful correspondent, though her correspondence dwindled as her illness progressed.

. . . .

## RICHARD P. FEYNMAN TO J. H. STEVENSON, MARCH 5, 1943

Mr. J. H. Stevenson
University of California
Berkeley, California

Dear Sir:

In your "first memorandum" you ask that questions concerning moving and living conditions be held until we receive the second memorandum. However, time is short, and your letter is not expected for several days. I am fairly sure your letter will not contain the answer to my question so I am taking the liberty of writing you now.

My wife has active pulmonary tuberculosis, and on this account has been required to rest in bed. This fact caused Prof. Oppenheimer and myself some concern when we were deciding whether I could come to the site or not, but Prof. Oppenheimer felt, finally, that almost certainly some way of taking care of her could be arranged.

There are two apparent possibilities. (a) She might stay at the post hospital. She has been receiving no special treatment, except bed rest, and no special equipment is necessary, except that it should be possible for her to get x-rayed once in a while, I suppose. (b) She might be accommodated in some nearby sanatorium. I should then want to be able to leave the post at least once a week.

Naturally I would not like to start moving with her until I know just where she will go and that she is expected. On the other hand I want to move out as nearly at the same time as the rest of the Princeton group as possible.

My wife and I would very much prefer to have her on the post ((a) above), and we hope it can be arranged that way. There is however no financial problem either way.

Sincerely yours,
Richard P. Feynman


## RICHARD P. FEYNMAN TO J. H. STEVENSON, MARCH 15, 1943

*Mr. Stevenson wrote back suggesting three places Arline could stay: (1) a small hospital 30 miles from the site, (2) a sanatorium in Albuquerque 100 miles*

*from the site, or (3) a dude ranch 20 miles from the site. Because construction was still underway, he encouraged Arline to stay away from the site until the latter part of the summer.*

*He concluded, "I am writing this at length to assure you of the certainty of making your wife comfortable and giving her good care at some point not too far from the site. There will be no question about your visiting her every week."*

Dr. J. H. Stevenson
University of California
Berkeley, California

Dear Dr. Stevenson:

Thank you very much for your nice letter of March 10, concerning the possible accommodations for my wife. We are both very pleased with the possibility you described and are very anxious to go out there.

Since Mrs. Feynman's case is an active one, the dude ranch is impossible (as is, in fact, any other place where there is no doctor in attendance). Either of the other two possibilities sounds very fine.

It occurred to me that we might save a little time, trouble, and money in the following way. When I come down to the site I could bring my wife with me to stay at one of the places you mentioned. I could then look around to see if there is a nicer place. If we find one we like better, we could move my wife. It will not be hard for her to move for a short distance like 50 miles.

If you do not think this plan is wise, please let me know, and I will come out alone as you originally suggested.

Otherwise, do you think you could make arrangements for us? We would be leaving on "the Chief", of the Santa-Fe line from Chicago at 12:01 (noon) on Tuesday, March 30. Could you tell us what station is best to get off at, and how we go from there to the hospital (we could go by car, taxi, train, ambulance, truck, etc., but preferably not by bus)? If it is necessary to make a deposit at the hospital, I can send it to you. If you cannot find a place which will accept her temporarily, you could, if necessary, guarantee that she will stay for as long as two months.

I think that Mrs. Olum would help you, if you wish.* If, however, you

---

* The wife of Paul Olum, friend of Richard's and likewise a former assistant of John Wheeler at Princeton.

cannot make the arrangements for any reason, I can come out alone as you have suggested.

I hope I am not causing too much trouble. My wife and I thank you very much for what you have already done.

Sincerely,

R. P. Feynman

## ARLINE FEYNMAN TO RICHARD P. FEYNMAN, MARCH 26, 1943

3 more days

Dearest Rich—darling if you only knew how happy you've made me with this train trip of ours—it's all I've wanted and dreamed about since we've been married. It will mean so very much to both of us—I adore you darling—I want to be all a wife should be to the husband she loves—now I have a chance—but the future holds even more promise dear—when we have our own home and family—it's worth waiting and fighting for—darling I'll see you tomorrow and I know it will be hard to keep these pent up emotions in check—with only one day left—I'm so excited and happy and bursting with joy—I think, eat, and sleep "you"—our life, our love, our marriage—the great future we are building—every day we live counts—all we do and say and think—I always want to be close to you sweetheart—a part of all your thoughts and desires and ambitions—you're that close to me—and more—I live for you and the day we can have "Donald"—and all the other little things we've planned to do together—hang curtains and camp out in a tent, have "teas" for your students, and a chess game before our open fireplace in winter, showers in the summer and Sunday morning in bed reading the funnies—dearest I could go on indefinitely—there is so much left in life for us to share and explore together—I want to be there every moment—and even then, with you, it will be too short—

Darling how can I show you how much you mean to me when we're miles apart! If only tomorrow would hurry and come—I want to feel your warm cheek pressed close to mine and your tender embrace—your nearness fills me with such content! I love you sweetheart—body and soul—I long to be near you again—the way we've been in the past—but now our love and our life is richer—we really belong to each other—for all the world to know—I'm so proud of you and proud and happy to be your

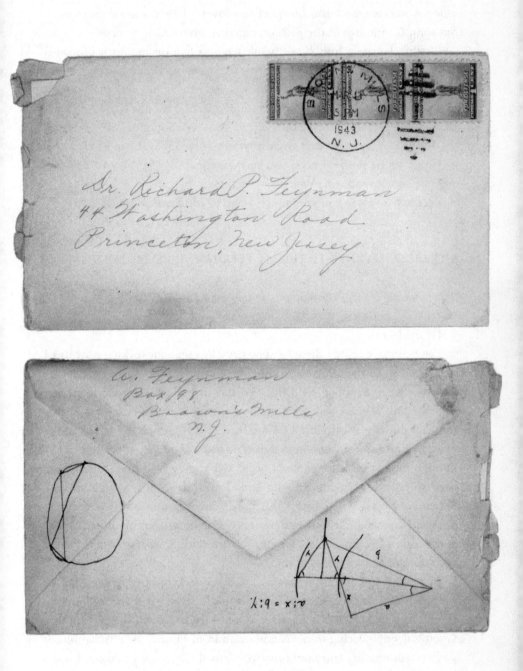

wife—you're a wonderful husband and lover—I have a good memory—but soon we can live in the present, together, even tho' it be brief.

Darling I have a hunch we won't have to live on memories long—I really believe (and I'll make it happen) we'll have our home soon now—we can do anything we really set our hearts and minds on—let's make a new gigantic effort together—or just continue the one we just started—I can't push alone dear—I need you behind me—encouraging me—because you're the only reason I care to do anything for—I must be in love!

Come to me soon—I need you and want you

Your adorable adoring wife and sweetheart

Putzie, who loves you very much—and so forth and so on


## RICHARD P. FEYNMAN TO LUCILLE FEYNMAN, JUNE 24, 1943

*The following letter is the first one written from Los Alamos.*

Dear Mom:

I've been sick in bed these last few days. I'm up and working now tho. I got a cold and I thought since a long drawn out cold lowers resistance and I have to keep in super condition, I'd stay in bed till I got rid of it, fast, which I did. While I was getting rid of the cold my ever-present gastronomical difficulties amplified themselves until I became conscious of the existence of my stomach—so I stayed in bed a little longer for that, and I haven't gotten rid of it yet, but I go to work anyway. It makes me a little tired so I sit down while solving equations instead of prancing around. When I got up (I was in bed 3 1/2 days) I discovered that 50% of the people were knocked out with some sort of dysentery from the water supply, and I'd been drinking tons of water for my cold! So that accounts for my being sick with two things at once.

It was fun, just as I always wanted to, I was sick and people brought me my meals and came to visit me and I took it easy in bed and had a radio (Putzie's—I had taken it here to fix it) and I read a book I once bought on Chemical Engineering from "Transportation of Fluids" to "Distillation." It was very interesting and now I'm a confirmed Chemical Engineer. I had a lot of visitors who brought me things among which were 3 oranges, one apple, crackers, jelly for same, pitcher for water, chocolate, Reader's Digest, book, news as to what was going on at the lab., etc. etc. It is funny, all the girls that came to visit brought stuff, but none of the fellows did.

I received your prune things during these times. Since I was not gastro-nomically on top of the world I have avoided eating them. When I get a lit-tle better inside and my intestines become more attuned to their responsibilities instead of taking their devil-may-care attitude, I'll eat them (the prunes I mean).

I have had several trips and climbs into canyons but nothing extra special to report.

I was thinking a little about Pop and his being sick and traveling all over the country. You said you would like him to retire and you told me all about the problem, except the details of how much you have saved up, etc. and how much scraping it would take if he retired.

I had an idea tho, which is not a product simply of my love of this coun-try. We have a department for purchasing here which has several faults (one being the guy in charge) and it would seem to me, although I don't know much about it, that they would like very much to have a guy like Pop. He would like it too, I'm sure. There would be no, or very little, traveling around, all the business being done by telephone and teletype. He would be partly out of the rush etc. of the business world and would be among aca-demic men to a great extent, which I'm sure he'd enjoy. On the other hand, it isn't all gravy. Purchasing these days is quite difficult, and everyone here is in a hell of a hurry for their stuff. Some of the things will be unfa-miliar to him, but that won't cause much difficulty. It will have its aggrava-tions, too, but it will be a damn important position in our project and scientific venture and he will be contributing more than I toward winning the war. What does he think? I'm not sure of the availability of the job, however.

RPF

.  .  .

## RICHARD P. FEYNMAN TO LUCILLE FEYNMAN, OCTOBER 27, 1943

*The project managers hoped to keep the exact location of the project a secret, and so no mailing address was given to the Los Alamos compound. Arline promptly ordered stationary with the cryptic Santa Fe post office box address used by everyone at the site.*

> DR. RICHARD P. FEYNMAN
> P.O.B. 1663
> SANTA FE, NEW MEXICO

Dear Mom,

I got a package from you of what, according to your letter, were brownies. I haven't opened them yet, but since the box is small I have become stingy and will open them in my room, rather than in my office. Thank you but I don't deserve them for I haven't written in such a long time.

You, who are a dietician, might be interested in the following story. In "Science" (the publication of the American Association for the Advancement thereof) was an article saying the Russians discovered that pine needles are a good source of ascorbic acid (Vitamin C—anti-scurvy). Well thereafter followed a whole series of articles to prove that the Russians weren't first. One of them said that in 1563 some Frenchman had his army all dying of scurvy, and asked the Indians about it and they told him to make a tea of pine needles and the men did that, using up a whole tree for the army and they all got better. Some years later another army had trouble but the Indians had all moved away and another tribe had moved in and they couldn't figure out which tree it was. Another article told about some guy back in 17 hundred and something who told sailors to take pine needles on ocean trips and there were stories about guys who more recently (18 something) were scientific about it, making the tea, describing and tasting it. All in all the Russians were late with their pine needle tea.

So, of course, I had to make pine needle tea. I did while visiting Putzie Sunday, not that either of us were short on Vitamin C (ascorbic acid) but my curiosity was aroused. Result—fair. Not too wonderful, not too bad. About as good as regular tea, but different. Crush the pine needles up—pour on boiling water and allow to steep a while. Serve hot, or with ice and lemon. Cheap. Remarkable fact is it tastes like pine needles.

R.P.F.

## ARLINE FEYNMAN TO RICHARD P. FEYNMAN, NOVEMBER 23, 1943

Tuesday nite

My darling husband it was so wonderful having you walk in today—it will always be that way dear—I'm glad you were a little lonesome for me and Snuggle*—it makes being together seem sweeter somehow—I felt lonesome for you too—and then when you arrived, it was heaven—and I was floating on a cloud—but really it was just your arms around me and you talking to me. I love you so very much sweetheart—I guess we share the same feeling so I don't have to try to describe it—I feel it most powerfully when we're lying close, resting, with my head on your shoulder—it makes me cry (like it did today). It seems too good—too much happiness you're such a good, kind husband—you're patient and understanding and so very good to dopey me—I'm smiling now and two big tears escaped—so you know I'm happy—I could rave about you endlessly dear, everything about you seems extra-special and nice to me—your legs are strong and muscular and you're tall and you can reach up and open that top window (above the door) without standing on a chair—you just reach up easily—then there is the baby talk you use sometimes when you love me—I like it all, the strong things and the silly ones—I like you to be a little lonesome for me and I like it too that my illness doesn't depress you—you're strong dear and you make me strong too—from all sides you're wonderful, my dearest husband. I like you strong dear, but it does feel nice to have you lean on me just a little (you ask me to take care of you and see that you go to the Dr. etc.) I feel necessary and I like to be able to help you—don't be too strong or too independent—miss me a little—once in a while—I love it. I adore you—

Your wife always

Putzie

## RICHARD P. FEYNMAN TO LUCILLE FEYNMAN, DECEMBER 10, 1943

Dear Mom,

We have just had a rather heavy snow around here. Everything looks swell in the snow. There are white mountains all around—to the west they are small mountains 8 or 10 miles away—to the east there are big moun-

---

* A toy stuffed elephant, pictured on page 28.

*Arline, Richard, and Snugglebun with their first Christmas tree,*
*December 1943.*

tains (e.g., Truchas, which we tried to climb) over 30 miles away. Last night
in the moonlight things looked good. Clouds were kind of low and just
skimmed the top of the west mountains, and you could see the east mountains sticking up thru them—you could see in places mountains both above
and below the clouds. Everything was lit by diffuse moonlight coming thru
the clouds. Here and there nearby there were spots lit up by houses or some
hidden street light. It looked much better than a X-mas card. See I'm getting an aesthetic sense.

The town council elections are coming up soon again. I hope I will
avoid—and will try to avoid—being reelected. It was a good experience,
etc. while it lasted but it takes a lot of time and trouble and I think it ought
to be somebody else's turn to do the job.

Putzie is preparing for X-mas and has all kind of things—a great big car-
ton full of stuff of all kinds. Tree decorations, presents, etc. I don't know
where she gets all the stuff—I don't give her a lot of money—poor girl.
According to plan, at least, we'll have a X-mas tree all decked out with
lights and stuff. We have a good time on X-mas.

I have to do shopping this weekend—according to orders from the boss.
Please don't give my address to anybody. There are Captains in the Army

who live up here who don't know what we are doing (even Majors). I'm not supposed to see anyone. I'd like to see Stappler when he comes and I'll try hard as hell to get permission to see him—and I might be successful;—but, don't give people my address because I might have to disappoint them.

LOVE,

RICH.

## RICHARD P. FEYNMAN TO LUCILLE FEYNMAN, FEBRUARY 7, 1944

Monday

Dear Mom,

I have instituted a new system. I was skidding in that Thursday routine so now I'll make it Monday. On Monday between supper and the town council meeting I have almost an hour (which this week is half eaten into already by business) which I will use to write in.

You asked twice about Arline. I didn't answer last time because I was waiting to see the Doc. I missed him Sunday and I won't wait again. The situation is:

1. Her coughing has reduced markedly—about 3 times a day—so she is being taken off codeine.

2. She is gaining weight—not however, uniformly, but about 1# or 1 1/2 a week on the average—some weeks none—others 2, etc.

3. She is feeling generally well—eats fair to good—digests everything O.K. etc.

4. Her sedimentation rate (a high one indicates infection) which used to be 18 in N.J. is 23 now. (The scale runs to 28, 7 is normal).

5. The sputum test (a high one means lots of germs) which used to be X is now I (the scale runs I to X—less than I means no germs—so called negative).

6. No x-rays have been taken recently.

7. Temperature has been essentially normal for a month. I don't know about pulse.

Except for 4. everything is better. Test 5 is unreliable however, it will have to be taken over, taking samples all during the day instead of just once. I will talk it over with Doc. to see what he says. Putzie quotes him as not understanding 4 but not taking it seriously, and generally thinking she has improved. I'll see him Sunday.

*Feynman, 1944.*

Mrs. G.* being here may have had something to do with it but is probably not the whole story because the improvement started somewhat before Mrs. G.'s arrival, and furthermore last time she was here Putzie was getting worse and was very bad when she left. Perhaps the gain in weight was due to her cooking good things to eat—I believe that the gain will continue.

T.B. has its ups and downs, however, and one needn't get discouraged when it gets bad—nor should one necessarily be encouraged when it gets better. Patience. (You see you always envied me when I didn't get all sorrowful and despondent about bad things—but on the other hand I don't get the fun of leaping around for joy when things seem to be turning out good. Don't get me wrong—I'm not sad, I'm just not wildly happy. I'll get that way, however, if this keeps up and looks permanent.)

Oh say, I'm sorry but it is discouraged that people come to see other people here. I got special permission to go to Albuquerque while Mrs. G. was there. I'd have trouble to see you if you were to come because it couldn't readily be argued that you came to see Arline primarily and it was none of their business. You can, of course, do that, but I don't think I could get to see you. It's kind of tough and I miss you all. We'll have to wait for the end of "hostilities."

So long,

RPF

## RICHARD P. FEYNMAN TO LUCILLE FEYNMAN, FEBRUARY 29, 1944

Dear Mom,

Don't feel so bad about your typing. It is OK and getting better all the time. Certain errors ought to be watched out for, like bixxard instead of buzzard. You almost shocked me for a while. It is pretty good now tho. In a little while you should start the practice of correcting each error where you make it, with an eraser, etc. after you get a little better. It slows one up a lot, but it is the only way to get a finished job.

I haven't worked on Pop's codes for the last week so there is no news on that at all.

Instead, I gave, last Thursday, a little talk. We have a math club up here which meets once in 2 weeks. I had to give the second talk in the series. My subject was, "Some interesting properties of numbers." It was all arithmetic—

---

* Mrs. Greenbaum, Arline's mother.

*The Feynman family, 1940s.*
*From left to right: Richard, Joan, Melville, and Lucille.*

nothing harder—just arithmetic. I had some nerve showing just arithmetic to all the mighty minds around here—but I swear that was all it was. Unfortunately I didn't have any problems whose answer was 7 oranges so you might not have enjoyed it. But Moose and Pop would have had a swell time.

Well, all the mighty minds were mighty impressed with my little feats of arithmetic. The marvelous things, apparently without explanation which I demonstrated numbers would do before their very eyes. Then, zip,—I'd reveal the explanation and all would be clear—they should have known it all the time—of course. I went pretty fast and didn't give them a hell of a lot of time to work out the reason for one fact before I was showing them another—still more amazing. I had a really wonderful time giving it. It was similar to a set of first lectures in algebra which I had conceived of once as the best way to teach that subject—start with arithmetic and work up. People told me later that they had a swell time. They'd come up to me in the halls with proofs of various things I had showed the night before, etc. Well sorry to bore you with all this but it still isn't seven oranges. Try this one; if oranges in one store are 2 for 5¢, and in another store are 3¢ each, and I buy some oranges in each store, and come out to have spent just 19¢ for oranges, how many did I buy? Right you are! (I won't ask you how many you bought in each store unless you ask Pop to help—then for Joan we ask, what is the largest number of ¢ that I could have said instead of 19 so that you could still be sure of how many were bought?) (No half-cents you know).

That ought to hold the family for 3 minutes while I tell them I love them all.

RPF


## RICHARD FEYNMAN TO ARLINE FEYNMAN, DATE UNKNOWN

*While at Los Alamos, Feynman became an expert on the safety of the facilities where active materials were stored, handled, and transported. He traveled to the Oak Ridge National Labratory in April 1944 for the first airplane flight in his life, with hidden documents strapped to his back, to report on the plant's safety standards. This letter is from a subsequent trip.*

Dearest Putzie:

I have, naturally, been somewhat busy—but not as busy as I have been the other times I came out here. They didn't have things planned out very well and I have to stay an extra day to do the work of 1 1/2 days so I am here 3 days.

I hope you haven't been too miserable while I'm away. Maybe your visitors were able to cheer you up somewhat. I love you and thought of you a lot on the plane trip—when I had time to think. I used to sit in the airport waiting for a connection, smiling to myself and thinking—Putzie is good.

To begin with, the airport forgot to call me at the hotel the way they promised—and so I missed the plane—and got re-routed to another plane route. We got to St. Louis and then I had to stay there a while so I went into town. I had forgotten what a big city was like. Cars, buses, big buildings—and too much noise. Not too pleasant. I went to the movies (and saw Dagwood and Blondie). I also ate supper in a very swanky restaurant—but it was hot so I couldn't take it and I took off my coat and vest and ate in my shirtsleeves. But nobody complained so it was O.K. I'm not used to all these "comforts of civilization."

I have seen the Stevensons and was to their house night before last—I am also going tonight.

I have a meeting this morning soon and I had better prepare what I am going to say at it.

To you, darling, I am going to say I love you.

I love you,

R.P.Feynman

## ARLINE FEYNMAN TO RICHARD P. FEYNMAN, AUGUST 22, 1944

Tuesday
RICHARD DARLING
I LOVE YOU
PUTZIE

Darling it would be better if you were twins—I think and I know, I need you too—I guess I've gone off the deep end again—in the morning I'll have an intravenous, among other things—I feel phooey dear and only you can change that—can you come darling—if it *doesn't* interfere with your work—I love you dear
   Your wife
   Putzie

It's good to know that even if the medicine fails, there is always your smile, and your hand—it's effective dearest.

## DRAFT OF LETTER FROM RICHARD P. FEYNMAN TO RICHARD GUBNER, M.D., DATE UNKNOWN

*In late August 1944, a doctor at Equitable Life Assurance Society of the United States wrote to Richard about his expressed interest in ongoing studies on the use of sulfabenamide in the treatment of tuberculosis. "The studies are in a very preliminary stage," he concluded, "and we have no assurance that the drug is of definite value." A subsequent letter expressed regret that he was unable to forward any of the drug to individuals, though he did assure them that it was non-toxic, and suggested Richard and Arline "wait for a period of two months."*

You wrote us recently to tell us that new information on sulfabenamide would be available in a few months—but that if our case seemed urgent enough there was a small possibility of the release of some of the drug. We were very encouraged by your interest and appreciate it very much.
   Since we didn't know exactly with what demands for the drug we would interfere, it was difficult to judge whether the urgency of the case was at all great enough to warrant asking for the drug at this time. After some discussion, we realized that you were the one in the best position to

make this judgement if we sent you the details about this case. So the doctor (William H. Thearle) is sending you very shortly the results of recent tests and other facts about my wife's case. Could you decide for us the reasonable thing to do? If you want additional information let us know.

We realize that we are asking you to do something that we should have done. Could we assume tho, that we are just asking you for advice, so that we would continue to assume the responsibility for having followed your advice? In that way you should feel as little uncomfortable as possible. If, for example, you suggest that we wait for a few months, that the drug shows promise at the end of that time, but that by waiting it has come too late to help us.

We do not want to interfere to any extent with your experiments, either by cutting into the drug supply, or by taking up your time now with these matters. The experiments are certainly more important than any individual case. We are grateful, therefore, for all the time you have been able to give us.

Sincerely,
Richard P. Feynman

*A supply of sulfabenamide was forwarded to the doctor supervising Arline's case in December 1944.*

## ARLINE FEYNMAN TO RICHARD P. FEYNMAN, JANUARY 31, 1945

My dearest darling I love you—you became such a part of my life in the last few days, I feel lost without you. I'm happy tho', thinking about you and knowing you'll be here soon. We'll be close together again, talking and reading and sharing lots of jokes about being married and its tribulations (like Nelson!). Snuggle is showing him the town—which gives me extra time. Darling we have such fun living together, being a man and his wife—it becomes more wonderful daily—and I mean wonder-full—it's glorious and deep and lasting emotion—I'm very much in love with Richard, the man, the husband, the lover, the father, the scientist, the dope (if you'll pardon the expression). I love you Richard—you reach into me, you fill every part of my mind and body. I didn't mean dope dear, I meant the foolish, fun-loving you (it was a poor choice of words). I'm in love with your honesty, your clean-cut reasoning, your straightforwardness and your strength—your belief in us and our enduring love.

We'll never have enough time to use up all the love we have.
Your wife and girlfriend
Putzie

P.S. Hope you got back O.K. Monday—it was cold at night—how about those long shorts?

## RICHARD P. FEYNMAN TO ARLINE FEYNMAN, DATE UNKNOWN

Thurs Morn

Dear Putzie:
I got your letter and I'll see you this weekend. I'll mail your letter to Lola too.
Last night I awoke at 3:45 AM for some screwy reason and couldn't go to sleep again so I washed socks. There were a million of them. It took two hours. And then I took a shower and went back to bed (at 6) and awoke this morning at 8.
The rest of the laundry I'll do some other time—they have washing machines here that cost 25¢ per hr. to use and maybe I will do that. Ironing shirts requires learning I guess.
How's everything?
I love you, darling.
RPF

## RICHARD P. FEYNMAN TO ARLINE FEYNMAN, FEBRUARY 1945

Tues. night

Hello Sweetheart:
I love you. It feels good to write to you. I have a problem which I can't handle and I'd like to discuss it with you. Maybe you would know what to do.
My wife and I thought that it might be a good idea to move her from Albuquerque to the site. Everything was OK at Albuquerque but we thought it would be better because we could see each other every day instead of once a week and she could meet the people I work with, etc. and all in all we would have more of a life together.

It didn't work out well however. She has been here 2 1/2 days and has cried almost all that time. She is unhappy because of a large number of things. For instance, when she coughs and needs an injection for it, it is some time after she rings the buzzer before the nurse comes—and even then there is an argument whether four hours are up since the last one and whether she really needs it. She is worried, and I think not too unjustly, that she will cough too much some time and ring for the nurse and she'll choke before the 15 or 20 min. it sometimes takes the nurse to come. Again, she would like certain changes in diet (although she hasn't talked this over with the dietician yet). The first day she was very unhappy about the visiting hours, because it appeared that I might not be able to see her all day Sunday. This has been fixed since. Other things that bother her are the noise from the nearby children's ward, the fact that her door is closed always, the oxygen tank runs out too quick, etc. Her general complaint is that the nurses don't know much about caring for TB and I have the general feeling although I may be all wet that she also doesn't like the general harshness of sticking to the rules attitudes which the army nurses have and is used to a more gentle view as taken by the more elderly and less busy (I think) nurse at her old sanatorium. (For instance, the nurses at the other san would argue, but more gently, (I guess) as to whether she ought to get a hypo—and they in this way tried to help her cut down the amount she took. She took of course no offense to this.)

Now in view of all these things she is quite unhappy, and wants to return immediately to Albuquerque. She thinks of all kinds of ways to speed up the process (she can't go immediately because room is not available presumably), as for example taking a room with less than optimum nursing care and having her mother come from New York to take care of her. She is very upset and equally insistent that something be done to move her right away. She wants immediate action.

In my eyes, however, most of the major difficulties might easily be overcome—just as the visiting hours was—and many of the others are so minor that they only require getting accustomed to. The reason I hesitate to move back to Albuquerque as soon as we possibly can is that I think there is a great deal to be gained in our mutual happiness by having us together if she could become acclimated to the changes, and I think that in fact in time she *would* be able to take an easier going less fault finding attitude—and a much more cooperative attitude toward the people which are trying to take care of her.

This last point is where she entirely disagrees with me and takes the atti-

tude that things here are hopelessly awry and incurably bad from the point of view of her being well taken care of.

Now the obvious (to my mind) solution to all this is to take it easy have patience and just wait and see whether things can be made to get better, and to wait and see if she really could get used to new things and be really happy here, or whether indeed, they are hopelessly awry.

But she thinks it a forgone conclusion that she can never possibly be happy here and that's that. But she is almost hysterically upset and picks on any difference no matter how slight to show the failings of this place. She is actually unreasonable, I think, in her criticisms and wild in her plans of what to do. (E.g., leave to go home to her mom—etc., etc.—things (I hope) she would never really do).

Should I give up trying to do what I think is the reasonable thing (i.e., take a calmer attitude of wait and see—2 1/2 days is a short time to make adjustments) and go hog-wild into trying to move her back to Albuquerque immediately as she wants now?

Put it one more way; she says she is weak and can't take the changes. She says she is weak and can't argue the point with me. She says she is weak and can't tell, for example, the dietician what she wants—it is too much effort. So weak, therefore, that there is no hope of solving the problems and she wants to go back.

But she has been weak before—and gradually became stronger as we talked things over. This time she says she is so weak physically and mentally that she won't get strong enough to deal with the problems sensibly.

All I want to wait for is for her to get and feel strong enough to explain her needs patiently as many times as necessary to those who take care of her—and to feel strong enough to explain the problem in a reasonable way to me so that I'll see that it is *reasonable* to return to Albuquerque.

I think I can *in time* get her strong enough mentally to face the problems in a reasonable way, and weigh the pros and cons and explain to me how it lives up.

On the other hand, if I fail, this time is all time during which she is unhappy—miserable would be more accurate—and why stall around with her unhappy just because I overestimate her potential strength and powers of adjustment?

Should I be the strong one in the family and try to pull her up to me and then do what looks best to both of us when we are both strong (which may of course be to go to Albuquerque) or should I give in to tears and weakness and do what appears as hysterically necessary to us if we are both

weak? The question I want to answer first is not whether we go back to Albuquerque or not, but rather in what frame of mind we should be when we decide to go back or not.

I honestly don't know—but I hope you could give me some advice. You see the problem from the other side. The real trouble is the urgency, because if I hesitate to think about it she remains, perhaps, unhappy that much longer. Yet all I'm asking for is time.

What do you think? I love you and respect your opinions. I love you.

R.P.F.

## RICHARD P. FEYNMAN TO ARLINE FEYNMAN, FEBRUARY 28, 1945

*After an unsuccessful attempt to relocate to the base at Los Alamos, Arline returned to the Albuquerque sanatorium.*

MRS. RICHARD P. FEYNMAN
BOX 1663
SANTA FE, NEW MEXICO

Hello Putzie:

Do you mind if I use my wife's stationary? It came today—and I think you said I should use it.

Today, I didn't get any mail so I have nothing to send to you but my love.

I went to the dentist today for a check up. I got my teeth cleaned—but my gums hurt a little now. I have a couple of cavities. My next appointment is for May 26! They are, apparently, quite busy. Should I maybe go to a dentist in Albuquerque? Don't go and make me an appointment now.

How is everything down there? I'm anxious to know whether you got over the excitement of the trips and stay here in a reasonable time. I hope you are feeling better, certainly safer.

Tonight we had the usual meeting, which I went to. But now I want to go to bed early in accordance with our plan.

It is only 10 o'clock so I can get 9 1/2 hours sleep.

I have several books and pictures and stuff of yours which I ought to bring down. I'm not sure but I think the six crates and ten cartons went off today. Pleasant dreams to you when you get it and each box has to be lifted over your bed to get them on the porch. Wow, what a mess. It might be eas-

ier to move your bed out to one side on the porch and then move it back when all the boxes are put in place. Poor Putzie.

I love you, sweetheart—even with the six crates and ten cartons.

I love you.

RPF

## RICHARD P. FEYNMAN TO ARLINE FEYNMAN, MARCH 2, 1945

Dearest Putzie:

I got two nice letters from you yesterday. One was a poem. Pretty good. Don't try to write if you feel so bad. I'll be down to see you Saturday. It is good you can always rely on your mom to take care of you if all else fails. Then you'll have someone day and night. Pretty good.

Whatever you worry about these days, don't worry about the expense. It really doesn't amount to much and we have saved lots of dough just for needs like this.

I was up till 5 AM last night (or this morning) working out a way to simplify operations and make them faster on the machines. I think I've worked out a real improvement. It was a lot of fun too. I expected to sleep real late this morning to compensate but instead I woke up at 9:30 and couldn't go back to sleep. Now it is 10 AM (and I love you).

I thought you'd feel too exhausted to write so I called Mrs. Masterson last night and told her that you were all set and thanked her very much, etc.

I have a few ideas I just got this morning that I'd like to try out, so I guess I'll be going to work.

So long, sweetheart. Everything will probably come out O.K. I love you. I'll see you tomorrow. I love you.

Husband

## RICHARD P. FEYNMAN TO ARLINE FEYNMAN, MARCH 5, 1945

Monday, 3:30 PM

Dearest Putzie:

You aren't going to like this too well—but I was up all night last night working on stuff. I got to bed at 8AM. I slept until now.

I think I will take it easy till suppertime and take a walk into the mountains for two or three hours. It will be fun. I haven't gone walking in the daytime for a long time. It is a little cold and there was snow on the ground when I went to bed, but it might be fun anyhow. I'll put on some extra warm socks and another sweater. Gee it is a good idea.

I met Thoma on the bus going back from Albuquerque. Thoma is the one who went with me and two girls that time two years ago—for a walk. Before his wife was up here.

His wife took their kid Teaberry and herself home around Christmas time for a vacation. While she was away he got a court order saying she was getting a divorce. He was completely surprised and bewildered, and still is. He still (although he spoke to her) doesn't understand why she left him. He knows it has absolutely nothing to do with other women. It has something to do with his being so busy and working so hard that she thought he wasn't a good enough father toward the kid—he thinks maybe.

I think that if a woman wants to leave a man after he has taken care of her for so long she should at least have the courtesy to tell him about it herself and talk it over as the trouble arises (instead of shocking him with a court order out of a blue sky etc.), but even more particularly it is her duty to see to it that the man understands clearly the reasons for her separating.

Never fear my love. I'm still trusting you to love me forever—and I know I will love you forever too.

R.P.F.

## RICHARD P. FEYNMAN TO ARLINE FEYNMAN, MARCH 7, 1945

Dearest Putzie:

I had a nice walk yesterday into the mountains. There are lots of places there to go and I think I will try to go more often.

Because I woke up so late yesterday I didn't want to go to bed too early last night—so I went to sleep at 3 AM and woke up at 11 AM today. So I lazed around until 12 and ate lunch and now I am writing to you.

I walked up to the base of the hills and said, "I'll climb this one," but when I got close I found I was separated from it by a big deep canyon so I kept going uphill along one side of the canyon until I got to the top of a ridge neighboring the hill I had wanted to climb. There the canyon disappeared and I could have made the intended hill—except my alarm clock—which I had taken with me to tell time said I was up 1 1/4 hours and so I'd

better go back if I wanted to get in in time for supper. So I went back down.

I had a piece of iron pipe about 5 feet long for a walking stick—so now my hand is stiff and it is hard to hold the pencil.

There was snow on the hills and I used the stick to prod it so I didn't go over any that was too deep. Actually none of it was too deep.

I worked after supper.

And how are you, darling?

I love you.

R.P.F.

## RICHARD P. FEYNMAN TO ARLINE FEYNMAN, MARCH 8, 1945

Hello Darling,

I'm glad you are feeling better. I got two letters from you yesterday. If you want to move into a bigger room, go ahead—it might be nicer for you. Only it better have a porch if I know you.

I took a shower this morning (I went to bed 11:30 and woke up 9 or 10 and lazed around). The shower has tin walls so I had a wonderful time making noise.

Now it is nearly 12 o'clock and I'm going to eat lunch soon. I sure am taking it easy, taking the morning off.

Last night I went for a walk—and discouraged the guard that guards the gate at night. I showed him an open gate in the fence not 50 yards from where he was checking passes. The gate was wide open with a road leading to and away from it and a car could easily have been driven through. So they closed it. Then I walked further and found (1) another big gate across a road wide open so a car could go thru, (2) a doorway cut in the fence so people could walk thru. I went thru the second (and passed a man going thru the other way) and came around to the guard again from the inside and he was very surprised. I explained it to him and the officer in charge—but I bet they don't do anything.

It is crazy anyway. They only guard that gate in the nighttime—so only nocturnal spies are kept out. If the army needs manpower they might as well take that guard off his post.

So long, sweetheart. I'll see you soon. I love you.

RPF

## RICHARD P. FEYNMAN TO ARLINE FEYNMAN, MARCH 10, 1945

Hello Putzie:

I suddenly realize that I have to figure out my income tax by next week. I better start soon. I think I'll bring what I can Saturday and fill out our monthly expense charts, and then I can go back and work on it Monday Night. It isn't as hard as everyone says, and it is fun because you can figure that every idea you get has just saved you so and so much money.

A girl, named Jane, from the hospital here called me to find out how you were, etc. She said she is coming this weekend, to Albuquerque. She is going to visit you either Saturday night or Sunday.

I didn't get to bed until 3 last night again, but I didn't wake up until 10:30 so I guess that is O.K., or isn't it.

Today is Friday so I will see you tomorrow.

I am going over to the boss' house (Hans).* I said I would watch their kid Henry, in exchange for reading the Encyclopædia Britannica. Only tonight they have also invited me to supper so really I get double pay for minding the kid.

That's all the news, darling, so I will say so long, and I love you.

R.P.F.


## RICHARD P. FEYNMAN TO ARLINE FEYNMAN, MARCH 13, 1945

*The mathematical symbols that Richard used for his epithet are his, not mine.*

Tues. Night

Dearest Putzie:

Lots of things happened today. First, I found out about the boxes. The x±<= people up here didn't send them out from there yet. Anyway, it has its compensations because I was able to get income tax dope out of the

---

* Parenthetical in the original. The reference is to Hans Bethe, the German physicist and longtime professor at Cornell. His three famous review articles on the state of nuclear physics in the 1930s had placed him at the top of his field. Bethe began his war work at the MIT Radiation Laboratory, and in 1943 Oppenheimer tapped him to head the theoretical division at Los Alamos.

boxes. (1) It was not in the big wooden crate. (2) It was, luckily, in the second crate I tried. (3) Not all of the stuff on finances was in the red and white striped box, as I find now—but only stuff for the last half year. However, I worked on the stuff tonight and found I had nearly all the dope and could figure everything sufficiently close with what I had. You will have to file a return again this year. I will mail it to you. All you have to do is sign it and mail it right out. Your return will be late therefore and so we are subject to some penalty, in principle, but I think since it will be a day or two late at most I don't think it will make any difference and the collector won't say anything. I'll send you all that tomorrow.

Right now I'm working and in between waiting for one machine or another to finish its job—I'm writing to you. Also in between, I am loving you.

The boxes should start out tomorrow.

I am not going home next week. I spoke to the boss and he said almost surely there would be some business in New York that I would be a good guy to send out on. And that means I can kill two birds with one stone and also get my carfare paid. So I'll wait.

I was surprised to hear about your Mom being so sick. She should try to go home as soon as possible. Tell her to take the train. (If she needs dough I'll pay her way.) It is too hard for a sick person to take the bus—and she'd feel better as soon as she went to lower altitude I bet.

So long sweetheart. It is 4 AM Wed. Morn.

I love you.

I didn't write yesterday.

I Love You.

RPF

## RICHARD P. FEYNMAN TO ARLINE FEYNMAN, MARCH 14, 1945

Wed. Night

Dearest Putzie:

Enclosed you will find an envelope. Pull it out, and remove what is in it—your income tax returns). As you remove it in front of you will appear two dotted lines—each marked with a penciled X (bottom of page 1 of return). One is marked (signature of taxpayer)—sign your name there—the other is (date)—put the date you do the signing there. Don't cheat on the

date because the return is late or some more obscure reason. Sign it and date it when you sign. That is all you do. Put it back in the envelope—seal it and mail it. Then sometime this next year you will receive a check for $203.06—just give it to me.

If you want to read the report before you sign, that is more honest. But don't forget to enclose the slip marked WITHHOLDING RECEIPT.

Want the story on our taxes? I had to pay $30.08 and you get a credit of $203.06 so we get a net credit of $173. Last year we had a $263 credit (we got paid $282 because it was 10 months late so they give us interest). However, the total tax last year was only $314 (they withhold dough from your pay—that is where you get the credit). This year it is (mine and yours) $489.43. They held back $662.40 from the salary tho so we make $173 credit.

The boxes went out today. I labeled them and helped pack them on the truck so I know, at least, they were sent on their way to Santa Fe for Santa Fe Trailways to be sent to you collect.

Have you got a nurse yet?

How is your mom?

Should I come to Albuquerque before Sat.?

I love you little darling. I love you.

RPF

P.S. Don't worry about the tax return being late. It cannot possibly amount to more than $6 (for 30 day late) and I doubt they will make any penalty for a few days. NO STRIPES FOR ME—I MEAN YOU. *YOUR RETURN IS THE LATE ONE—MINE IS ON TIME!!* Ha Ha.

## RICHARD P. FEYNMAN TO ARLINE FEYNMAN, MARCH 18, 1945

Dearest Putzie:

I think I've been working a little too hard again this week. I got enough sleep usually but I went to bed at 3 and got up at 11 most of the time. The trouble is that it busts up my schedule so that I am always too tired to write to you and I put it off till morning or something and then after a while a day's gone and you have missed a letter.

I went to the house of a couple that you don't know last night. The guy used to be an instructor at Cornell. The girl used to work in a zoo for a while or something—anyway she likes animals. They have two dogs—one

of which is due to have pups soon. She took the latter to the veterinarian and he inspected the dog and said, "Nice nipple development" at which she looked down modestly and said, "Oh, thank you."—much to the doctor's embarrassment.

There was another fellow there too—so this guy played the UKE while the husband played the clarinet and I banged on the furniture, etc. with various objects. After that we saw pictures (Kodachrome) of Ithaca, N.Y., of New Mexico and of his wife in revealing poses (pictures I said).

Today I got up early for a meeting but something was haywire with the clock and I am 1/2 hour early.

Also today I am going to see you. I love you very much. So long, sweetheart.

R.P.F.

## RICHARD P. FEYNMAN TO ARLINE FEYNMAN, MARCH 22, 1945

Thurs. Morn.

Dearest Putzie:

I got a very good letter from you yesterday. I'm glad I can straighten you out and make you happy so easily. Only I rather wish it would take and last longer—not because I mind talking to you, but because you wouldn't have those periods of unhappiness before I can get to talk to you.

In the letter was a little note—"Things to Remember"—one item of which was emphasized with green ink. Namely "Rest and Relaxation." Altho it was only 3 in the afternoon (Wed.), after reading that I went to bed immediately—and slept for 17 hrs. until 8 this morning (Except for a 1 hour intermission because some louse named Julius was practicing playing the recorder—an infernally popular wooden tube for making noises bearing a one-one correspondence to black dots on a piece of paper—in imitation to music—producing a weird cat-call like effect).

In order to explain this unusual behavior of mine might I say (1) The group leader of the group I am taking care of broke his leg skiing. (2) We had to prepare things to start a new problem. (3) We run 24 hours shifts. (4) Nobody knows what's going on except me and the guy with the busted leg—and I don't know it so good. Hence I was up from 8 AM Tues. morn to 3 PM Wed. night, 31 hrs. and by that time the new problem was running smoothly.

And where were we when I woke up? Exactly where we were when I went to sleep! Somebody made an error 15 min. after I left and we had to go back, correct it, and start over from there. That is why I have to work so hard. I'll have to think of another way of administering things so that I personally don't have to be there to keep things going right.

And I love you, thru it all.

RPF

## RICHARD P. FEYNMAN TO ARLINE FEYNMAN, MARCH 27, 1945

Dearest Putzie:

My upper lid is kind of swollen today so I can't open it fully. I am going to the doc to see what it is at ten this morning. If this writing is cockeyed it is because I see it only thru one eye. I suppose it is just a big sty, but it is so damn big that I want to check up.

Sunday night I went to bed at 11:30 to 7:30 and last night 12 to 8 so I am back at normal hours. I get three meals a day and 8 hours sleep and I'm going to try to keep it up. Last night after supper I went for a short ride in Claus' new car—he just got it.* We went to where some Indian caves were and crawled around in those a while—then it was dark so we came back to work.

Things have calmed down here now for a while, and, except for minor disturbances the fellows seem to be able to take care of all the problems almost by themselves.

I am writing this Tuesday Morning. I love you Tuesday Morning. But that is just a symptom of a far more extensive ailment. I love you always. You are a nice wife. I like to come to visit you—and I wish it were Sat. already—which it nearly is.

How did you make out this week?

I love you, little Putzie.

So long,

RPF

. . . .

---

* Claus, or Klaus, Fuchs was later revealed to be a spy for Russia.

## RICHARD P. FEYNMAN TO ARLINE FEYNMAN, APRIL 3, 1945

Tuesday Morning, 10 AM

Dearest Putzie

You will be interested in two things. First, yesterday I got everything going smoothly and so now I won't be working late hours anymore (I worked till 12 last night), and second, I took a shower. I slept late this morning just for fun and I'm beginning to take a more relaxed view of life (I even read a book for a 1/2 hour before I went to sleep). I think the worst is over and now I can take it easy.

There is a third thing you will be interested in. I love you. You are a strong and beautiful woman. You are not always as strong as other times but it rises and falls like the flow of a mountain stream. I feel I am a reservoir for your strength—without you I would be empty and weak like I was before I knew you—but your moments of strength make me strong and thus I am able to comfort you with your own strength when your stream is low.

I find it much harder these days to write these things to you—there isn't quite the personal intimate contact that I used to get out of letters. I will come Sunday and tell them to you—I will love you Sunday.

No news from here. Oh yes, we have a regular gestapo up here. They took a guy for over an hour in a smoke filled room with men sitting around in the dark—just like in the movies—firing questions at him to prove he was a Communist. They didn't succeed—because he wasn't. The poor guy couldn't work good the next day because they got him out of bed the night before. They claim they are trying to keep spies out of this place. It is dopey, because they leave the gates open at night often by mistake. Don't get scared tho they haven't found out that I am a relativist yet!

I love you sweetheart,

RPF

## RICHARD P. FEYNMAN TO ARLINE FEYNMAN, APRIL 4, 1945

Wed Morn

Hello Putzie:

I did work a little late last night (1 AM) but I slept until 10 this morning.

It was cold as negative hell around here yesterday because it was snowing and windy and etc.

Yesterday afternoon just after lunch a man I hardly knew, living at Fuller Lodge, asked me to pick a lock to a storeroom. It seems the keepers of Fuller Lodge mislaid the key to the gate lock to a storeroom that he had to put his stuff in. So I collected 2 paperclips, one screwdriver, a nail and a few other miscellaneous objects to impress him, and opened the lock in about 2 minutes with one of the paper clips and the screwdriver. He was quite impressed, and happy—as was I because I'm not too good at it and often fail. I used to be much better, but I lost the touch a little I think. The other night I picked a Yale lock to the place where I sleep in the tech area. Altho I had the key I wanted to win a bet. Did I tell you how I broke into a file cabinet to get the contract for the ski-tow association when they were going to have a big meeting and needed it? I was sitting in Julius' room and these two guys come running up the stairs panting and saying, "Oh there he is, thank god!" After that I couldn't help but get the thing out of the locked file for them. I still haven't figured out a way to get into my own steel safe without knowing the combination. That will surely be a triumph if I ever do that.

The key to my interest in all this is probably because I like puzzles so much. Each lock is just like a puzzle you have to open without forcing it. But combination locks have me buffaloed.

You do too, sometimes, but eventually I figure out you. I love you, too.
RPF


## RICHARD P. FEYNMAN TO ARLINE FEYNMAN, APRIL 12, 1945

Thursday Morning

Hello Putzie:

I've got to write to you more often. Give me a little bawling out Sunday. I've been busy these last few days, as usual—but I've gotten enough sleep. The worst was Tues. night when I went to bed at 2:30—but I didn't get up till 12. Last night I went to bed at a reasonable time (11:30). By a great exertion of will-power, I left work at 11:15 just after the graveyard shift came on. Yesterday we found an error which I made and which made us start one of the problems all over again—putting us back to Saturday or so—but, by putting all hands on deck in *3 hours* we

had the problem started over and ready for the machines—and all in all we lost only about 1 day. Except for that debacle I had expected things to be relatively calm the latter half of this week—and I still think they will be. The first thing I'll do when things calm down is go for a walk or something for an hour. I haven't taken over an hour off (Except the hour I drove around with Claus in his new car) for three weeks except the weekends when I visit you. Those are good. They set me up for another week.

I love you, little Putzie. I'm sorry I'm working so hard I don't have time to think about us and write to you like I used to.

I love you, darling. Feeling better?

RPF


## RICHARD P. FEYNMAN TO ARLINE FEYNMAN, APRIL 19, 1945

Thursday Night

Dearest Putzie:

Julius borrowed my clock so I don't know exactly what time it is, but is before 12 I think.

I almost got home at 11:30 but just as I got to the door the fire siren went off so I went off to help with the fire. By the time I got there—which was in just a minute or two—the fire was out. Just a flare up of chemicals that was gotten under control right away. Some guy driving in his pajamas gave me a hitch over there. When I got there my group of fellows who were on shift were milling around wanting to know if they could help—so I took them along too—but there was nothing to do.

I haven't heard from you—but don't write if you don't feel like it. I'll be down Saturday at the usual time of 8:30 or so, unless I miss a bus again. If I can get a good hitch I might even make it sooner—maybe.

Two of the guys on one of the shifts got sick at the same time—damn it. And I want to start a new job tomorrow! The fellow that broke his leg isn't back yet.

Pretty soon I'll be the only one left to work on the job.

Keep working darling—the stakes are high. I'll see you Sat. How's it going? Drink a glass of milk.

I love you, sweetheart.

RPF

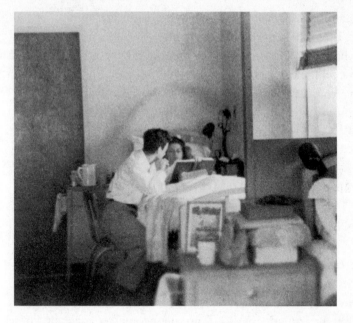

*At the sanatorium, 1945.*

## RICHARD P. FEYNMAN TO ARLINE FEYNMAN, APRIL 21, 1945

Sat. Morn.

Dearest Putzie:

I got your postcard. I'll come and give you strength this afternoon.

I went to bed at 12 last night—but didn't write you a letter because I was too tired. I didn't shave either. I just shaved this morning. The last time I shaved was when I was in Albuquerque—Sunday night. Five days. I'd just as soon quit shaving altogether, but you look so damn dopey and it isn't too comfortable (I mean, I look so damn dopey—you look dopey whether I shave or not—I love you).

I've been working everyday so there is no news. I go to bed OK now—but still I work from 8:30 AM to 11:30 PM with 2 hours off for lunch and supper, or 13 hours a day. I remember my $20/week (or was it a month?) job at the Arnold Hotel which was 11 hrs. one day, 13 the next, alternately. That was harder than this because it wasn't as interesting and the hours weren't voluntary. In 3 days I nearly do a 40 hr. week—in four days over 48, so if I took off all of Saturday and Monday (and Sunday) each week, I'd

still be doing enough to earn my paycheck (in the opinion of those who pay me, I mean—between us, I don't know).

Maybe I'll get a chance today to find out how to go about getting a sensible doc. to look at you.

I love you, little Putzie.

Drink a glass of milk.

RPF

## RICHARD P. FEYNMAN TO ARLINE FEYNMAN, APRIL 24, 1945

Tues. Morn.

Dearest Putzie:

I love you.

There was no excitement on the trip back up here.

Say maybe you got a notice from the income tax collector. Forget it, it's OK—it just means they are transferring records from Camden, N.J. to Albuquerque, New Mexico.

About finances. Per month: in round figures

| | |
|---|---|
| Dr. Therale | 10– |
| Nurses | 300– |
| Room and Oxygen | 200– |
| Cash To Me | 50– |
| My Rent | 40– |
| | 600– |
| | |
| Income | 300 |
| Loss/Month | 300– |

We lose $300 per month, and have over $3300 so we can last 10 months (assuming no extra costs for operations, etc.). Do you think it necessary to sell the ring and piano now? It is up to you. Should I go back to eating at the mess hall—in ten months, it will save $150.

Say, looking at the list of expenses I see something that looks like lack of balance. The doctor is only $10 and all the rest is $500. I think we do not have sufficient medical supervision and are being overcharged for things at the San. I don't know quite what to do about it tho'. What do you think?

I slept OK last night. I'm going to bed regular hours—getting enough sleep, etc. Things have calmed down a lot.

I'll see you soon again.

Drink a glass of milk. Is your temp too high—well, remember and drink it for me later.

I love you.

RPF

## RICHARD P. FEYNMAN TO ARLINE FEYNMAN, APRIL 25, 1945

Wed Morn.

Hello Putzie:

I got to bed a little later than usual—2 PM and slept till 10. I won't do that anymore. I love you—and I see how bad everything would get if on top of all else I were to get sick. I will work hard to stay well from here on in.

I got a postcard from you with the good news that you are trying hard. Keep it up. How about a bottle of milk now? The card also implied that a watch was being sent by you—I haven't gotten it yet. Why didn't you leave it there and I'd work on it Sunday. I don't have much time up here—and fewer tools, no oil.

I just read over the chapter on repair of watches in the book on "Time and Timekeepers." There are more and more elaborate ways of cleaning a watch as the watch gets more expensive. I'd just as soon work only on cheap watches—but I suppose I'll get an expensive one someday. Perhaps I better practice on my good watch which is in a hell of a state because I fooled around with it so much.

I just got it out to look at it. It seems to be OK except that the minute hand is missing. It works OK otherwise tho. I found another dollar watch too which has a bent hairspring. Should I fix that up and give the watch to you? Do you want it?

In the old days watches were so inaccurate that they only had hour hands—the minute hand, if one, would be off so far that it would be useless. I see that with just the hour hand on my watch I can easily tell time to five minutes or less without the minute hand. I think I'll carry it around and try it.

I love you, little one.

I love you.

RPF

## RICHARD P. FEYNMAN TO ARLINE FEYNMAN, MAY 2, 1945

Dearest Putzie:

The prof. under whom I worked at Princeton, John, is coming out here today and I have to meet him at the train. Isn't that nice. I'll get all the news of Jeanette and Tita and Jamey or whatever their names are.

I haven't seen him in a long time—it will be real good to see him again. It is nice that they send me out as the welcoming committee. That is because I was his student, and also because some of the things he wants to know, I know. This is all a result of stuff I told that Colonel the time I had to leave you all of a sudden.

Drink some milk!

You are a nice girl. Every time I think about you, I feel good. It must be love. It sounds like a definition of love. It is love. I love you.

I'll see you in two days.

RPF

## RICHARD P. FEYNMAN TO ARLINE FEYNMAN, MAY 3, 1945

*For a short time Richard and Arline believed she was pregnant. The "Friedman test" refers to a pregnancy test that involves injecting some of the woman's urine into an unmated female rabbit. The ovaries of the rabbit are then examined to determine whether the woman is pregnant.*

Dearest Putzie:

I got your letter about the negative test. I saw the Doc. up here. He says that they can make the test at the Van Atta Laboratories in Albuquerque because (he thinks) they have rabbits there. He says he doesn't take too much stock in the Friedman test (this was the obstetrician here) and has seen it come out wrong either way. He says feeling the growth in the uterus is the best way. Maybe we can get that doc to poke around again in a month.

I'll get the x-ray plates for you. I'll be down Saturday early—around 12:30 if all goes well.

The doc came around special to tell me of a mold growth, streptomycin, which really seems to cure TB in guinea pigs—it has been tried on humans—fair results except it is very dangerous as it plugs up the kidneys or something—and some have nearly been killed by it. He says he thinks they may soon lick that—and if it works it will become available rapidly. I

wonder if Therale could watch the progress of these experiments so if they look good we can know it as soon as possible.

Keep hanging on tho—as I say there is always a chance something will turn up. Nothing is certain. We lead a charmed life.

I love you, sweetheart.

RPF

## RICHARD P. FEYNMAN TO ARLINE FEYNMAN, MAY 3, 1945

JULIUS ASHKIN
P.O.B. 1663
SANTA FE, NEW MEXICO

Arline,

Here's a note from Richard.

Julius Ashkin

Hiya Putzie:

The trouble is I'm in the office and didn't have any stationary so I borrowed this from Julius. It is lucky you sent it to him in the first place. He wrote that at the top so you wouldn't get a nervous breakdown and figure he was in love with you, when I write: I love you. Because I do love you and my name is not Julius (or Nick) (which reminds me the baby better not be born with a mustache, or I'll know who!) but Richard, your loving husband.

I only worked till 11 last night so I got to bed early for a change.

Fred's wife is going to have a baby real soon, she's at the hospital now and he just left to go over there—I just wished him luck saying, "I hope everything comes out all right." He has the cigars, and candy (a custom up here, for frail men who don't smoke cigars, they have candy—I smoke cigars). I get about one free cigar a week up here.

They drained the water out of our pond and have bulldozers and diggers pushing the clay around on the bottom. I don't know what they are doing, but I'll bet anyone (I just bet T.A.) even money it doesn't hold water.

That's all the news, sweetheart.

I love you.

R.P.

*Richard and Arline, 1945.*

## RICHARD P. FEYNMAN TO ARLINE FEYNMAN, MAY 9, 1945

Hello Sweetheart:

I don't think I'll get drunk anymore. I didn't do anything to regret—it just wasn't as much fun as being sober. Last night around 9:30 I was working when the husband of the woman who talks so loud asked me over to their house to celebrate VE day. So I went, and drank more wine than I ever did before. And I got drunker than I ever did before too. I didn't even think I could act sober if I wanted to—I was making a lot of noise (they got me my drum), etc. I don't like it because I know I didn't drum good or make good jokes and I wasn't all there to appreciate other people's jokes—I sort of got "individual" and found it hard to pay attention to other people. We went all over the town singing and beating drums, pots and pans, etc. It sounds like a lot of fun but I know I would have enjoyed it better were I more sober.

Late in the evening I went to sleep and then they woke me up and I went home with Claus.

I don't feel bad this morning, not much hangover or anything. I took a shower and am all fixed up.

I guess every guy has to get pretty drunk once in his life in order to be sure he doesn't like drink.

Same with smoking, after cigarettes, pipe and cigars I gave the whole thing up. I'm getting moraller and moraller as I get older—that's bad.

I think of you a lot—even when I was drunk. I love you very much. I love you. I'll see you soon, darling.

RPF

## RICHARD P. FEYNMAN TO ARLINE FEYNMAN, MAY 10, 1945

*For Richard's birthday on May 11, Arline sent away for fake newspapers announcing his birthday. They were delivered to the base at Los Alamos and widely distributed.*

Dearest Putzie:

I love you.

Last night I ate over at T.A.'s house. We had spaghetti and meat balls. Three big meat balls each. We all got too full and had lots of trouble eating the strawberry shortcake for dessert.

This place is flooded with newspapers—the *Herald* I believe! "Entire nation celebrates birth of R.P. Feynman!" Gee whiz, from all the newspapers it is almost true. They have copies tacked to the wall where I work and people borrowed copies to take home to their wives. Big news. Maybe I'll get some presents out of it—or maybe they'll just give me 27 socks on the back. Anyway, I would have forgotten that my birthday is so close if I hadn't been so forcibly reminded.

I love you, thru it all tho. You are a nice wife.

RPF

## RICHARD P. FEYNMAN TO ARLINE FEYNMAN, MAY 11, 1945

Friday

Dearest Putzie:

Lots of people are wanting copies of that newspaper to show to other people, etc. News sure gets around. I suppose this is my birthday today.

How are you feeling on my birthday?

Hans got a newspaper too and made some comment about you being wonderful. I took it to be sarcasm of course, even if he didn't sound like he meant it that way (I think, not for publication, that you are wonderful—but how your trying to embarrass your husband proves that, I don't know).

I think maybe I'll take a few hours off and try to dream up a scheme for embarrassing you.

Moose wrote me a letter which I haven't answered yet, wondering whether "Mom" knows about the baby—she didn't know whether to mention it in a letter to her "Mom."

Paul said when he was in NY he spoke to a doc. about us—the doc says he thinks there is no trouble at all to an abortion. If Therale is worried about anesthetic, because you need oxygen, make it a spinal anesthetic which won't interfere with breathing at all.

I haven't heard any news yet. Is the Friedman + or –? If + and the doc can feel something next time he comes, do you think we should do it?

Don't worry darling.

I love you, darling.

RPF

*The sudden discussion of abortion must be related to the two letters Arline received from a concerned doctor at the Deborah Sanatorium, where she lived*

*after she and Feynman were first married. The doctor strongly urged her to "interrupt" any pregnancy immediately: "Do not wait one day longer." It turned out she was not expecting. Her periods had ceased, probably due to reasons alluded to in the following letter.*

## RICHARD P. FEYNMAN TO ARLINE FEYNMAN, MAY 15, 1945

Tues Morn.

Hello Sweetheart,

I didn't write you last night because I was busy—but I'll write you this morning. If I mail it by noon it will get out in the same mail—I think. I'll have to find out about that.

I worked till 12:30 last night—because when the new shift came on at 11 only one man came—the other couldn't get thru the gate because his pass expired Sunday. So I did a lot of telephoning here and there, etc. The result of it all being that I got him in an hour later. We usually have 3 men on a shift, but the third one was sick.

I was just thinking last night, about you as usual. You are getting very thin and show all the symptoms of what looks like starvation. Yet I'm sure, that tho you don't eat a great deal, you certainly eat enough that you shouldn't be starving. Why isn't the food being assimilated? Is your intestinal system on the blink, or is it some other difficulty such as lack of air (although I don't see how shortage of air could act that way)? If it is the former, would not a direct infusion of food into the blood be a good idea? How about intravenous feeding of sugars and possibly other substances. It might be worth a try. Ask Therale about this and see what he says. Ask him why you keep losing weight so fast on what you eat. If he says you don't eat enough, then intravenous should help that much more—if you eat as much as you can and take sugar in intravenously too, it's just that much more. Does a test of your blood show that there is sufficient food being carried around by the blood, but the cells don't pick it up—in which case intravenous wouldn't help—or does it show that the blood isn't getting any food from the digestive system—in which case intravenous would do the trick? Where is the difficulty, at the "digestive system to blood transfer" or at the "blood to cells" transfer? Ask Therale.

I love you, darling.

R.P.F.

## RICHARD P. FEYNMAN TO ARLINE FEYNMAN, MAY 17, 1945

Thurs Morn.

Dearest Putzie:

I didn't write you yesterday.

I got a package from home like I said, containing six shirts from Wender and Goldstein. This is a good thing.

I didn't find out how to darn my socks. Some woman in the hospital, who has to stay there quietly a while asked me if I had any buttons to sew on or socks to darn—when she found out I had, however, she seemed very disturbed so I won't bother her. Don't get worried, she's the wife of a friend of mine and is in there because of complications connected with pregnancy. You should stop worrying about me and other women. Everything is under control—and I love you only.

Last night I read the encyclopædia: Tuberculosis, Tuff, Tuleremia, Tunicata, Tumor, Turkey and others between which I couldn't remember. I'd read the TB article before, but there wasn't much in it. Tuff is the kind of volcanic ash we have around here. The next is a disease of rats and rabbits, a man got the first case in Utah in 1913 and it was on the increase after that. Tunicata is a group of very strange microscopic animals that contain a cellulose (like plants), and have the rare element Vanadium in their blood (where we have Iron, insects have Cu, plants have a similar substance with magnesium (chlorophyll)). Tumors you already know about, and Turkey, the country, also—I didn't finish the last article.

I was taking care of Henry, the son of Hans. He has grown a lot, and now he can walk. He is a good guy. Didn't cry the whole time, although he fell on his bottom several times while walking around the house.

That is all the news.

I love you.

RPF

*Feynman would babysit Henry Bethe in exchange for access to the* Encyclopaedia Brittanica, *his favorite book. Decades later, he could recite the subject divisions on the volume covers rapid-fire.*

·   ·   ·   ·

## RICHARD P. FEYNMAN TO ARLINE FEYNMAN, MAY 22, 1945

Dearest Putzie:

I didn't tell you about my trip back here Sunday night. Everything was OK until we got to Española—and there I saw a Ferris wheel and lights and stuff of a carnival so I got out of the bus right there and gave no further thought to how I was going to get back up the hill.

I took a ride on the Ferris wheel and then went on a thing that whirled "chairs" hung by chains around in a circle. They had various things like hoops to throw on nails—baseball tosses etc. to win prizes—which were statues of Christ, etc. I didn't play those because I didn't see much in the prizes.

I gave three little kids rides on a little airplane gadget that they were looking at.

It was a very small carnival, but it was fun.

I then started to hitchhike back—the first car by (after about 1/2 min.) stopped and gave me a ride. In fact, I drove it, because the driver was tired. The car had three girls in it. But they were kind of ugly so I remained faithful without even having the fun of exerting will power to do it.

I love you little darling. I thought of you a lot at the carnival. We used to have fun at carnivals. Get better and we'll do it again.

I love you.

RPF

## RICHARD P. FEYNMAN TO ARLINE FEYNMAN, MAY 23, 1945

Dearest Putzie:

I went to a town council meeting last night, there were lots and lots of people there—and lots of excitement.

What had happened was that they had put restrictions on some of the girls' and men's dorms—they put MP's to guard the girls' dorms and to see that no men were in the girls' rooms at night, and only men with visitors could be taken into the "dayroom" (a sort of common living room in each dorm) and the lights are to be on in this room all during the night, etc. All dopey restrictions. In all dorms (including mine, for instance) a matron was to be around 24 hrs a day—purpose not clear, obviously to see that there wasn't too much noise and etc.

Since these restrictions had to do with the homes of the dorm residents

and since they were put on without notification or discussion—the MP's just appeared—people were rightly sore. They said they could take care of such problems themselves—and some of the dorms organized committees to make rules for quiet etc. for their dorm—a committee to which members who didn't like something could complain.

I got sore along with the other people and finally made a suggestion that the Town Council be the one to review evidence that people in a dorm can't get along together and that they suggest remedies, etc.

So they are going to ask for authority to do this. And they are asking that the matrons be there only during cleaning hours in the daytime. We'll see what comes of it.

They also made up a new one about people on busses needing statements from their foreman to leave here (to check absenteeism) on a weekday. Some freedom.

One guy got up and said if we were prisoners he demanded to know at least which branch of the army captured us.

Quite exciting.

I love you darling. Also I haven't been in a girls' dorm for over a year as far as I can remember.

Love,
RPF


## RICHARD P. FEYNMAN TO ARLINE FEYNMAN, MAY 24, 1945

Thursday Night

Dearest Putzie:

I miss your letters. Maybe you could get Pop or a nurse to write a postal for you once in a while—telling how you are and that you love me, etc.

We had another meeting of Town Council with dormitory representatives, one representative from each dorm. I had been chosen (last night) (at a meeting I didn't attend) as representative of my dorm. We hemmed and hawed a while but finally I got an idea and suggested that we all sign a piece of paper saying we wanted to run our own lives, etc. That is what we finally did and two dozen or so names were signed each by a representative of a different dorm (each dorm had chosen a rep.) saying we could run our own problems and lives, if we got into trouble we wanted the Town Council to be authorized to handle it, and that we thought the guys in the

Mrs. R. P. Feynman

Presbyterian Sanatorium

Albuquerque, New Mexico

housing administration etc. were a bunch of lunkheads! Or words to that effect. It was good. Maybe we'll get what we are after.

Last night I went to eat supper at somebody's house who you don't know too well. A metallurgist also there was the Italian man I admire so much—who I think is so smart. We just talked about junk after supper. The food was good and so was the company. As usual I was the only one without a coat or tie. I'm getting good at that. Now I can't go to anybody's house with a coat because everybody will be insulted to whose house I go without a coat.

I love you, little darling. How are you? I'll see you soon again—just 2 days to Saturday.

I love you.

RPF

## RICHARD P. FEYNMAN TO ARLINE FEYNMAN, MAY 31, 1945

Wed. Morn

Dearest Putzie:

There is a big forest fire quite a way (about 10 miles) north of here. It

has been going for 2 days now. I can see the smoke out of the window now, and the flames at night.

They asked for volunteers yesterday. So I went out to fight the fire. Unfortunately, it was not very well organized so our work was for naught. We made (170 men) a ditch 18 inches wide thru the forest two miles long about. We had to go by brush and fallen trees and stuff. Quite a job. Only they didn't leave patrols to patrol it—but told everyone to come down from the mountains just when the fire was coming up toward the ditch. On the way down I saw 4 places where sparks had started smoldering on the other side of the ditch. If patrols were left they could shovel the sparks back or put them out, etc.

Now they have just asked for volunteers again. But I don't think I'll go again. All yesterday's effort was wasted. Anyway I'm busy because I didn't work yesterday. We were out from 3 PM to 3 AM yesterday.

News: I got a raise. Quite a sizeable one. I used to get $380, minus a lot of taxes and stuff ~ $300. Now I am going to get $450! Minus a lot of taxes and stuff. I don't know what it will amount to exactly after they subtract taxes but I'll let you know as soon as I do. It looks like I did a good job on the new job. Well, now that I got the raise I'll start changing jobs again.

A charmed life as you always say—as our expenses go—so goes our income.

I love you, little sweetheart.

RPF

## RICHARD P. FEYNMAN TO ARLINE FEYNMAN, JUNE 6, 1945

Wed. Night

My Wife:
I am always too slow. I always make you miserable by not understanding soon enough. I understand now. I'll make you happy now.

I understand at last how sick you are. I understand that this is not the time to ask you to make any effort to be less of a bother to others. It is not the time to ask any effort at all from you. It is a time to comfort you as you wish to be comforted, not as I think you should wish to be comforted. It is a time to love you in any way that you wish. Whether it be by not seeing you, or by holding your hand, or whatever.

This time will pass—you will get better. You don't believe it, but I do. So

I will bide my time and yell at you later—now I am your lover, devoted to serving you in your hardest moments. I am your husband, call on me for help—or tell me to go—as you prefer. I will understand everything. I want to comfort you.

I will come this week and if you don't want to bother to see me just tell the nurse. I will understand darling, I will. I will understand everything because I know now that you are too sick to explain anything. I need no explanations. I love you, I adore you, I shall serve you without question, but with understanding.

I am sorry to have failed you, not to have provided the pillar you need to lean upon. Now, I am a man upon whom you can rely, have trust, faith, that I will not make you unhappy any longer when you are so sick. Use me as you will. I am your husband.

I adore a great and patient woman. Forgive me for my slowness to understand. I am your husband. I love you.

*Arline Feynman died on June 16, 1945.*

## RICHARD P. FEYNMAN TO LUCILLE FEYNMAN, AUGUST 9, 1945

*The world's first atomic bomb was detonated in the New Mexico desert on July 16, 1945.*

Dear Mom—

Now I am in Cincinnati, waiting for a plane out. Am I dumb. You must have all gotten a good laugh at the absent-minded guy that thinks his sister goes to school in the summer time. I realized my error about a half hour after I sent Joan the telegram.

There is lots in the newspaper about the atomic bomb now, so I know some things I can tell you about. Remember, I left Saturday night on the plane. I got in Sunday near noon (Albuquerque) and was met by an army car and taken to the site, arriving three o'clock. I went right to the boss's house. His wife had made sandwiches, etc. We were all scheduled to leave on busses at 5 P.M. (so I made it by two hours) to go south, about 100 miles south of Albuquerque because we were to witness an experimental trial of our bomb. It was scheduled for 4 A.M. Monday morning—weather permitting. There were (now I'm flying) three busloads of anxious scientists. Interesting events en route were, first three busloads of anxious scientists

stopped, waited at the side of the road while one especially anxious scientist (not me) got off and went into the bushes for a while. Second— Albuquerque, the largest town in New Mexico was swamped—all drug stores, cafes, etc. were full of the same guys when we stopped for refreshments on the way down, showing how small Alb. really is. We eventually arrived at our vantage point on a ridge overlooking a great bowl of desert at the center of which, 20 miles away, was our gadget. It was mounted on a hundred-foot steel tower but we couldn't see that. We knew where to look because of searchlights, which were shining on it and alternately on the clouds—the weather looked bad.

Dark glasses were distributed (welder's glasses). I looked at my flashlight through them and could hardly see it. Then everyone sat down to eat and wait for 4 A.M. There I appreciated the efforts of the boss's wife. Roast chicken and lemonade and chocolate.

We had two radios—one like a police radio to listen to and talk into a ground station—and one to listen to reports from a plane in the air, which was to fly over, take pictures, drop measuring instruments and see how it felt from the air. I ran the radio to the plane. One of the two radios didn't work—mine. I sweated blood on it. I got reports from some of the men who knew as to what frequency the plane was to transmit. I tuned all around there, changed the position of the antennae, switched every switch on the thing, but no success. Meantime, we had learned that because of the weather, it was to be postponed. All I could get near the right frequency was a San Francisco short wave station playing music. But this enabled me to tune everything as well as possible (there were literally ten dials and no one knew what they were all for—but I found out by experimenting with San Francisco what they all did). Finally I tuned to as accurately as possible where I thought the thing would be as everyone was anxious as hell to find out what was what and they weren't getting much dope from the other set because the guys at the other end were too busy to answer our questions. I went over to the other radio, hoping to ask them to contact the other station to check what frequency the plane was transmitting on. They were too busy, but when I got back, one of the radio electronics experts among the scientists we had there was happily jumping up and down in front of the set and it was saying as clear as you please, O.K. we see your searchlights over. I felt dumb, I suppose those electronics boys know how to tune radios. I asked him how he did it. He said he didn't—he just walked past and the voice came out. They had just started to transmit. The reason I didn't get them before was that they weren't saying anything.

In a few minutes of listening (around 5 A.M.) to them I heard them say "the shot will be at 5:30, it is now minus thirty minutes." Everyone set their watches and crowded around the radio. "Minus 10 minutes"—then "Minus 3 minutes." People scattered over the hill so they wouldn't be in each other's way. They took out their dark glasses. Some even put on suntan oil. A bunch of crazy optimists, I thought. I had helped to figure out how powerful the bomb should be. I knew how many things had to go just right to get a really big blast and I wanted a full solid experience if it did go—so I was going to look at it directly—no dark glasses for me. I did get behind the windshield of the weapons carrier which had the radio on it, just so the ultraviolet light, if any, wouldn't hurt my eyes. I heard a voice of the man at my right—"it ought to be in 15 seconds." I got behind my glass, stared at the spot. Would it go—would everything go right?

I was blinded by a terrific silver white flash—I had to look away. Wherever I looked an enormous purple splotch appeared—it was just as bright when I closed my eyes. "That," said my scientific brain to my befuddled one, "is an after-image caused by looking at a bright light—it is not the bomb you are looking at." So I turned back to look at the bomb. The sky was lit up with a bright yellow light—the earth appeared white. The yellow gradually became darker, turning gradually to orange. In the sky I saw white clouds from above the gadget caused by the sudden expansion following the blast wave—the expansion cools the air and fog-clouds form—we had expected this. Some thing—creates clouds. The orange got deeper, but where the gadget was it was still bright, a bright orange flaming ball-like mass. This started to rise, leaving a column of smoke behind, below looking much like the stem of a mushroom. The orange mass continued to rise, the orange to fade and flicker. A great ball of smoke and flame *three miles across* it was, like a great oil fire billowing and churning, now black smoke, now orange flame. Soon the orange died out and only churning smoke, but this was enveloped in a wonderful purple glow. Another after—image I thought, but on closing my eyes it did disappear and appeared on opening them again. Others said they saw it too, probably caused by ionized air produced in the great heat. Gradually this disappeared, the ball of smoke rising majestically slowly upward, leaving a trail of dust and smoke behind it.

Then suddenly there was a sharp loud crack followed by resounding thunder. "What was that?" cried the man at my left, a war department representative. "That is the thing," I yelled back. He had forgotten that sound takes much longer than light to travel, and what we had seen so far was a silent picture—the sound track for which was one minute and forty sec-

onds late. I knew then that the bomb was a success—big as it appeared at twenty miles, I was still more impressed with the solid sound of the thunder echoing in the hills.

We jumped up and down, we screamed, we ran around slapping each other on the back, shaking hands, congratulating each other, guessing at the energy released—it had worked as well as anyone could have dared to expect. Everything was perfect but the aim—the next one would be aimed for Japan, not New Mexico. (continued on way back in Knoxville airport waiting for plane to Cinn.) We finally got into the buses and started home. We asked one of the bus drivers on the way what his impression of the explosion was. "Well, I don't know—you see I never had an opportunity to see one of these things go off before."

Later pictures and observations showed that an area almost one mile in diameter was covered by a green glasslike glaze formed by melting the sand at the surface. The sand is brown, the glaze is bright green. It is a wonderful sight from the air to see the green area with the crater at the center, in the brown desert.

Well, when we got back I had the fun of telling lots of people about it. The fellows working for me all gathered in the hall with open mouths, while I told them. They were all proud as hell of what they had done. Maybe we can end the war soon. It was too much to hope. We went back to work.

Some expeditions went out to the mountains around Alb. and saw the sky light up so brightly and worried for a moment that we had miscalculated and all the experimenters six miles away were cooked. It was seen in three states—over two hundred miles in all directions. The head of the Alamogordo Air Base had to put out a statement that they accidentally blew up an ammunition dump.

So long.

## RICHARD P. FEYNMAN TO ARLINE FEYNMAN, OCTOBER 17, 1946

*This letter is well worn—much more so than others—and it appears as though he reread it often.*

Thursday, Oct. 17, '46
D'Arline,
I adore you, sweetheart.

I know how much you like to hear that—but I don't only write it because you like it—I write it because it makes me warm all over inside to write it to you.

It is such a terribly long time since I last wrote to you—almost two years but I know you'll excuse me because you understand how I am, stubborn and realistic; and I thought there was no sense to writing.

But now I know my darling wife that it is right to do what I have delayed in doing, and what I have done so much in the past. I want to tell you I love you. I want to love you—I always will love you.

I find it hard to understand in my mind what it means to love you after you are dead—but I still want to comfort and take care of you—and I want you to love me and care for me. I want to have problems to discuss with you—I want to do little projects with you. I never thought until just now that we can do that together. What should we do. We started to learn to make clothes together—or learn Chinese—or getting a movie projector. Can't I do something now. No. I am alone without you and you were the "idea-woman" and general instigator of all our wild adventures.

When you were sick you worried because you could not give me something that you wanted to and thought I needed. You needn't have worried. Just as I told you then there was no real need because I loved you in so many ways so much. And now it is clearly even more true—you can give me nothing now yet I love you so that you stand in my way of loving anyone else—but I want to stand there. You, dead, are so much better than anyone else alive.

I know you will assure me that I am foolish and that you want me to have full happiness and don't want to be in my way. I'll bet that you are surprised that I don't even have a girlfriend (except you, sweetheart) after two years. But you can't help it, darling, nor can I—I don't understand it, for I have met many girls and very nice ones and I don't want to remain alone—but in two or three meetings they all seem ashes. You only are left to me. You are real.

My darling wife, I do adore you.

I love my wife. My wife is dead.

Rich.


P.S. Please excuse my not mailing this—but I don't know your new address.

# 1946–1959

After the war, Feynman decided not to take the position that the University of Wisconsin had been holding for him and instead opted for a post at Cornell. His father died in October 1946, little more than a year after Arline's death. It was a bleak time in Richard's life, and this is reflected in the relatively listless tone of many of the letters written during this period. As he later described his state of mind at the time:

> I had a very strong reaction after the war of a peculiar nature. It may be from the bomb itself, and it may be for some other psychological reason. . . . Already it appeared to me, very early, earlier than to others who were more optimistic, that international relations and the way people were behaving was no different than it had ever been before, and it was just going to turn out the same way as any other thing, and I was sure that it was going therefore to be used very soon. . . . This was before we knew that the Russians were quickly developing one, but there was no doubt in my mind that they could develop one. What one fool can do, another can.*

Professionally, however, he was coming into his own. His attendance of the Shelter Island conference in 1947—Edward Teller, Hans Bethe, Abraham Pais, Isidor Rabi, John von Neumann, John Wheeler, Julian Schwinger, Linus Pauling, Willis Lamb, and Robert Oppenheimer were

---

* From Christopher Sykes, *No Ordinary Genius: The Illustrated Richard Feynman* (New York: W. W. Norton, 1994).

among the twenty-four attendees—helped solidify his position among the leaders in his field. Feynman's papers on quantum electrodynamics, the work that eventually won him the Nobel Prize, also date back to this period.

In 1950 he accepted a position at Caltech—spending his first year on sabbatical in Brazil. Much of the correspondence from the following years revolves around purely professional concerns: asking for help in finding errors in an academic paper and sending his regrets when asked to return to Los Alamos. His efforts and achievements were recognized with the Albert Einstein Award in 1954. The arrival of Murray Gell-Mann at Caltech soon thereafter subsequently generated a fruitful—and now legendary—collaboration and rivalry.

In 1958, my mother, Gweneth Howarth, came to America at my father's urging. Unfortunately, little of the correspondence between them has survived, though it is clear from his letter of May 29 of that year that Gweneth was an adventurous woman—daring enough to move across the Atlantic to be his housekeeper after the briefest of relationships.

· · · ·

## RICHARD P. FEYNMAN TO R. C. GIBBS, OCTOBER 24, 1945

Prof. R. C. Gibbs
Department of Physics
Cornell University
Ithaca, New York

Dear Professor Gibbs:

When I heard, a few weeks ago, that Mr. Bethe had nearly decided to go to Columbia, I was very disturbed and did my best to try to get him to stay at Cornell. The reason I chose to go to Cornell a year ago was that I wanted to go to a school where there was an active experimental group doing research in nuclear physics. Only in this way could I keep abreast of progress by means of the theoretical problems and questions which could arise in connection with the experimental works. At that time, I was looking forward to working with Dr. Bethe and having an experimental group of such men as Bacher, Rossi, Parratt, Greisen and we were thinking of McDaniel and Baker also. If, however, Dr. Bethe did not go to Cornell then Bacher and Greisen would not go (Rossi had already decided not to

return) and I didn't see how we could attract other young men with so little to offer. I decided to come on November first as planned anyhow because the date was so late and you had been counting on it. But I did intend to tell you that I wanted to stay as short a time as possible.

I know many young men here whom we should want at Cornell and have spoken to them about the situation. I have had very little success when it was assigned that neither Dr. Bethe, Bacher nor Greisen would be there. On the other hand, if we assumed that these men would be there and described the program we had in mind, they were very interested. Unfortunately, they also have other offers and are being pressed so that they are impatient with the uncertainty at Cornell. We have already lost one very good electronics man in this way.

So it seems to me there are just two possibilities. Either Dr. Bethe (and therefore, Bacher, Greisen and other young men) go to Cornell, and the department is one of the best in the country, or else the Physics Department will find itself in such a poor state as to be unable to attract the abler of the young physicists who are now being released from war work.

I am, therefore, in favor of anything which will result in the first alternative. This means, I believe, that Professor Bethe would be chairman of the department after your retirement. From the point of view of administration of the department and the control of policy, I think this would be a very good thing. Dr. Bethe has done a wonderful job as leader of the Theoretical Division here. He managed this with remarkable facility. Everyone felt free to work on whatever he wished, yet all the work was coordinated and the job was done. And you can understand that the policy of the entire project very often depended on Theoretical conclusions. There should be absolutely no doubt as to his abilities as an administrator.

On the other hand, it would be unfortunate for physics, indeed, if he were to spend a very large fraction of his time away from research. Therefore, I think he should have a vice-chairman who could take up as much as possible of the purely administrative duties.

I do hope that you can find some such arrangement which is satisfactory to all. It is important that this be done as quickly as possible. I am looking forward to becoming an active member of an active Physics Department.

Sincerely,

R. P. Feynman

. . . .

## RICHARD P. FEYNMAN TO ROBERT OPPENHEIMER, NOVEMBER 5, 1946

Professor J. R. Oppenheimer
Department of Physics
University of California
Berkeley, California

Dear Oppy:
It looks black for my proposed visit to California.
Cornell is all loaded up with graduate students. There are many more than ever before, and we have all we can do to handle them. The depart-

ment was, therefore, very reluctant to let me go for the spring semester, as I was needed to help handle the load.

There is another personal reason. My father has died and I do not want to go too far away from my mother in New York, at least for a while.

I was looking forward very much to my visiting you at Berkeley, as you know. I am sorry to have to disappoint both myself and you. I had been hoping to see many friends again, but that is the way things go.

Sincerely,

R. P. Feynman

## RICHARD P. FEYNMAN TO R. D. RICHTMEYER, APRIL 15, 1947

*After receiving an inquiry as to whether he would be interested in returning to Los Alamos as a consultant for ten weeks during the summer and whether he was interested in attending a nuclear physics conference still in the planning stage, Feynman wrote the following response.*

Mr. R. D. Richtmeyer
Los Alamos Scientific Laboratory
P.O. Box 1663
Santa Fe, New Mexico

Dear Bob:

My plans for the summer are not very definite. I expect to loaf around a great deal and I do not know whether I will be through New Mexico.

However, I do not think that I will have time to do any work at Los Alamos in the near future so I would just as soon not bother about filling out all the blanks for the contract until there is some definite reason for it.

Sincerely yours,

R. P. Feynman

. . . .

*Teaching at Cornell, 1948.*

## RICHARD P. FEYNMAN TO E. O. LAWRENCE, JULY 15, 1947

Professor E. O. Lawrence
Radiation Laboratory
University of California
Berkeley, California

Dear Professor Lawrence:
I have just written to Professor Birge to tell him that I will not be in California next year.

It was really an awfully difficult decision to make but everything seemed to balance out except the weather and the fact that I was already settled at Cornell. Neither of these seemed like a very important consideration so I have had a lot of trouble in finally making up my mind. When I heard that Weisskopf was not going to be at Berkeley, I finally decided that I would stay at Cornell next year.

I don't know how to thank you for the really glorious time I had in California. Probably, it was just the kind of time everyone always has in California but it seemed to me to be especially good. Please wish my best to Serbeis and MacMillans as well as to your wife and kids and thank them

very much for making my stay there so enjoyable. No doubt somehow, we will all see each other again.

Sincerely yours,

R. P. Feynman

## JACK WILLIAMSON TO RICHARD P. FEYNMAN, MARCH 25, 1949

*Jack Williamson, who published his first story in 1928, was in the early stages of a long and illustrious career in science fiction.*

Dear Professor Feynman:

This is to ask your opinion of a notion about the nuclear binding forces—that they may not be actual forces, but instead a consequence of the nature of space-time.

If space-time is an effect of the mass it contains, might it not be composed of ultimate units somehow reflecting the quantum nature of individual particles? And might those minimum units or packets of space-time be of such a nature that the disruptive force of electrostatic repulsion cannot act inside any packet, but only between different packets?

And might the atomic nucleus then be composed of one such unit of space, shaped by the particles it contains?

It would follow that energy is required for the formation of these units of space—a different amount of energy for each number and arrangement of contained particles, the mechanism being such as to account for the packing fraction curve and all the complicated phenomena of nuclear masses and energies.

One helium nucleus, for the simplest example, requires only one unit of space-time in which to exist, so that the formation of a helium nucleus from four hydrogen nuclei liberates the energy of the three surplus packets. (Neglecting the amount of energy involved in altering that one packet to hold four particles instead of one.)

The stability of such a nucleus would not be due, then, to any actual force binding the parts of it together, but instead to the two conditions that, first, the repulsive force cannot act inside the unit of space-time containing it, and, second, the parts cannot separate without acquiring energy to form additional units of space-time to contain the fragments.

The instability, on the other hand, of such massive nuclei as those of radium and uranium would be due to the circumstance that the packets of

space-time containing them have become overfilled, to the extent that energy is available, from that spent in enlarging them, to form separate packets for an alpha particle or the fragments of fission.

If really successful, this idea of course ought to account for such other entities as electrons and mesons. An electron might be an empty or substantially empty packet, therefore representing much less mass and energy—it might be the actual quantum of space-time. I don't, however, see any simple answer to the problem of the meson.

The mathematical relationship of such units of space-time to their sum or effect in the more familiar space-time of the macrocosmic universe— that relationship, it seems possible to me, might help to bridge the gap which now exists between quantum mechanics and the relativistic physics of the macrocosmos.

However, I am not a mathematician—which of course means that I am unable to develop this idea or to assay its value, if any. I am not even sufficiently familiar with the technical literature to be sure that it is new. It does seem to me to have a certain plausibility, as well as the logical advantage of avoiding the need of any separate kind of force to contain nuclear energy.

I should be very grateful for any comment which you might have the time and inclination to make on this suggestion. A stamped envelope is enclosed.

Yours sincerely,

Jack Williamson

## RICHARD P. FEYNMAN TO JACK WILLIAMSON, MAY 30, 1949

Mr. Jack Williamson
Portales, New Mexico

Dear Mr. Williamson:

I was interested in your idea concerning the binding of nuclear particles but I did not find it stated precisely enough to be able to understand it. What I mean is that I do not see how one would explain the mathematical relations of the units of space-time, etc.

As you know, a theory in physics is not useful unless it is able to predict underlined effects which we would not otherwise expect. I do not think that your idea is sufficiently developed to enable that to be possible.

As an example you may explain the instability of massive nuclei as those of radium and uranium due to circumstance "that the packets of space-time

containing them have become overfilled." The question is would you have expected that they become overfilled when we get to nuclei as massive as radium and uranium? Why are they not overfilled for atoms of copper or iron? The question is a quantitative one in order to determine exactly which atoms are massive enough to be unstable.

I hope I do not discourage you too much in thinking about these things but I would suggest that you try to make the ideas as definite and precise as possible.

Sincerely yours,

R. P. Feynman

## RICHARD P. FEYNMAN TO T. A. WELTON, NOVEMBER 16, 1949

Professor T. A. Welton
Department of Physics
University of Pennsylvania
Philadelphia, Pennsylvania

Dear Ted:

Right now I don't feel much like giving a colloquium at all. I am trying to write up some more of my stuff and I would like to have time just to stay in one place and work. On the other hand I would like to see you again some time so I don't know what to say. Why don't you try me again sometime next semester when I am sick and tired of working?

I am enclosing reprints of my papers. I gather from your letter that you did not try to read then because if you had I assure you would find them very simple, at least if you don't try to prove that all the things I say are correct. You know how I work so most of it is just a good guess. All the mathematical proofs were later discoveries that I don't thoroughly understand but the physical ideas I think are very simple. Start with the one about positrons. I wish you luck.

Sincerely yours,

R. P. Feynman

.   .   .   .

## RICHARD P. FEYNMAN TO JULIUS ASHKIN, JUNE 5, 1950

Professor Julius Ashkin
Department of Physics
University of Rochester
Rochester, New York

Dear Ash:

I am sending you a manuscript of my next paper. I was just hoping that you would have time to read it and to discover all the errors as you were so kind to do on the other two papers that I wrote. I and the printer have the only other two copies so it is a very valuable thing. However, if you do not want to study it so carefully you may still have it as a prize for being so good to me last time.

On the other hand, if you have no time to read it at all and are not interested, would you please send it back because Professor Bethe wants a copy for a course he is going to teach this summer and all I can give him is the typist's manuscript.

I would appreciate any comments you have to make on it. I will send you a copy of my next paper from California.

On cleaning my desk I discovered a little note telling me to send a bill for my expenses to the seminar at Rochester. I don't have the courage to send the bill so late so I am sending it to you with a mild hope that I can get $22.00. If, however, it is too late and causes a lot of confusion, please do not worry about it because I am well paid in California. Thanks very much.

Sincerely,
R. P. Feynman


## RICHARD P. FEYNMAN TO M. L. OLIPHANT, DECEMBER 12, 1950

Prof. M. L. Oliphant
Australian National University
Canberra, Australia

Dear Professor Oliphant:
I have received your kind letter and that of Ernest Titterton telling me of

*Feynman, 1950.*

the fine opportunity in Australia. I have thought it over and decided not to accept.

I am very interested in attempts to begin research in other parts of the world. I want to wish you great success. I am spending next year in Brazil in connection with an analogous development. The present concentration of research facilities and scientific universities has obvious dangers.

Sincerely,

R. P. Feynman

## RICHARD P. FEYNMAN TO JERROLD R. ZACHARIAS, JANUARY 18, 1951

Dr. Jerrold R. Zacharias, Director

Laboratory for Nuclear Science and Engineering

Massachusetts Institute of Technology

Cambridge, Massachusetts

Dear Zach:

I am writing this letter to save you a telephone call in answer to your proposition to use my nasty mind with Intelligence.

I have decided against doing this. The reason is I do not believe that my talent is sufficiently unique in this direction. I think my abilities and training as a physicist probably has a more direct use in some other (but as yet unknown) way. I suspect that there are plenty of ingenious people able to deal, as well as I, with the problems that you mention, who are at present employed as anything from sales-manager to criminal.

Thank you for considering me for the job, however. Frankly my position would be stronger if I saw something I could actually do directly with physics. Maybe physics in its own right has some value since the national emergency is not yet complete war.

Sincerely,

Richard P. Feynman

## JOHN WHEELER TO RICHARD P. FEYNMAN, MARCH 29, 1951

Professor Richard Feynman

Norman Bridge Laboratory

Calif. Institute of Tech.

Pasadena, California

Dear Dick:

I know you plan to spend next year in Brazil. I hope world conditions will permit. They may not. My personal rough guess is at least 40 percent chance of war by September, and you undoubtedly have your own probability estimate. You may be doing some thinking about what you will do if the emergency becomes acute. Will you consider the possibility of getting in behind a full scale program of thermonuclear work at Princeton through at least to September 1952?

Los Alamos has asked Princeton to pitch in on this business. I am returning there full time the end of May to push it. Spitzer, Schwartzschild, Ford and Toll are going to give half to full time. Others are going to give part time. Spitzer, Hamilton and I are engaged in active recruiting. The university has made available a large separate building. Los Alamos has prepared a draft contract. John von Neumann, Goldstein, and Richtmeyer will be actively concerned with the work, especially as involves the Princeton MANIAC. The Princeton development far from representing a downgrading of the Los Alamos effort has been requested to help Los Alamos get even more done. I can't discuss feasibility in this letter, nor a number of exciting new ideas which have been churning about here the last weeks. They are along a line rather different from that which Bob Christy plans to follow. Both Edward Teller and I would like to describe them to you in person to see if you don't think it is urgent for the defense of this country that most promising of these schemes be developed as soon as possible.

The following reasons make me think you might wish to give serious consideration to this request for your help:

(1) Already without the benefit of thermonuclear oomph, atomic bombs form a major part of this country's war potential. At peak production during War II we turned out about 4 kilotons a day in conventional high explosives. In the crude and highly arbitrary measure of total energy release this output means one fifth conventional atomic bomb per day, or in 700 days, 140 old fashioned atomic bombs. For comparison take any newspaper guess as to atomic bomb output. Then ask if there is any justification for the hair-shirt philosophy of many nuclear physicists—"Nuclear physics is interesting; therefore we mustn't work on it in case of war; it's better to forget physics and tell the admirals and generals how to do tactical and strategic this-and-that." Clearly there's a lot to be done in such directions. One may even achieve a factor of two gain. But what business has the country's best physicists fooling around with a factor two when factors of five and twenty are at stake? If they do, the

country may feel in return it ought to give atomic weapon development to the generals!

(2) Princeton's job is to be idea factory and do primordial design, Los Alamos to work in these fields, too, but also to carry things through all the practical stages right to the end. The shortage of people on the idea assessment and primordial design end is to me terrifying. You would make percentage-wise more difference there than anywhere else in the national picture.

(3) We intend to get together at Princeton, a group of supercritical size that will really get somewhere.

(4) I'm planning to work full time and I would hope you might consider that, too, owing to the urgency of the international situation. However, there would be an alternative opportunity for as large a fraction of pure academic connection as you might feel appropriate if you don't think the emergency has got to the full time stage yet.

Summarizing—

(1) It would be enormously helpful if you could come now, either to Los Alamos or to Princeton.

(2) If you feel the emergency isn't yet at that stage, but may possibly soon get there, it would be very encouraging to us all if you would say that you will consider seriously pushing the thermonuclear business.

(3) If your answer on (2) is affirmative, would you be willing to fill out the enclosed clearance forms to keep open the degrees of freedom for early participation?

(4) And would you care to ring me collect at Los Alamos 2–2776, or write me, to let me know what your feeling is?

Best wishes,
John Wheeler

## RICHARD P. FEYNMAN TO JOHN WHEELER, APRIL 5, 1951

Professor John Wheeler
P.O. Box 1663
Los Alamos, New Mexico

Dear John:
As you know, I was planning to spend my sabbatical leave in Brazil. I am uncomfortably aware of the very large chance that I will be unable to go.

Until that situation becomes definite, however, I do not wish to make any commitment for work next year.

Best wishes,

Richard P. Feynman

## RICHARD P. FEYNMAN TO LUCILLE FEYNMAN, AUGUST 30, 1954

August 30, 1954

My Dear Mom;

You have nothing. A small room in a hotel. Stuffy and no home with friends and family in it. A job that gives no enlightenment or has no further aim than to be done each day, building nothing for yourself. No easy transportation but to be jostled by the crowds. Nor fancy meals, nor luxurious trips, nor fame nor wealth. Children who rarely write. You have nothing.

So say your friends, but they are wrong. Wealth is not happiness nor is swimming pools and villas. Nor is great work alone reward, or fame. Foreign places visited themselves give nothing. It is only you who bring to the places your heart, or in your great work feeling, or in your large house place. If you do this there is happiness. But your heart can be as easily brought to Samarkand as to the Hudson river. Peace is as difficult to achieve in a large house as in a small one. Feeling can be brought to any work. Your friends of wealth have nothing because of it that they would lose, if with more modest means.

In the sea of material desire that is our country you have found an inlet and a harbor. You are far from perfectly happy, but are as contentful as you can be, with your make-up in the world that is. That is a great achievement, or a great woman.

Why do I write this? Because you have told me these things many times, and I have nodded, vaguely understanding. But you mention them again and again, so perhaps you think I do not understand. For so few understand, each friend questions you, each relative hounds you with the query, how can you live in such a tiny place, how can you work in that unbearable shop with those horrible sales girls? You know how. They could never do it, nor can they live as contentedly in any other way, for they do not possess your inner strength and greatness. A greatness which has come to realize itself thru the knowledge that, beyond poverty, beyond the point that the material needs are reasonably satisfied, only from within is peace.

*Richard and Lucille, mid 1950s.*

I offer you all my resources of wealth. What do you want, what will you take? You can have anything $10,000 could bring you. I have offered many times. Not $10 worth can you think you need that you will let me give you. I bother you no more. I will never say it again, but you must always know that I will give you any material thing of wealth you could desire. Now or in my ability in the future. You have no insecurity. And tho you wrack your brains to think of something—not the smallest item suggests itself to you. No man is rich who is unsatisfied, but who wants nothing possess his heart's desire. No need to concern yourself with friends' attempts to help. You are not forced to live as you do. Your son's offer proves that. It is your choice, your life, your simplicity, your peace and your contentment. It needs no further justification.

And I can offer all I own, even if I were selfishly doing so, for I know you want none of it.

When I offer it, what do you ask? You ask that I write to you. What can I give more easily, and am yet more stingy about? Tho I know your strength now requires nothing for its self-confidence,—tho I know you could live

without my writing by accepting such a fact and living with it,—I do not desire to test your strength or to make your burden more heavy. What son has a mother who in such circumstances asks less of him!

My duty is clear, right action obvious. May I have the strength of resolve that this be the beginning of a more regular correspondence. I hope that the lesson of your strength in life will inspire me more often to try to add a bit you really want. I offer no more fans. If you want them ask. I hope I can write more often to a most deserving and inspiring woman. I love you.

Your Son.

*The lack of correspondence from her son would eventually cease to be an issue. Lucille Feynman moved from New York to Pasadena, California, in 1959 to be closer to Richard.*

. . . .

## RICHARD P. FEYNMAN TO THE DEPARTMENT OF STATE, JANUARY 14, 1955

Department of State
Washington, D.C.

Gentlemen:

Yesterday I received a letter from Mr. Zaroubin, the Soviet Ambassador, inviting me to take part in a scientific conference in Moscow! A copy of the letter is enclosed.

This comes to me as a great surprise and I do not know what to do. Although the letter indicates that the conference is purely scientific, in the present relations of our country to Russia, such a visit by me obviously could have considerable non-scientific repercussions and be of interest to the Department of State. I would be very grateful to you if you could give me any advice. I should like to cooperate with your desires in this matter.

Does the Soviet invitation of foreign scientists represent a partial change of the iron curtain policy? Can we hope that this is a step back to more normal scientific relations with that country? Should we take advantage of such a change to find out what the scientists are thinking about and how healthy the scientific community is in Russia? Is there danger that I would be kept there and not permitted to return? I worked on atomic energy during the war, but not afterward. I am an expert on quantum electro-

*Drumming, mid 1950s.*

dynamics and the theory of elementary particles, scientific fields which have no apparent military application at present, so it is not unreasonable that they would invite me if the invitation is above board. A further possibility is that it is partly as a consequence of a letter I wrote to the scientist Landau, commenting on some reprints he sent me. I enclose a copy of that. There was no answer. Are you aware of any other scientists being invited to this meeting?

I am willing to proceed in any way that seems to you to be in the best interests of the country, even if it should mean some personal danger.

Sincerely,

R. P. Feynman

## RICHARD P. FEYNMAN TO THE ATOMIC ENERGY COMMISSION, JANUARY 14, 1955

Atomic Energy Commission
Washington, D.C.

Gentlemen:

Yesterday I received a letter from the Soviet ambassador inviting me to a scientific conference in Moscow! A copy of his letter is enclosed. I don't know what to do about it, and have written to the State Department for advice.

I thought you would be interested because I was connected to the Los Alamos project during the war, so the danger that I might not be able to return, or the attitude of public opinion must be considered. Any suggestions you could make would be appreciated.

Sincerely yours,

Richard P. Feynman

## TELEGRAM
## RICHARD P. FEYNMAN TO THE DEPARTMENT OF STATE, FEBRUARY 17, 1955

Department of State
Washington, D.C.

Refer to my letter of January 14 asking your advice about invitation I received from Soviet Academy Sciences to visit Moscow for scientific con-

*Classroom at Caltech, 1955.*

ference. Zaroubin today says Soviet Academy will also pay travel expenses from USA to Moscow and back. I should like to answer as soon as possible to leave time for various arrangements. Could you please tell me your sentiments in this matter.

Richard P. Feynman

## RICHARD P. FEYNMAN TO WALTER RUDOLPH, FEBRUARY 24, 1955

Mr. Walter Rudolph
Department of State
Washington, D.C.

Dear Sir:

Over a month ago I wrote a letter to the Department (but not directly to your office) informing them that I was invited to a conference in Moscow on Quantum Electrodynamics and Elementary Particles. The invitation came through the Soviet Ambassador from the Soviet Academy of Science. It seemed to me that it may be of interest to the country and to science to encourage such exchanges and conferences. On the other hand, there might be good reasons, involving the relation of our country to Russia, why this may be a bad idea. I asked the Department for their view. I received no answer. Likewise a night letter sent last week went unanswered.

It is possible that these letters went astray, or were delayed because they were addressed to no one in particular, but to the Department in general. Dr. Koepfli, here, suggested that it would be much better if I wrote directly to you.

There is very little time left until the meeting. The meeting is March 31, and abstracts of papers to be presented should go in on March 1st. Furthermore, I would have to arrange air transportation and my passport and visa. Since I received no answer and some decision had to be made, I decided to accept the invitation. If there are any objections I hope you could register them immediately.

The passport I have is not valid for travel in Russia. Mr. Wallace Atwood, of the National Academy of Sciences, will soon bring it to the Department to get permission for such travel. I hope it doesn't take too long.

I should like to emphasize again that I would like to cooperate fully with the Department. I would be perfectly satisfied with a refusal of per-

mission to go, without feeling that any rights were violated. The situation between the countries is delicate and I am sure that you are far better at judging the repercussions of such a move than am I. On the other hand, if there are no objections I would like to go to the conference.

Could you please acknowledge this letter, even if no decision has yet been made, so I know at least whether this and the others have been received.

Sincerely,

Richard P. Feynman

## RICHARD P. FEYNMAN TO WALLACE W. ATWOOD, JR., FEBRUARY 24, 1955

Mr. Wallace W. Atwood, Jr.
National Academy of Sciences
Washington, D.C.

Dear Mr. Atwood:

Thank you very much for the interest you have shown in my problem of getting to the conference in Moscow. I am especially grateful for your telephone conversation. I still haven't heard from the State Department, however.

You were very kind to offer to see about my passport. I wish there were some other way to do it so that I wouldn't bother you with it, but I am afraid to take the risk of getting it tied up. I called the passport division here in Los Angeles. They just laughed and said, "You can't get a passport to go to Russia." When I insisted they said to write to the Department of State directly.

So, I hope you don't mind that I am asking a favor of you. My passport is enclosed. Could you take it to the Department of State and ask them to O.K. for a trip to Moscow to attend the conference there. I will also write a letter today to them and tell them that, since they have given me no other advice, I am applying to have my passport O.K'd for travel in Russia, and that you will bring it over. I am also writing the Soviet Ambassador, telling him I would like to go to the meeting, but that my passport has not yet been O.K'd for travel in Russia.

If the passport is O.K'd, could you get the visa from the Soviet Embassy and send it back to me.

I feel embarrassed in having to ask these favors of you, when I do not even know you personally. I am grateful to you for your help.

Sincerely,

Richard P. Feynman

## RICHARD P. FEYNMAN TO A. N. NESMEYARROV, FEBRUARY 25, 1955

*A February 16 letter from the Academy of Sciences of the Soviet Union informed Feynman that his expenses to and from Moscow would be paid.*

Mr. A. N. Nesmeyarrov

President

Academy of Sciences of U.S.S.R.

Moscow, Russia

Dear Mr. Nesmeyarrov:

I should like to thank the Academy of Sciences for their kind invitation to attend the conference on Quantum Electrodynamics and Elementary Particles to be held March 31 to April 6. Their generosity in paying for all my traveling expenses makes it financially possible for me to come.

I shall accept the invitation, but there remains one uncertainty. My passport at present is not valid for travel in the USSR and I have applied to the U.S. Department of State for the necessary extension. If it is granted I shall surely come. I am embarrassed to have still to give you such an indefinite reply even after the long delay which I have made in replying to you. I thank you for your patience in this matter.

The invitation asked that I send you, at this time, the thesis of whatever communication I may wish to present. I assume that means a summary, rather than the complete paper. Unfortunately, I have not done much successful original work in the field which is not already published. I enclose a summary of work in a closely related problem. Perhaps you will decide that this is not suitable for the conference because the subject is somewhat afield. Please do not hesitate to tell me if this is the case. I am also preparing two other papers, assuming that one or the other will be of interest. One is on the present situation with regard to the accuracy of the comparison of quantum electrodynamics to experiments, and a discussion of the theoreti-

cal problems remaining to be solved to make this comparison more complete. Another is this: There was recently a conference at Rochester, N.Y. on high energy physics. I could give a summary of this conference, particularly the latest experiments of interest with mesons which have not yet been published. These things are not original work of course, and perhaps others intend to report on them. If there is any desire for it I could give a more general introductory survey talk on the present situation in theoretical physics of elementary particles. Or I could be more specific and describe some incomplete original work on the effects of closed loop diagrams in the pseudoscalar meson theory.

I would appreciate it if you could tell me which of these subjects would fit best into the plan of the conference so that I could concentrate my attention on preparing them.

I have done some work on the theory of liquid helium which may be of interest particularly to Professor Landau who has done so much work on this problem. I am looking forward to discussing it informally with Professor Landau. But if there is any opportunity, outside the conference, at which he would like me to give a more formal talk on this subject, I shall be glad to.

Thank you again for the kindness and generosity of your invitation. I hope I can come, but in any event, I am sure it will be a very successful conference.

Sincerely,

R. P. Feynman

## K. D. NICHOLS TO RICHARD P. FEYNMAN, FEBRUARY 28, 1955

Dear Professor Feynman:

This is in reply to your letter dated January 14, 1955, which requested advice concerning the invitation received by you from the Russian Embassy to take part in the Conference which will be held by the Academy of Sciences in Moscow from March 31 to April 6, 1955.

In our review of foreign travel on the part of employees or former employees of the United States atomic energy program, it is our policy not to interfere with such travel, unless it is of such nature as to indicate an undue risk to security or involve the personal safety of the traveler.

Because of the highly classified information to which you have had access during your association with the atomic energy program, it is our view that your attendance at this Conference would constitute an

unwarranted risk and it is strongly recommended that you decline the invitation.

Your interest in notifying the AEC of this matter is appreciated.

Sincerely yours,

K. D. Nichols

General Manager

## RICHARD P. FEYNMAN TO A. N. NESMEYARROV, MARCH 14, 1955

Mr. A. N. Nesmeyarrov
President,
Academy of Sciences of U.S.S.R.
Moscow, Russia

Dear Mr. Nesmeyarrov:

In my last letter I accepted the invitation to the conference on Quantum Electrodynamics provided a passport would be granted by the State Department. This matter is still pending but in the meantime circumstances have arisen which make it impossible for me to attend. I hope that my vacillations have not caused you too great inconvenience.

I should again like to thank the Soviet Academy for their kindness and generosity in inviting me. I am sure that it will be a very successful conference, and wish you the best of luck with it.

Sincerely,

Richard P. Feynman

## RICHARD P. FEYNMAN TO WALTER M. RUDOLPH, MARCH 14, 1955

Mr. Walter M. Rudolph
Assistant to the Science Adviser
Department of State
Washington, D.C.

Dear Sir:

Thank you for your letter of March 3, acknowledging mine of February 24.

Since I wrote you, K. D. Nichols, of the Atomic Energy Commission,

has written to me recommending that I do not go, as such a trip would constitute an unwarranted risk in view of the fact that I had access to highly classified stuff during the war.

I have decided to take his advice. Therefore I withdraw my request that my passport be OK'd for travel in Russia. Mr. Wallace Atwood will pick it up.

As you know, I had already accepted the invitation pending the OK of the passport. I have now written to the Soviet Ambassador that the passport matter is still pending, but that circumstances have arisen which make it impossible for me to attend. This letter and others involved are enclosed.

Does it strike you that prompt replies to requests for their State Department's views might make it easier for citizens to act in such a way as to protect the department from unnecessary embarrassment?

Sincerely,

Richard P. Feynman

## WALTER J. STOESSEL, JR., TO RICHARD P. FEYNMAN, MARCH 15, 1955

Dear Dr. Feynman:

I refer to the letter which you addressed to the Department on January 14, 1955 and to subsequent communications with the Science Advisor's Office and the National Academy of Sciences concerning the invitation which you received to take part in the All-Union Conference on the Quantum Theory of Electrodynamics and the Theory of Elementary Particles to be held in Moscow, U.S.S.R., March 31 to April 6, 1955 under the sponsorship of the Academy of Sciences of the U.S.S.R. The long delay in reaching a Department position concerning your response to this invitation has been occasioned by the necessity of consulting with a number of interested offices and agencies including those which are aware of considerations stemming from your war-time experience in the field of atomic energy. It now appears that these considerations make it highly undesirable for you to visit the Soviet Union and, consequently, we urge that you decline the invitation.

It is too early to say whether a tendency on the part of the Soviet Government to encourage somewhat larger numbers of Western scientists to visit the Soviet Union and, conversely, to send a few more Soviet scientists abroad represents any basic change in the Soviet attitude toward the

international exchange of scientific information. However, there are strong indications that the Soviet Government in so doing is primarily motivated by the prospect of propaganda gains in the international political field and has little intention of establishing more normal scientific relations which would involve greater exchange of mutually beneficial scientific information. In this connection, it might be noted that the presence of Western scientists at purely internal Soviet conferences and meetings is more susceptible to effective Soviet propaganda exploitation than contacts between Soviet and Western scientists at meetings or conferences of recognized international organizations or groups.

Sincerely yours,

Walter J. Stoessel, Jr.

Officer in Charge

USSR Affairs

## RICHARD P. FEYNMAN TO MR. WALTER J. STOESSEL, JR., APRIL 4, 1955

Mr. Walter J. Stoessel, Jr.

Department of State

Washington, D.C.

Dear Mr. Stoessel:

Thank you for your letter of March 15. In view of a letter from K. D. Nichols, of the Atomic Energy Commission, voicing similar views to those in your letter, I have declined the invitation to attend the Soviet Conference of Quantum Electrodynamics.

Sincerely yours,

R. P. Feynman

## RICHARD P. FEYNMAN TO RALPH BOWN, MARCH 7, 1958

*A member of the "Advisory Board in Connection with Programs on Science" wrote Feynman to discuss the formalities of a pending working relationship. Feynman had agreed to be an advisor for a television program that Warner Brothers was producing for the Bell System Series. Among other formal legal stipulations, the letter stated that "it is provided that for each program an indi-*

*vidual be named 'the designee' from whom Warner may accept authoritative comment, advice and recommendations in the name of the Scientific Advisory Board."*

Mr. Ralph Bown
Advisory Board in Connection with Programs on Science
New York, New York

Dear Mr. Bown:
Thank you for your formidable letter describing the legal interrelations. Who is the "designee"? Is that me or am I an advisor, or what the hell? Put it in plain clear one-syllable words, please.

Anyway the Warner guys have an author named Marcus. He has come to my office on two occasions each for about half a day (so you owe me one day's pay more). The purpose was to get more complete detailed explanation of some of the scientific matters in the report I wrote (like simultaneity in relativity, how short times are measured, etc., etc.). He is very intelligent and I was successful in explaining a great deal to him.

Although the gimmicks, etc. were not on the agenda, he told me about them, and left a document describing his plans. I made no comment on these ideas, telling him they are not my business.

(On the other hand, my hair stood on end as I read the "ideas" for presenting the material. But I kept my hat on and it wasn't noticed. It will relieve me a little if I can say a word to somebody so I can let out steam. So please don't consider the following as a valid or official opinion. It is just me letting off unofficial views and is to be kept safely within these parentheses).

(The idea that the movie people know how to present this stuff, because they are entertainment-wise and the scientists aren't is wrong. They have no experience in explaining ideas, witness all movies, and I do. I am a successful lecturer in physics for popular audiences. The real entertainment gimmick is the excitement, drama and mystery of the *subject matter*. People love to learn something, they are "entertained" enormously by being allowed to understand a little bit of something they never understood before. One must have faith in the subject and in people's interest in it. Otherwise just use a Western to sell telephones! The faith in the value of the subject matter must be sincere and show through clearly. All gimmicks, etc. should be subservient to this. They should help in explaining and describing the subject, and not in entertaining. Entertainment will be an automatic byproduct.)

Don't worry, I'm keeping my hat on and will limit myself to scientific advice only.

Sincerely,

R. P. Feynman

## RICHARD P. FEYNMAN TO MIMI PHILLIPS, JUNE 1958

*Feynman's influential work on liquid helium started in 1953, and it continued to occupy him (and his collaborator Mike Cohen) over the next five years. This letter was written to the young daughter of his cousin in June 1958 while on the way to the Conference on Low Temperature Physics, in Leiden, the Netherlands. It was later published in the Phillips's local newspaper.*

Flying Over England

Dear Mimi:

Am I not terrible—not answering your two letters and cards! You write very good letters—and you are right that I should answer them and write to my Mom—after I finish this one I will write to Mom.

I am on my way to Europe and am flying over England. I am going to land in Holland, in Amsterdam. I have to give a talk at a conference. It is about how liquefied helium behaves.

It is a very strange liquid indeed—it can flow coasting through even the finest cracks without you having to push it. You know how water will very slowly seep through a piece of cloth, or dirt, say? Well, liquid helium flows right through very easily.

It has other crazy properties and physicists have been trying a long time to understand all about it, by doing lots of experiments and lots of thinking. The biggest thinking step forward was by a man named Landau in Russia in 1941 (and the second big thinking step was by me, and now we understand it pretty well). So he was to be honored by giving the first big lecture at this conference (which is about all kinds of strange things that happen at very low temperatures).

But he can't come (we all suspect it is because the Russians won't trust him to leave the country, maybe he would run away). So I have to do it. It is day after tomorrow and I haven't figured out what to say yet! They just told me I would have to do it.

After that I go to another conference in Geneva, Switzerland. This is about all the strange new particles we get when we hit two atoms together very, very hard (it is called a conference on high energy physics).

Atoms are complicated, maybe like watches are—but they are so small, that all we can do is smash them together and see all the funny pieces (gears, wheels and springs) which fly out. Then we have to guess how the watch is put together. In the last few years we've been having enough trouble trying to distinguish one gear wheel from another and to count them. Now it looks like we know most of the parts that go in—but nobody knows how they fit together.

How long will it take for us to figure that out? Five years, ten years? Will I be able to help?—I'll try. I'll have to think very hard and imagine all kinds of possibilities. Why do you think we want to bother to figure out what atoms are made of and how they are put together?

When I come back I am not going out West right away, but I'll stay in Ithaca, N.Y., at Cornell University until Christmas. Maybe I'll see you then. Thank you very much for writing. Why didn't I answer sooner? Because I am bad, bad. Most people are bad, you know, in one way or another—but they are not always bad, and they have other good points to compensate. So if you think I'm bad for not writing—see, I'm not always bad—today I'm good—and I have some compensations because I remember we had a very, very good time together in Connecticut.

Best of luck and regards to your Mom and Pop.

Dick Feynman

P.S. How goes the piano lessons?

## RICHARD P. FEYNMAN TO BILL WHITLEY, MAY 14, 1959

Mr. Bill Whitley
Public Affairs KNXT
Hollywood, California

Dear Mr. Whitley:
On May first you recorded an interview of me by Bill Stout for expected use on your program, "Viewpoint," for May 10.* The tape was not used, and

---

* See Appendix I for the transcript.

you have asked me to record another interview. No very clear reason was given for this request. It was said at one time that my views might antagonize people, at another time that the fault was entirely with Mr. Stout (that by his questions he intimated too strongly that he agreed with me).

Yesterday I heard an audio tape recording of the interview. I found that I had ample opportunity to express my views, that these views were expressed honestly and sincerely and in a calm logical and undogmatic manner. I cannot conceive that antagonism could result from the way I expressed myself, but only perhaps from the fact that I did express myself. It is clearly stated that my views are my own personal opinion, and that not all scientists agree. The viewpoints expressed, or others very close to them, are held by a very large number, albeit perhaps a minority, of very intelligent people in this country. There is no reason why they should not find some expression on our public communication channels such as television.

Mr. Stout conducted the interview with very considerable skill. The questions were clear and unambiguous and so designed to permit me to develop and express my ideas in the clearest way possible. His remarks were solely in the form of questions; neutral questions which in no way implied agreement or disagreement with the ideas I was developing.

The television industry can be proud to be part of the tradition of freedom of expression of this country. And the program bearing the proud name "Viewpoint" makes an important contribution in discussing the controversial issues of our time. Nevertheless, I consider your refusal to utilize the program recorded with me as a direct censorship of the expression of my views.

I see no reason to make a new recording. I will not change my views, and I would not want to change my manner of expressing them.

If you are still concerned about the position of Mr. Stout, please feel free to make, or have him make, an announcement to the effect that the station, or he, does not agree with my views.

In view of these considerations, could I ask you to please reconsider your decision?

Awaiting an early reply, I am

Yours sincerely,

R. P. Feynman

*Feynman did not give another interview, and the station ran the program. However, the program was shown earlier in the day than its advertised time, thus guaranteeing a much smaller audience.*

*Richard and Gweneth, 1959.*

## RICHARD P. FEYNMAN TO GWENETH HOWARTH, MAY 29, 1959

Dear Gweneth,

Well, at last!

I was overjoyed to hear that you are coming at last. We've been waiting so long! What did you finally do to the embassy to wake them up? I need you more than ever—I'm getting sick of my own meals—all I can cook is steak, lamb chops, and pork chops with peas, corn or lima beans and there is not much variety in that. I'm looking forward to being much happier—and now in 3 weeks I'll be calling for you at the airport!

But please write right away anything you know about the flight, like TWA flight number, or exact time of arrival in Los Angeles—or exact time it leaves N.Y. (you said arrives at Los Angeles at 11 AM—but TWA says they have none that do—only flight 5 leaving N.Y. at 9:30 AM, arriving Los Angeles 11:35 AM. Is that the one?) The more details I have the better I can check upon things if something seems to be mixed up. Also give me the time of arrival of the Stadenhaus and whether and where you stay over night in N.Y. or what. I have to take care of you too, you know. As soon as you arrive here you are a responsibility of mine to see you are happy and not scared—so I'd like to know where you are supposed to be to call you to see if everything is all right. In case of any trouble at all please call me—put in 10¢ and tell the operator you want to make a collect call station to station to Los Angeles, SYCAMORE 7–XXXX. If I don't answer try the school where I work—this time tell the operator you want to make a collect call person-to-person to Richard Feynman, telephone number SYCAMORE 5–XXXX. If you can't get me and are positively desperate, try calling collect to Matthew Sands—person-to-person—same number SY–5–XXXX. I'll tell him you may call, and you can tell him your troubles and he will straighten it all out. By the way, he is glad to hear you are coming at last, because I promised he and his wife they would be the first dinner guest after you come. He says he wants pheasant! What should we give him?

Anyway, don't be too scared—America is a good place and the people all speak English—after a fashion—so just ask what you want to know.

If you came 2 weeks earlier I'd sure have a lot for you to do—I'm going to be on television, in an interview with a news commentator on June 7th and there may be a lot of letters to answer.

Leave your ski things there. We do have skiing here in the mountains 60 min. away, but I have never skied—so save the weight. We may go once or

twice if you want to try to teach me how—but we can borrow (or rent) things from friends for the few times, if ever, we go. They are too heavy to bring along just for that.

The weather is hot in summer—warm in winter—I have never seen it freeze. Usually 60°–65° or so. Cold nights. A heavy coat is not necessary. I never wear a coat at all except on coldish rainy days (very few) when I wear a rain-coat. But a light coat will come in very handy.

I went to buy tires for your car—but they are very expensive and I'll have to shop around—so I decided to wait and let you do the shopping around until you find a reasonable price—there is no sense in putting fancy expensive tires on an old car. Don't worry about the driver's license—we can get that here. Anyway you'll need a lesson or two on the right side to get familiar with our streets and the car.

It is the season for the beach here too—but I haven't gone yet. Last week I went on a camping trip with friends into the desert for 2 nights. It's fun.

O.K. I'm managing poorly without you. Come quick.

Yours,
Richard

# 1960–1970

## The National Academy of Sciences

It all started with a letter enclosed with a statement dated November 1959: "According to the records in the business office, your dues have not been paid for the two previous fiscal years." What follows is an exchange about the value of honors—the merit of merits, if you will—that would crop up many times over the course of Feynman's life. Though the National Academy of Sciences had been helpful in the quest to attend the Moscow conference a few years earlier, the distinction of membership now held little appeal.

The exchange between Feynman and the various concerned parties at the Academy forms a strangely comic testimony not only to Feynman's deep-seated distaste for clubs whose value rests primarily on their exclusivity, but also to his very keen persistence.

· · · ·

### B. L. KROPP TO RICHARD P. FEYNMAN, NOVEMBER 1959

Dear Dr. Feynman,

Enclosed is a statement of your dues as a member of the National Academy of Sciences for the year beginning July 1, 1959.

According to the records in the business office, your dues have not been paid for the two previous fiscal years. Perhaps there is an error in our

accounts and we would appreciate your advising us when payments were made. If on the other hand your records are in agreement with ours, please send a check for $30.00 which will pay your dues to June 30, 1960.

Yours very truly,

B. L. Kropp

Assistant Business Manager

## RICHARD P. FEYNMAN TO B. L. KROPP, NOVEMBER 9, 1960

Mr. B. L. Kropp

Assistant Business Manager

National Academy of Sciences

Washington, D.C.

Dear Mr. Kropp:

Enclosed is my check for $40 to pay for my dues to the National Academy of Sciences.

I have found that I have little interest in the activities of the Academy, so would you please accept my resignation as a member.

Sincerely yours,

R. P. Feynman

## RICHARD P. FEYNMAN TO THE NATIONAL ACADEMY OF SCIENCES, FEBRUARY 20, 1961

Dear Sirs:

I should like to resign my membership in the Academy.

I do not find myself with enough time or interest to actively participate.

Sincerely,

Richard P. Feynman

.   .   .   .

## DETLEV W. BRONK TO RICHARD P. FEYNMAN, JULY 15, 1961

Dear Dr. Feynman,

The Home Secretary of the Academy has told me of your letter of resignation. I have hoped that I might see you in person and then, because of my high regard and great admiration for you, urge you to reconsider your action.

I felt particularly sorry personally that you were written to by our Business Office regarding Academy dues because I myself have felt strongly that it is not appropriate for the Academy to assess its members. When I suggested abolishing dues at an Academy meeting several years ago, there were differences of opinion; some thought as I do, some that dues should be increased, the majority were for continuation of the present policy. I do not understand the reason for dues nor their justification. The income received therefrom is a trivial few thousand dollars compared with an annual expenditure of more than $15,000,000. I do not like the idea of requesting payments from our members who serve science freely through the Academy in so many ways and who did not request election in the first place. I realize that this was not your motive in sending your resignation, but I wish to express my regret that you were written to as you were. I will endeavor to see that it does not occur again.

Regarding the more significant matter of your membership, I hope that you will permit us to keep your name on the lists of the Academy. Your election was an expression of our high regard for your achievements. Your presence in the membership adds distinction to the Academy and that, in turn, aids us in the fulfillment of our many undertakings in the furtherance of science. I realize that to many members the Academy's activities are of relatively little interest at any one time, but I hope that with its rapidly broadening scope you may from time to time find it an organization of some significance to you.

With warm personal regards, I am

Yours sincerely,

Detlev W. Bronk

President

·  ·  ·  ·

# RICHARD P. FEYNMAN TO DETLEV W. BRONK, AUGUST 10, 1961

Dr. Detlev W. Bronk, President
National Academy of Sciences
Washington, D.C.

Dear Dr. Bronk:

I am sorry that you had to be bothered by this matter of my wanting to resign my membership in the Academy. It must be quite a job worrying about all the peculiar whims of all of the strange birds that make up your flock.

It was impolite of me to scratch a small note on the dues bill; I should have written a more formal and clarifying letter. As you guessed, the dues have nothing to do with it at all.

My desire to resign is merely a personal one; it is not meant as a protest of any kind, or a criticism of the Academy or its activities. Perhaps it is just that I enjoy being peculiar. My peculiarity is this: I find it psychologically very distasteful to judge people's "merit." So I cannot participate in the main activity of selecting people for membership. To be a member of a group, of which an important activity is to choose others deemed worthy of membership in that self-esteemed group, bothers me. The care with which we select "those worthy of the honor" of joining the Academy feels to me like a form of self-praise. How can we say only the best must be allowed in to join those who are already in, without loudly proclaiming to our inner selves that we who are in must be very good indeed. Of course I believe I am very good indeed, but that is a private matter and I cannot *publicly* admit that I do so, to such an extent that I have the nerve to decide that this man, or that, is not worthy of joining my elite club.

Maybe I don't explain it very well, but suffice to say that I am not happy as a member of a self-perpetuating honorary society. For that reason I have not made any recommendations for memberships in the past (except during my first year) and have always wanted to resign, but didn't get around to it until one day when I was sending in my dues.

Therefore if it is not a matter of extreme seriousness, or one that will cause considerable embarrassment; if it is a matter that can be done quietly and easily without talk, discussion and concern of the entire membership, I would prefer if you would accept my resignation and take my name from the Academy lists.

Please appreciate that it is not a matter of deep principle to me, so that if my resignation causes you any major difficulties, feel free to disregard my preferences. But your flock will contain one very peculiar, sad and reluctant bird.

May I take this opportunity, while I am writing to you on another matter, to add a personal note. Without your knowing it, you have been honored and respected by me from my earliest student days twenty years ago. At the Princeton graduate college some biology student friends suggested I listen in on a course on cell-physiology given by E. Newton Harvey. He would not hear of it unless I did the work along with the others. It consisted of reading original papers and reporting on them and I was assigned Adrian and Bronk's paper on the single nerve impulse. What a fundamental discovery, and what a good experience to read it in the original! (Incidentally, they are probably still laughing at the Princeton Biology Library at the story of the graduate student who asked the librarian if she had "a map of the cat." "You mean a zoological chart!" she cried indignantly, but I got it and located the tibicus anticus and all the flexors and extensors, so I could understand your paper.) Not long afterwards, at a meeting of some kind to which Harvey took us, I saw in person the great Bronk, himself.

It was exciting to me because I personally knew your work. In physics we did not have that close contact with original papers—the teaching methods were different—and it was to be a few more years before I was to be impressed by meeting the great scientists in my own field, under circumstances when I knew so well the reason for their greatness.

So, with kind personal regards, I am

Sincerely yours,

Richard P. Feynman

## DETLEV W. BRONK TO RICHARD P. FEYNMAN, OCTOBER 26, 1961

Dear Dr. Feynman,

When I receive a letter that gives me great pleasure, I put it aside to re-read. Because I treasure it, I tell myself I must answer it when I have the time and spirit to answer it as it deserves to be answered so that I may show my gratitude. The consequences of these habits are that the letters which I prize the most go unanswered the longest. Such was your letter of last summer.

Last night Ted Berlin, Keffer Hartline and I talked late. The conversation turned to Dick Feynman, his work on general relativity, quantum electrodynamics and massless particles, how everything he does is made interesting to many by the rare quality of his mind. But I was mortified then to recall that I had still to answer your letter. Thank you for it.

And thank you for your willingness to continue as a member of the Academy. I find the elections distasteful as do you; I have done my best to reduce the emphasis on the "honor" of election and make the Academy as much as possible an agency for furthering science. Because of my great regard for you, I am grateful that you will continue as a member at least during my last year as President.

If you ever return to your old haunts in the East, it would give us much pleasure if you would stay with us at The Rockefeller Institute. A room and a warm welcome from many will always be awaiting you.

With kind personal regards, I am

Yours sincerely,

Detlev W. Bronk

## RICHARD P. FEYNMAN TO THE NATIONAL ACADEMY OF SCIENCES, JULY 1, 1968

Gentlemen:

Please accept my resignation as a member of the National Academy of Science. My reasons are entirely personal and reflect in no way on my opinion of the Academy.

Sincerely yours,

Richard P. Feynman

## RICHARD P. FEYNMAN TO FREDERICK SEITZ, JUNE 12, 1969

Dr. Frederick Seitz, President
National Academy of Sciences
Washington, D.C.

Dear Fred:

When your telegram and letter came, I was out so I did not answer.

Bromley called and I told him no, because it is not the kind of thing I enjoy doing.

On another entirely different subject, that has absolutely no relation to your recent request: I should like to resign from the Academy for personal reasons that have nothing to do with its activities on behalf of the government, nor your administration of it. I have been trying for years to do it quietly without making any waves, or any political point, for it has no more significance than my personal, rather childish, likes and dislikes. Please accept my resignation.

Sincerely yours,

Richard P. Feynman


## RICHARD P. FEYNMAN TO PHILIP HANDLER, JULY 15, 1969

Professor Philip Handler
Biochemistry Department
Duke University Medical Center
Durham, North Carolina

Dear Professor Handler:

My request for resignation from the National Academy of Sciences is based entirely on personal psychological quirks. It represents in no way any implied or explicit criticism of the Academy, other than those characteristics that flow from the fact that most of the membership consider their installation as a significant honor.

Sincerely yours,

Richard P. Feynman


## PHILIP HANDLER TO RICHARD P. FEYNMAN, JULY 31, 1969

Dear Dr. Feynman:

I have your somewhat cryptic note of July 15. I am troubled by the reasoning which led to your decision, particularly in view of the fact that the Council and I are determined to transform the Academy in at least some part into a working body which will fulfill the obligations of our noble charter. We are seeking to increase the meaningful roles of the Academy

and mechanisms whereby the membership can participate in a rewarding way. Wouldn't you rather join us in that effort?

Sincerely yours,

Philip Handler

President


## RICHARD P. FEYNMAN TO PHILIP HANDLER, AUGUST 14, 1969

Dr. Philip Handler, President

National Academy of Sciences

Washington, D.C.


Dear Dr. Handler:

Thank you for your letter of July 31.

My note of July 15 still stands. Please accept my resignation from the Academy.

Sincerely yours,

Richard P. Feynman


## RICHARD P. FEYNMAN TO FRANCIS D. MURNAGHAN, JANUARY 22, 1970

*In a December 21, 1969, letter to Feynman asking why he had resigned from the Academy, Dr. Murnaghan wrote, "I learned some days ago of your resignation from the National Academy and have been worrying about it ever since." He went on to say that the resignation was a loss to the Academy and may indeed have done it harm. He wanted to know if there was anything he and others could do to improve the organization.*

Dr. Francis D. Murnaghan

Baltimore, Maryland


Dear Dr. Murnaghan:

My resignation from the Academy was for purely personal reasons and was not meant to imply any criticism of the Academy, nor to damage it in any way. It is just the result of a peculiar quirk in my personality which arises in my reaction to "honors."

I am sorry to hear that such a minor thing caused you any worry.

Best regards,

Richard P. Feynman


## RICHARD P. FEYNMAN TO J. NEYMAN, APRIL 28, 1970

*An April 21 letter from Dr. Neyman inquired whether the rumor—that Feynman had resigned from the National Academy of Sciences—was true and, further, whether he had a statement about the reason.*

Dr. J. Neyman
Department of Statistics
University of California
Berkeley, California

Dear Dr. Neyman:

My resignation from the National Academy is a fact. The reason is purely personal and should reflect in no way on the Academy.

Sincerely yours,

Richard P. Feynman

# 1960–1965

After flirting with the idea for a long time, Feynman spent a summer, and then a sabbatical year, doing research in the area of molecular biology at Caltech. While working with Bob Edgar and others on viruses that attacked bacteria, he discovered several cases of "back mutations" that occurred close together—perhaps giving a clue to the sequence of DNA. Edgar urged him to publish his findings, and at the invitation of James Watson, Feynman later gave a talk on the subject to the biology department of Harvard University. He then returned to his first love, physics.

Gweneth became his wife on September 24, 1960. He soon began to boast of his newfound domesticity, and perhaps it provided a buffer from the demands of his growing renown.

He was increasingly turning his attention to popularizing physics, and in 1961 he worked as the scientific advisor for the film "About Time," which NBC aired on prime-time television. Part of the Bell System Science Series, the hour-long program brought Feynman's name and personality to an audience far beyond his colleagues and fellow scientists. The letters from perfect strangers—students, laymen, and the occasional fan—soon followed.

My brother, Carl, was born in 1962, the day before the presentation to Feynman of the E. O. Lawrence Award, a prize given by the Department of Energy for contributions in the field of atomic energy. The local newspaper ran a picture of Richard and Gweneth at the hospital, he holding his medal and she holding infant Carl.

More significantly for his reputation as a teacher, this same period saw the birth of The Feynman Lectures on Physics. Based upon a series of fresh-

man and sophomore lectures given at Caltech, and edited by Robert B. Leighton and Matthew Sands, the books have become a unique and timeless portrait of physics—a work of art, in the eyes of many. During the 1962–1963 academic year, Feynman also gave his now-legendary Lectures on Gravitation. Soon thereafter Feynman's lectures on more advanced topics were also published. A 1963 letter to a colleague at Cornell discusses another burgeoning enterprise: Feynman lectures on film.

. . . .

## RICHARD P. FEYNMAN TO WILLIAM H. MCLELLAN, NOVEMBER 15, 1960

*In 1959 Feynman realized that it should be possible to reduce the physical size of stored words, even machines, to a scale of a few tens or hundreds of atoms in each dimension—that is, he envisioned a new field of science now called* nanotechnology. *In a famous lecture titled, "There Is Plenty of Room at the Bottom,"\* he expounded this view and promoted it among scientists and engineers by offering prizes of $1,000 each for two challenges: (1) to "take the information on the page of a book and put it on an area 1/25,000 smaller in linear scale in such a manner that it can be read by an electron microscope"; and (2) to build "a rotating electric motor which can be controlled from the outside and, not counting the lead-in wires, is only 1/64 inch cube." The motor challenge was met almost immediately, but it was achieved with craftsmanship only—no new technology was needed, which is what he had hoped to encourage.*

Mr. William H. McLellan
Electro-Optical Systems, Inc.
Pasadena, California

Dear Mr. McLellan:
I can't get my mind off the fascinating motor you showed me Saturday. How could it be made so small?

Before you showed it to me I told you I hadn't formally set up that prize I mentioned in my *Engineering and Science* article. The reason I delayed was

---

\* Delivered to a meeting of the American Physical Society at Caltech on December 29, 1959; paper published in the February 1960 issue of Caltech's *Engineering and Science*.

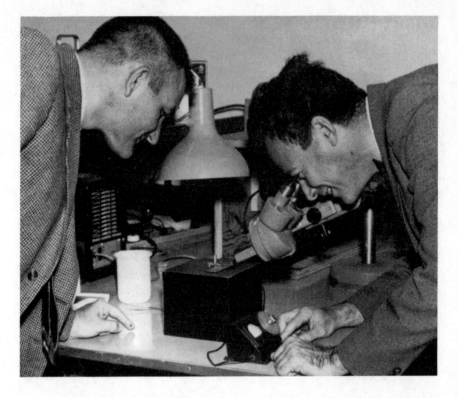

*Feynman and William McLellan with micromotor, 1960.*

to try to formulate it to avoid legal arguments (such as showing a pulsing mercury drop controlled by magnetic fields outside and calling it a motor), to try to find some organization that would act as judges for me, to straighten out any tax questions, etc. But I kept putting it off and never did get around to it.

But what you showed me was exactly what I had had in mind when I wrote the article, and you are the first to show me anything like it. So, I would like to give you the enclosed prize. You certainly deserve it.

I am only slightly disappointed that no major new technique needed to be developed to make the motor. I was sure I had it small enough that you couldn't do it directly, but you did. Congratulations!

Now don't start writing small.

I don't intend to make good on the other one. Since writing the article I've gotten married and bought a house!

Sincerely yours,

Richard P. Feynman

## RICHARD P. FEYNMAN TO RONNIE KERNAGHAN, FEBRUARY 20, 1961

*An eighth-grader writing a job survey needed some information about the life of a theoretical physicist—the courses needed to become one, as well as the potential job opportunities, working conditions, and pay rates.*

Mr. Ronnie Kernaghan
Paso Robles, California


Dear Mr. Kernaghan:

I am sorry but I have no information on any of the questions you asked me.

If you asked me is it an adventuresome and exciting life trying to find out about how nature works, I could answer that. It is, and it is a lot of fun. But make sure you have talent for it.

Sincerely,

R. P. Feynman


## RICHARD P. FEYNMAN TO EARL UBELL, FEBRUARY 21, 1961

Mr. Earl Ubell
New York Herald Tribune
New York, New York


Dear Earl:

A Miss Draper from your paper kindly sent me a copy of your article "Gravity, the Disrespectful."* Very good, I thought, as I read it. Usually science reporting is no damn good at all. But the last line—ah, there is a man that understands, and says it better than I can!

I have a vague recollection of being short or impatient with you somewhere in the past. At least that is how I have become toward the newspaper science press generally. I apologize; to you, but not to the others.

Sincerely,

Richard P. Feynman


---

* See Appendix II.

## RICHARD P. FEYNMAN TO FLOYD GOLD, APRIL 5, 1961

*A letter from an old friend, Floyd Gold, dated March 25, 1961, expressed concern about his son's future education. "He is fifteen years of age. . . and is at present constructing a digital computer as his entry in the Westinghouse Science Talent competition." Gold went on to say that his son was completely self-taught except for a seven-week course in the field of computer logics. The assistant principal of Far Rockaway High School (Feynman's alma mater) suggested that Gold write Feynman and ask for a list of courses and schools to which his son might aspire. Mr. Gold was also curious about Feynman's feelings on engineering versus a pure science approach to his son's education.*

Mr. Floyd O. Gold
Far Rockaway, New York

Dear Floyd:

It is very exciting to hear from someone from the old days whose name I remember. I married Arline Greenbaum from Cedarhurst. She died in 1946 and I am now beginning my third marriage, to an English girl. No kids yet, but I hope to soon.

My advice to your son is this. It is very lucky he is interested in something and gets delight out of doing something. He should be encouraged to do exactly as he wants—I don't mean in the future—I mean day to day, without some grand plan. At his stage, engineering and physics education is nearly the same, and will be for several years. Many is the man who has changed from one to the other even after graduating from college (but not as late as after graduate college) without any great difficulty.

So let him play with computers and ideas as hard as he can. His math background will develop as he needs it to understand his circuits, etc. He must have freedom to pursue his delight now, and when he becomes a great expert in something, he will find it easy to pick up other related subjects.

If, on the other hand (contrary to what your letter implies), he is average at *everything* and gets no "charge" out of doing anything in particular (or gets plenty of "charge" out of everything, but goes from one thing to another) then I don't know what to advise.

Sincerely,

R. P. Feynman

## RICHARD P. FEYNMAN TO FREDERICH HIPP, APRIL 5, 1961

*Frederich Hipp, a high school student, was fascinated by physics ("atomic theory and quantum mechanics in particular") and had built a cloud chamber for his science project. He was concerned, however, that he had little aptitude for math. His question to Feynman: "Can a person of normal mathematical ability master enough math to do work on some professional level in this field?"*

Mr. Frederich Hipp
New Milford, Connecticut

Dear Sir:

To do any important work in physics a very good mathematical ability and aptitude are required. Some work in applications can be done without this, but it will not be very inspired.

If you must satisfy your "personal curiosity concerning the mysteries of nature" what will happen if these mysteries turn out to be laws expressed in mathematical terms (as they do turn out to be)? You cannot understand the physical world in any deep or satisfying way without using mathematical reasoning with facility.

How do you know you don't have an aptitude for math? Perhaps you disliked your teacher, or it was presented wrong for your type of mind.

What do I advise? Forget it all. Don't be afraid. Do what you get the greatest pleasure from. Is it to build a cloud chamber? Then go on doing things like that. Develop your talents wherever they may lead. Damn the torpedoes—full speed ahead!

What about the math? Maybe (1) you might find it interesting later when you need it to design a new apparatus, or (2) you may not go on with your present ambition to understand everything, but instead find yourself a leader in some other direction, such as building the most ingenious rocket-ship control devices, or (3) biological problems may ultimately absorb all your interest and talent for doing experiments and learning about nature, etc.

If you have any talent, or any occupation that delights you, do it, and do it to the hilt. Don't ask why, or what difficulties you may get into.

If you are an average student in *everything* and no intellectual pursuit gives you real delight, then I don't know how to advise you. You will have

*The Feynman Lectures on Physics, 1962.*

to discuss it with someone else. It is a problem that I have not thought about very hard.

Sincerely,

R. P. Feynman

## RICHARD P. FEYNMAN TO HELEN CHOAT, JULY 26, 1961

*The "About Time" film was by this time over two years in the making. Warner Brothers produced the footage, and N. W. Ayer and Son was chosen to publish the accompanying materials and workbooks.*

Miss Helen Choat
N. W. Ayer and Son
Rockefeller Center
New York, New York

Dear Miss Choat:

You asked me to review the "Biographical Material" about me. The second sentence of the first paragraph is wrong. It should read something like

"He started work in connection with the atomic bomb at Princeton in 1941 and continued with the project until its successful conclusion at Los Alamos in 1945." I didn't work with Einstein, Einstein didn't work on the atomic bomb, and Einstein is not the father of the atomic bomb.

Finally, please omit the reference to the National Academy of Sciences. Also I am not sure if I am a member of the American Association for the Advancement of Science—I don't remember.

You also wanted in a few lines how I became attracted to science: "My father, a business man, had a great interest in science. He told me fascinating things about the stars, numbers, electricity, etc. Wherever we went there were always new wonders to hear about; the mountains, the forests, the sea. Before I could talk he was already interesting me in mathematical designs made with blocks. So I have always been a scientist. I have always enjoyed it, and thank him for this great gift to me."

Sincerely,

Richard P. Feynman

## RICHARD P. FEYNMAN TO GWENETH FEYNMAN, OCTOBER 11, 1961

*The following letter was written from the Solway Conference in Brussels. Gweneth was pregnant with Carl at the time.*

October 11, 1961

Hotel Amigo, Brussels

Hello, my sweetheart,

Murray and I kept each other awake arguing until we could stand it no longer. We woke up over Greenland which was even better than last time because we went right over part of it. In London we met other physicists and came to Brussels together. One of them was worried—in his guidebook the Hotel Amigo was not even mentioned. Another had a newer guide—five stars! and rumored to be the best hotel in Europe!

It is very nice indeed. All the furniture is dark red polished wood, in perfect condition; the bathroom is grand, etc. It is really too bad you didn't come to this conference instead of the other one.

At the meeting next day things started slowly. I was to talk in the afternoon. That is what I did, but I didn't really have enough time. We had to

## Hôtel Amigo

Hello, my sweetheart;

Murrays I kept each other awake arguing until we could stand it no longer. We woke up over Greenland which was even better than last time because we went right over part of it. In London we met other physicists and came to Brussels together. One was worried — in his guide the hotel Amigo was not even mentioned. Another had a newer guide — five stars! and rumored to be the best hotel in Europe!

It is very nice indeed. All the furniture is dark red polished wood in perfect condition the bathroom is grand, etc. It is really too bad you didn't come to this conference instead of the other one.

At the meeting next day things started slowly. I was to talk in the afternoon. That is what I did but I didn't really have enough time for we had to stop at 4PM because of a reception scheduled for that night. I think

*Richard and Gweneth, 1961.*

stop at 4 pm because of a reception scheduled for that night. I think my talk was OK tho—what I left out was in the written version anyway.

So that evening we went to the palace to meet the king and queen. Taxis waited for us at the hotel—long black ones—and off we went at 5 pm, arriving through the palace gates with a guard on each side, and driving under an arch where men in red coats and white stockings with a black band and gold tassel under each knee opened the doors. More guards at the entrance, in the hallway, along the stairs, and up into a ballroom, sort of. These guards stand very straight, dark grey Russian-type hats with a chin strap, dark coats, white pants, and shiny black leather boots, each holding a sword straight up.

In the "ballroom" we had to wait perhaps 20 minutes. It has inlaid parquet floors, and L in each square (Leopold—the present king's name is Baudoin, or something). The gilded walls are 18th century and on the ceiling are pictures of naked women riding chariots among the clouds or something. Lots of mirrors and gilded chairs with red cushions around the outside edge of the room—just like so many of the palaces we have seen, but this time it was alive, no museum, everything clean and shining and in perfect condition. Several palace officials were milling around among us. One had a list and told me where to stand but I didn't do it right and was out of place later.

The doors at the end of the hall open—guards are there, and the king and queen so we all enter slowly and are introduced one by one to the king and queen. The king has a young semi-dopey face and a strong handshake, the queen is very pretty. (I think her name is Fabriola—a Spanish countess she was.) We exit into another room on the left where there are lots of chairs arranged like in a theatre, with two in front, also facing forward, for K & Q later, and a table at the front with six seats is for illustrious scientists (Niels Bohr, J. Perrin (a Frenchman), J. R. Oppenheimer etc.).

It turns out the king wants to know what we are doing, so the old boys give a set of six dull lectures—all very solemn—no jokes. I had great difficulty sitting in my seat because I had a very stiff and uncomfortable back from sleeping on the plane.

That done, the K & Q pass thru the room where we met them and into a room on right (marked R). All these rooms are very big, gilded, Victorian, fancy, etc. In R are many kinds of uniforms, guards at door, red coats, white coat sort of waiters to serve drinks and hors d'oeuvres, military khaki and medals, black coat—undertaker's type (palace officials).

On the way out of L into R, I am last because I walk slowly from stiff back and find myself talking to a palace official—nice man—teaches math part time at Louvain University, but his main job is secretary to the queen. He had also tutored K when K was young and has been in palace work 23 years. At least I have somebody to talk to, some others are talking to K or to Q; everybody standing up. After a while the professor who is head of conference (Prof. Bragg) grabs me and says K wants to talk to me. I pull boner #1 by wanting to shake hands again when Bragg says, "K, this is Feynman"; apparently wrong— no hand reaches up, but after an embarrassed pause K saves day by shaking my hand. K makes polite remarks on how smart we all must be and how hard it must be to think. I answer, making jokes (having been instructed to do so by Bragg, but what does he know?)—apparently error #2. Anyway, strain is relieved when Bragg brings over some other professor—Heisenberg, I think. K forgets F and F slinks off to resume conversation with Sec'y of Q.

After considerable time—several orange juices and many very very good hors d'oeuvres later—a military uniform with medals comes over to me and says, "Talk to the queen!" Nothing I should like to do better (pretty girl, but don't worry, she's married). F arrives at scene: Q is sitting at table surrounded by three other occupied chairs—no room for F. There are several low coughs, slight confusion, etc. and lo! one of the chairs has been reluctantly vacated. Other two chairs contain one lady and one Priest in Full Regalia (who is also a physicist) named LeMaître.

We have quite a conversation (I listen, but hear no low coughs, and am not evacuated from seat) for perhaps 15 minutes. Sample:

Q: "It must be very hard work thinking about those difficult problems."

F: "No, we all do it for the fun of it."

Q: "It must be hard to learn to change all your ideas" (a thing she got from the six lectures).

F: "No, all those guys who gave you those lectures are old fogeys—all that stuff was in 1926, when I was only eight, so when I learned physics I only had to learn the new ideas. Big problem now is, will we have to change them again?"

Q: "You must feel good, working for peace like that."

F: "No, never enters my head, whether it is for peace or otherwise we don't know."

Q: "Things certainly change fast—many things have changed in the last hundred years."

F: "~~Not in this palace.~~" (I thought it, but controlled myself.)

F: "Yes," and then launched into lecture on what was known in 1861 and what we found out since—adding at end, laughingly, "Can't help giving a lecture, I guess—I'm a professor, you see. Ha, ha."

Q in desperation, turns to lady on her other side and begins pleasant conversation with same.

After a few moments K comes over, whispers something to Q who stands up and they quietly go out. F returns to Sec'y of Q who personally escorts him out of palace past guards, etc.

I'm so terribly sorry you missed it. I don't know when we'll find another king for you to meet.

I was paged in the hotel this morning just before leaving with the others.

Phone call—I returned to the others and announced, "Gentlemen, that call was from the queen's secretary." All are awestruck, for it did not go unnoticed that F talked longer and harder to Q than seemed proper. I didn't tell them, however, that it was about a meeting we arranged—he was inviting me to his home to meet his wife and two (of four) of his daughters, and see his house. I had invited him to visit us in Pasadena when he came to America and this was his response.

His wife and daughters are very nice and his house was positively beautiful. You would have enjoyed that even more than visiting the palace. He planned and built his house in a Belgian style, somewhat after an old farmhouse style, but done just right. He has many old cabinets and tables inside, right beside newer stuff, very well combined. It is much easier for them to

*Richard and Kiwi, or "Snork," 1961.*

find antiques in Belgium than for you in Los Angeles as there are so many old farms, etc. He has large grounds and a vegetable garden—and a dog—from Washington—somebody gave the king and the K gave to him. The dog has a personality somewhat like Kiwi because I think he is equally loved.* He even has a bench in his garden hidden under trees that he made for himself to go and sit on and look at the surrounding countryside. The house is slightly bigger than ours and the grounds are much bigger but not yet landscaped.

I told him I had a queen in a little castle in Pasadena I would like him to see—and he said he hoped he would be able to come to America and see us. He would come if the Q ever visits America again.

I am enclosing a picture of his house, and his card so I don't lose it.

I know you must feel terrible being left out this time—but I'll make it

---

* Kiwi was my parents' much-loved dog. My father was very fond of his dogs over the years—lounging on the floor with them, teaching them the usual tricks of begging, shaking, getting the paper. He once taught one to stick out her tongue on command, though we never found a good use for that skill.

up someday somehow. Don't forget I love you very much and am proud of my family that is and my family that is to be. The secretary and his wife send their best wishes to you and our future.

I wish you were here, or next best thing, that I were there. Kiss SNORK and tell Mom all my adventures and I will be home sooner than you think.

Your husband loves you.

Your husband.

*Gweneth herself would plan a number of adventures over the years. Trips to areas as remote as the tribal lands of the Tarahumara Indians in Mexico, and many camping trips to the middle of nowhere, were the family norm.*

## RICHARD P. FEYNMAN TO VOLTA TORREY, NOVEMBER 15, 1961

*Feynman gave a talk titled "The Future of Physics" at the MIT Centennial Convocation in April 1961. The centennial authorities considered publishing a book of the speeches made that day, but the project was eventually dropped. The Technology Review had a high opinion of Feynman's talk and wanted to publish it. Mr. Torrey wrote Feynman to obtain permission and enclosed the proposed manuscript.*

Mr. Volta Torrey, Editor
Massachusetts Institute of Technology
Cambridge, Massachusetts

Dear Editor Torrey:

There is some confusion about the manuscript of my talk at the MIT Centennial. First I gave the talk and a tape recording was made (call it Version A). Then somebody related to the Centennial book sent me a highly edited and rewritten version (call it Version B) which is similar to your story and is the manuscript they gave you which you sent me. But I didn't like version B and sent back asking for the original tape, (Version A), which I corrected and returned as Version C.*

Try to get Version C from the book boys. I like it much better than version B which you want to print. Wouldn't you rather print version C? I have no copy of C or I wouldn't give you the run-around.

---

* See Appendix III.

Thanks for your patience.

Sincerely,

Richard P. Feynman

## RICHARD P. FEYNMAN TO ALLADI RAMAKRISHNAN, JANUARY 30, 1962

*The following letter is in response to a request to publish some unpublished work on a model of strong interactions. Mr. Ramakrishnan had already received permission to include some of Feynman's notes on electrodynamics in a book, and he hoped to include the unpublished work as well. He wrote, "If you do not mind its inclusion, I will send it on to the press; otherwise I have to delete it and this may cause considerable confusion at this stage for the printers." Mr. Ramakrishnan also offered to include any comments as a footnote.*

Mr. Alladi Ramakrishnan, Director

The Institute of Mathematical Sciences

Madras, India

Dear Mr. Ramakrishnan:

I was sorry to see that you wanted to publish the model of strong interactions I once wrote about. The cover page of this work was marked "Publication of this work is not planned at present." I didn't publish it for a reason. I wasn't sure that it was right (I mean, that Nature was really being described correctly). Now I am even less sure. Almost certainly it is wrong.

It is my policy to try to keep up the standard of my own published work—at least so that each article should have some relevance to Nature and her laws. Speculations about how it might come out are not satisfactory unless I feel that with high probability the speculation is in fact correct (unless I am writing a review article on present-day speculations, etc.).

So you embarrass me by taking seriously something that I think is not true. I would prefer that you leave it out.

However, since this might cause considerable confusion for the printers, as a second choice add to what you have, the following footnote:

"Professor Feynman feels that the physicists' problem is not just to make speculative models, but to really find out what is going on in nature. Since, in his opinion, this model is a speculation and almost

certainly does not describe the observed reality, he does not think it of sufficient value to warrant publication."

Then you can, if you wish, go on to say something like "Nevertheless, the author thought that this model would be of sufficient interest to his readers as an example of the kind of thing being tried that he has included it here" or some such thing. Whatever you want. But if you do publish it, please add at least the paragraph preceding this one.

Sincerely,

R. P. Feynman

## RICHARD P. FEYNMAN TO MR. Y, FEBRUARY 8, 1962

*The television program "About Time" finally aired on the NBC network on February 5, 1962. A viewer whom I shall call Mr. Y, and who had complained previously to KRCA in Los Angeles about its presentation of the "orthodox" viewpoint on relativity, wrote the station to complain once again.\* He copied his letter to Feynman and four other scientists plus three organizations. In the letter, Mr. Y attacked the film's presentation of the* twin paradox, *charging that it was "propaganda" aimed at propagating incorrect orthodoxy, and asserted that his own correct viewpoint was being suppressed.*

Dear Mr. Y:

You sent me a copy of a letter to KRCA about the program "Time." I was scientific adviser to that program and am responsible for the inclusion of the traveling twin story. I included it because I believe it is correct. Please be assured that I am not interested in propaganda or persuading people. We agree that we should be interested only in what is right in science.

But if we are interested only in what is right, must we not conclude that the phenomenon suggested by the traveling twin is right? I honestly thought so, and I believed that the observed long lifetime of moving mu mesons confirmed it. You seem to think otherwise. I should be glad to learn of my error, if, indeed, it is an error. I have used these ideas in an essential way in my work in physics for many years and know of no direct violation of them—in fact, they seem to be very successful in predicting new phe-

---

\* I cannot call him Mr. X because that was Feynman's pseudonym, widely known among his fellow physicists.

nomena. I would be surprised, but very pleased, to learn that there is another point of view which successfully does two things:

(1) Predicts correctly all those effects which have been carefully observed experimentally so far;

(2) Predicts a different result for the traveling twin.

Sincerely yours,

R. P. Feynman

## RICHARD P. FEYNMAN TO MR. Y, MARCH 14, 1962

*Mr. Y replied with several successive, emotional letters that attacked Feynman for joining with the science establishment in suppressing Mr. Y's views but never explained clearly what those views were.*

Dear Mr. Y:

I have received a number of letters from you which I do not understand. As you say, the question is not so much a matter of philosophy but of what phenomena occur. In order therefore to see if we really disagree on the phenomena associated with the clock paradox, I should like to ask what, in your opinion, would be the result of the following experiment:

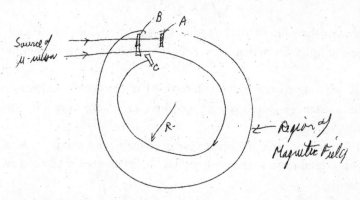

A weak beam of negative $\mu$-mesons of velocity v coming from a source pass through a thin counter B so the time of arrival of each can be noted. Half of them impinge on a piece of matter, A, thick enough to stop them, and in such circumstances that their subsequent decay can be detected. The mean time between entrance at B and decay in A (minus the small bit of time estimated to get from B to A, which can be made very small indeed) is measured. Let us call this $\tau_0$ (as you know, it comes out to be 2.2 x 10⁻⁶

seconds). In particular the probability/sec. of a meson in A to decay at time t is

$$e^{-t/t_0}.$$

In addition, those that miss A go around in a magnetic field through an angle close to 360° to a counter C. (Since C cannot be put exactly at 360° imagine a small correction is made to the data for the small time needed to go from C to the true 360° point.) How many counts do you expect will arrive at C? Supposing that the radius of the orbit is R, so the time to go around the circle is $T = 2\pi R/v$.

I presume your answer is one of two. Either (a) the lifetime to decay does not depend on the speed so the number in C is the same as in A after the same delay, namely

$$e^{-T/\tau_0}.$$

Or (b) the lifetime of a moving meson appears longer in the sense that the number arriving at C is

$$e^{-T/\tau}$$

where $\tau$ is not $\tau_0$, the value of rest, but rather

$$\tau = \tau_0 / \sqrt{1 - v^2/c^2}$$

where v is the velocity of the mesons going around the circle, and C is the speed of light. Or perhaps you expect a result different than either (a) or (b).

I was not able to tell by your letters what you would expect under these circumstances. Could you let me know, so we can discuss the matter further.

Will you accept my apologies for not answering your letters sooner. I am busy with several matters and do not take care of all my letters right away.

Sincerely,

R. P. Feynman

## RICHARD P. FEYNMAN TO MR. Y, APRIL 3, 1962

*Mr. Y, obviously pleased by Feynman having responded a second time and in such detail, sent a polite response that did not address the issue Feynman had raised. Feynman wrote back.*

Dear Mr. Y:

My letter of March 14 was only to see whether we agree or disagree on what we would expect to happen in a given experimental situation. Of course, v is less than c and need not approach c. (For example, v might be ⅔ c). What would you expect to happen?

If I don't know what your opinion is in this matter I don't see much use to our writing letters back and forth because I don't know whether my views disagree with yours or not. You speak of our "opposite viewpoints" and "my arguments" but I still don't know whether we have opposite viewpoints or any argument on the results of this experiment. In short, I don't know what your viewpoint is.

Therefore I hope to hear your opinion clearly and simply stated in your next letter. If you have any questions about the conditions of the experiment I shall be glad to explain them more fully.

Sincerely yours,

Richard P. Feynman

## RICHARD P. FEYNMAN TO MR. Y, APRIL 10, 1962

*Mr. Y responded by questioning the premises of Feynman's thought experiment when the meson speed gets extremely close to the speed of light, and he raised a number of other qualitative objections to the theory of relativity. Feynman tried once more.*

Dear Mr. Y:

Thank you for your letter of April 7. After reading it I still don't know what you would expect to happen in the experiment in question. The purpose of scientific thought is to predict what will happen in given experimental circumstances. All the philosophical discussion is an evasion of the point. The mesons do not go at the speed of light. To be specific, take the case of mesons of such momentum that the radius of curvature in a field of 10,000 gauss is 44 cm. (corresponding, in the usual method of calculation, to velocity equal to 0.8 of the velocity of light). Suppose the counter C is placed 10 cm. in front of the 360° point—there is surely plenty of room for that. My question is now specific: "The field is 10,000 gauss, the mesons are selected (by a previous magnet and slits) to have a radius of curvature of 44 cm. The counter C is 10 cm. in front of the 360° point. What will be the relation of the number of counts in A and in C?"

This is not an artificial and arbitrary question meant to trap you. It is

designed to see if we differ in our predictive understanding of Nature. We obviously differ in our way of analyzing and thinking, but the real question is whether we differ in our expectations of what will happen in given circumstances. If we do not, then there is no scientific question involved, it is only a matter of philosophic argument, and I would not be able to make a choice of who is right. If we do differ in our expectation then it is very easy to decide who is right. We do the experiment and see.

Please, [Mr. Y.], answer the question. Don't go on telling me I ignore relativity in "my thesis," or speaking of "my philosophy" or that I ignore the fact that a meson is a particle, etc., etc. All that may or may not be true. Are you trying to tell me I can't set up the counters in the arrangement indicated or that a magnetic field of approximately 10,000 gauss cannot be made, or that there are no mu mesons? If not, and you agree I can set up the apparatus in question—then tell me what you expect the counting rate in the counters will be.

Sincerely yours,

Richard P. Feynman

*Feynman's archived papers do not contain any further response from Mr. Y.*

## RICHARD P. FEYNMAN TO DOUGLAS M. FOWLE, SEPTEMBER 4, 1962

*A magazine article on Feynman and Caltech caught Mr. Fowle's eye, and he wrote to discuss Caltech's startling attrition rate of 35 percent. Mr. Fowle pointed out that Caltech accepted only 180 students out of 1,500 of the best and brightest applicants. He went on to say, "the force-out of so much as 35% entails a misdirection of faculty teaching time and talent as well as shameless cruelty to scores of fine young men."*

Mr. Douglas M. Fowle
Milwaukee, Wisconsin

Dear Sir:

I do not think that the methods of selection of freshmen at Cal Tech are good. I have never paid a great deal of attention to it, being "busy" with other things. This is not an excuse but an admission of irresponsibility. Thank you for your letter; perhaps it will help prod me into paying more attention to the problem.

There is today, in my opinion, no science capable of adequately selecting or judging people. So I doubt that any intelligent method is known.

I am not sure, but it is even possible that a large drop-out rate is not a result of selection but of what happens to the poor guys when they get here. For example, a student that has been at the very top of his class for all his previous schooling, finding himself below average at Cal Tech may have a 2:1 chance to get discouraged and drop out, for psychological reasons. No matter how we select them, half the students are below average when they get here.

These human problems are very difficult, and I don't know much about them—but I believe nobody really knows very much about them either.

Thank you for your letter—I surely will be more attentive when such matters are discussed in the future.

Sincerely,

Richard P. Feynman

## RICHARD P. FEYNMAN TO GWENETH FEYNMAN, 1962

Grand Hotel

Warsaw

Dearest Gweneth,

To begin with, I love you.

Also I miss you and the baby and Kiwi, and really wish I were home.

I am now in the restaurant of the Grand Hotel. I was warned by friends that the service is slow, so I went back for pens and paper to work on my talk for tomorrow—but what could be better than to write to my darling instead?

What is Poland like? My strongest impression—and the one which gives me such a surprise—is that it is almost exactly as I pictured it (except for one detail)—not only in how it looks, but also in the people, how they feel, what they say and think about the government, etc. Apparently we are well informed in the U.S. and the magazines such as Time and Atlas are not so bad. The detail is that I had forgotten how completely destroyed Warsaw was during the war and therefore that with few exceptions (which are easily identified by the bullet holes all over them), all the buildings are built since the war. In fact it is a rather considerable accomplishment—there are

very many new buildings—it is a big city, all rebuilt (from 7/8 destruction). Of course, as you know the genius of builders here is to be able to build old buildings. And so it is—there are buildings with facings falling off—walls covered with concrete with patches of worn brick showing thru—rusted window bars with streaks of rust running down the building, etc. Further, the architecture is old—decorations sort of 1927 but heavier—nothing interesting to look at (except one building).

The hotel room is very small, with cheap furniture, very high ceiling (15 feet), old water spots on the walls, plaster showing through where bed rubs wall, etc. It reminds me of an old "Grand Hotel" in New York—their faded cotton bedspread covering bumpy bed, etc. But the bathroom fixtures (faucets etc.) are bright and shiny—which confused me—for they seemed relatively new in this old hotel. I finally found out—the hotel is only three years old—I had forgotten about their ability to build old things. (No attention at all yet from waiter—so I broke down and asked one who was passing for service—a confused look—he called another over—Net result: I am told there is no service at that table and am asked to move to another. I make angry noises—action?—I am put at another table (given a menu and 15 seconds to make up my mind) and order SZNYCEL PO WIEDI-ENSKU (Wiener Schnitzel).

On the question of whether the room is bugged. I look for covers of old sockets (like the one in the ceiling of our shower). There are five of them, all near the ceiling—15 ft—I need a ladder and decide not to investigate them—but there is one similar large square plate in lower corner of room near the telephone—I pull it back a little (one screw was loose). I have rarely seen so many wires—like the back of a radio—what is it—who knows—I didn't see any microphones—the ends of the wires were taped (like connections or outlets no longer in use). Maybe in the tape is the microphone—well I haven't a screwdriver and don't take the plate off to investigate further. In short—if it isn't bugged they are wasting an awful lot of wires.

The Polish people are nice, poor, at least have medium style in (soup arrives!) clothes, etc., so there are nice places to dance to bands, etc. so it is not very heavy and dull—as one hears Moscow is. On the other hand you meet at every turn with that kind of dull stupid backwardness characteristic of government—you know like the fact that change of $20 is not available when you want to get your card renewed at the American Immigration Office downtown. Example, I lost my pencil and wanted to buy a new one at the "Kiosk" here. Pen is $1.10—No, I want a pencil—wooden with graphite. No, only $1.10 pens. O.K. how many Zlotys is that? You can't buy

it in Zlotys only for $1.10. Why—who knows. So I have to go upstairs for American money. Give $1.25—No, cannot give change—must go to the cashier of the hotel—the bill being written in quadruplicate of which the clerk keeps one, the cashier one, and I get two copies. What shall I do with them? On the back it says I should keep it to avoid paying customs duties. It is a Paper-mate made in the USA (the soup dish is removed).

The real question of government versus private enterprise is argued on too philosophical and abstract a basis. Theoretically planning may be good etc—but nobody has ever figured out the cause of government stupidity—and until they do and find the cure all ideal plans will fall into quicksand.

I didn't guess the palace in which the meetings are held right. I imagined an old forbidding large room—from 16th century or so. I forgot that Poland has been so thoroughly destroyed. The palace was brand new—we are in a round room with white walls with gilt decorations—and a balcony and ceiling painted with a blue sky and clouds—(The main course comes—I eat it—it is very good—I order dessert (pastries with pineapple 125 g)—incidentally the menu is very precise—the 125 g is the weight— 125 grams! Things like "filet of herring 144g" etc. I haven't seen anybody checking with a scale that they are not cheated—I didn't check if the Schnitzel was the claimed 100 grams). I am not getting anything out of the meeting—I am learning nothing. This field (because there are no experiments) is not an active one, so few of the best men are doing work in it. The result is that there are hosts (126) of dopes here—and it is not good for my blood pressure—such inane things are said and seriously discussed— and I get into arguments outside the formal sessions—say at lunch—whenever anyone asks me a question or starts to tell me about his "work." It is always either—(1) completely un-understandable, or (2) vague and indefinite, or (3) something correct that is obvious and self-evident worked out by a long and difficult analysis and presented as an important discovery, or (4) a claim, based on the stupidity of the author that some obvious and correct thing accepted and checked for years is, in fact, false (these are the worst—no argument will convince the idiot), (5) an attempt to do something probably impossible, but certainly of no utility, which, it is finally revealed, at the end, fails (dessert arrives—is eaten), or (6) just plain wrong. There is a great deal of "activity in the field" these days—but this "activity" is mainly in showing that the previous "activity" of somebody else resulted in an error or in nothing useful or in something promising, etc.—like a lot of worms trying to get out of a bottle by crawling all over each other. It is

not that the subject is hard—it is just that the good men are occupied else-where. Remind me not to come to anymore gravity conferences.

I went one evening to the home of one of the Polish professors (young—with a young wife). People are allowed seven square yards per person in apartments—he is lucky and has 21 for he and his wife—living room, kitchen, bathroom. He was a little nervous with his guests—(myself, Professor and Mrs. Wheeler and another) and seemed apologetic that his house was so small (I ask for the check—all this time the waiter has between two and three active tables including mine), but his wife was very relaxed and kissed her Siamese cat "Boobosh" just like you do with Kiwi. She did a wonderful job—the table for eating had to be taken from the kitchen—a trick requiring the bathroom door to be first removed from its hinges. (There are only four active tables in the whole restaurant now—four waiters). The food was very good and we all enjoyed it.

Oh, I mentioned that one building in Warsaw is interesting to look at—it is the largest building in Poland—the "Palace of Culture and Science" given as a gift by the Soviet Union—designed by Soviet architects, etc. Darling, it is unbelievable! I cannot describe it—it is the craziest monstrosity on land! (The check comes—brought by a different waiter—I await the change).

This must be the end, I hope I don't wait too long for the change. I skipped coffee because I thought it would take too long. Even so, see what a long letter I can write while eating Sunday dinner at the Grand Hotel.

I say again—I love you, and wish you were here—or better I were there. Home is good. (The change has come. It is slightly wrong (by 0.55 zloty = 15¢) but I let it go.)

Good bye for now.

Richard.

## RICHARD P. FEYNMAN TO THOMAS K. MOYLAN, MARCH 25, 1963

*Feynman was appointed to the California Curriculum Commission on March 14, 1963, and soon began the long process of evaluating textbooks used in the state's public schools.*

Mr. Thomas K. Moylan
Manager, Pacific Division
Silver Burdett Company
San Francisco, California

Dear Mr. Moylan:

Thank you for your letter of congratulations. In it you submitted some literature on your books and suggested that we meet. I have not read your material. I would feel uncomfortable to discuss the textbooks with any one publisher without doing the same for all. But to do the latter would take more time than I have available.

I expect to spend most of what time I have in going carefully over each of the textbooks submitted. I hope you do not feel I am being unfair to you. I am sure your books will speak for themselves.

Sincerely yours,

Richard P. Feynman

*There were similar responses to other publishers, including Scott, Foresman and Company, and Laidlaw Brothers.*

## ERNEST D. RIGGSBY TO RICHARD P. FEYNMAN, APRIL 17, 1963

Dear Dr. Feynman:

I am presently engaged in a doctoral research project concerning the nature of scientific methods as interpreted by active research scientists and philosophers of science. Realizing that the bibliographic resources often do not contain critical listings, I am forced to ask for certain assistance directly from certain scientists. If you could give some assistance to this effort, it would be greatly appreciated.

If you are willing, please briefly comment on your interpretation of the scientific method or scientific methods.

We have searched the standard bibliographic sources for statements by eminent scientists, such as yourself, on this topic. It is believed however, that some up-to-date statements directly from these scientists would greatly facilitate our program.

This request for assistance is not meant to infringe on the valuable time of anyone. If you find it inconvenient or undesirable to assist this effort, we shall, of course, understand.

We note that a sizable number of our science textbooks at all grade levels present the scientific method as more rigid than many practicing scientists consider it to be. We further trust that this letter of inquiry will help us to find certain more realistic description of this topic.

If you do not have time or do not wish to make a written response to

*Richard and Carl, 1963.*

this request, perhaps it will be possible for you to direct us to sources in which you have already made statements concerning the scientific methods—often such statements are included in articles which would not be indexed under the heading: scientific method.

Permission to quote you is respectfully requested.

Very truly yours,

Ernest D. Riggsby

Associate Professor

## RICHARD P. FEYNMAN TO ERNEST D. RIGGSBY, APRIL 30, 1963

Professor Ernest D. Riggsby

Troy State College

Troy, Alabama

Dear Professor Riggsby:

I am sorry that I have no published statements on the "scientific method." I have just given three lectures (called John Danz Lectures) at the University of Washington, the first and third of which contain remarks on the subject. How soon they will be published (from recorded tapes), I do not know.

Briefly, however, the only principle is that experiment and observation is the sole and ultimate judge of the truth of an idea. All other so-called principles of the scientific method are by-products of the above which depend on the nature of the material and what is found out. There are, furthermore, a number of tricks (like reasoning from analogy, or choosing the "apparently simplest" explanation) which have been found to increase the ease with which we cook up new ideas to subject to the test of experience.

Sincerely,

R. P. Feynman

## L. G. PARRATT TO RICHARD P. FEYNMAN, AUGUST 30, 1963

Dear Dick:

Did you ever write that letter to me so that I may transmit it to the Messenger Lectures Committee so that they can ask our President (of Cornell, that is) to extend to you the official, formal invitation?

Incidentally, a recent letter from the Commission on College Physics

*Coffee hour at Caltech, June 1964.*

asks if they could put your Messenger Lectures here on video tape for a larger audience. I don't know what the Messenger Lectures Committee might say about this, but, first, what do you think?

Sincerely yours,

L. G. Parratt

Chairman

*The Messenger Lectures have taken place annually since 1924, when Cornell grad and professor Hiram Messenger established a fund "to provide a course of lectures on the evolution of civilization for the special purpose of raising the moral standard of our political, business, and social life."*

## RICHARD P. FEYNMAN TO DR. L. G. PARRATT, SEPTEMBER 6, 1963

Dr. L. G. Parratt

Chairman, Department of Physics

Cornell University

Ithaca, New York

Dear Lyman:

I don't know what you want in your letter of August 30. If I am invited to give the Messenger Lectures I will be happy to accept. My subject would be "The Character of Physical Law"—the title of each individual lecture (how many are there—six?) will be hard to decide until the last minute (unless you insist strongly enough).

With regard to the video tape, I don't know what it is like to do that. All I care about is that it should not be distracting in any way. No strong special lights—no nearby camera in the front of the audience—no mess of cables all over the floor, etc. A simple camera in the projection booth or in the back is O.K. If it can be done so it is not obvious that is being done, so the audience is not disturbed by technicians jumping around and moving lights or cameras around, I see no objection, unless the committee feels it isn't a good idea.

I hope this letter is what you asked for.

Sincerely,

Richard P. Feynman

*Feynman gave a series of seven Messenger Lectures in November 1964.*

## RICHARD P. FEYNMAN TO D. BLOKHINTSEV, JUNE 25, 1964

D. Blokhintsev, Chairman
Organizing Committee
Joint Institute for Nuclear Research
Moscow, U.S.S.R.

Dear Sir:

Thank you very much for your invitation to the Dubna Conference. I have thought a good deal about the matter and would have liked to go. However, I believe I would feel uncomfortable at a scientific conference in a country whose government respects neither freedom of opinion of science, nor the value of objectivity, nor the desire of many of its scientist citizens to visit scientists in other countries.

Sincerely yours,

Richard P. Feynman

. . . .

## D. S. SAXON TO RICHARD P. FEYNMAN, OCTOBER 15, 1964

Dear Professor Feynman,

Dr. Marvin Chester is presently under consideration for promotion to the Associate Professorship in our department. I would be very grateful for a letter from you evaluating his research contributions and his stature as a physicist.

May I thank you in advance for your cooperation in this matter.

Sincerely,

D. S. Saxon

Chairman

Dick:

Sorry to bother you but, unfortunately, we really need this sort of thing.

David S.

## RICHARD P. FEYNMAN TO D. S. SAXON, OCTOBER 20, 1964

Dr. D. S. Saxon, Chairman

Department of Physics

University of California, Los Angeles

Los Angeles, California

Dear David:

This is in answer to your request for a letter evaluating Dr. Marvin Chester's research contributions and his stature as a physicist.

What's the matter with you fellows, he has been right there the past few years—can't you "evaluate" him best yourself? I can't do much better than the first time you asked me, a few years ago when he was working here, because I haven't followed his research in detail. At that time, I was very much impressed with his originality, his ability to carry a theoretical argument to its practical, experimental conclusions, and to design and perform the key experiments. Rarely have I met that combination in such good balance in a student. Was I wrong? How has he been making out?

Sincerely yours,

R. P. Feynman

*The above letter stands out in the files of recommendations. After this time, any request for a recommendation by the facility where the scientist was working was refused.*

## R. C. FOX TO RICHARD P. FEYNMAN, OCTOBER 26, 1964

Dear Prof. Feynman,

I can't tell you how much I enjoyed reading and studying Vol. I of your "Lectures in Physics." There is so much one can say about that book and really so little time to say it. So many books I have hammered and tonged at only to finally get lost before I reached the goal besides being so dry and dusty between cases. True, the rewards gained were worth the effort but so difficult are the ways. So many "obvious" steps are left out, nothing left but the bones, desiccated and almost uninteresting.

Why is it so many lousy books are written, lousy for me anyway and why does a gem like yours come by so rarely?

Surely your conversational like style adds so much. It reminds me of the dialogues of Galileo which started the whole thing. I know that the tendency toward brevity is appreciated by some but not by me.

And the way you handle the interesting little sidelights which I and others have certainly wondered about, but have been afraid of raising the stupid question so necessary to get the answer that you have given us.

Talk about wine; man that book of yours is really intoxicating. I can hardly wait for Vol. II. I'm sure I'm expecting too much but I sure wish there were going to be a Vol. III, IV, V, etc. all built on one another. If you need any encouragement I surely would like to furnish some. Boy would I like to get the lowdown on such things as Tensors, Group Theory, Quantum Mechanics etc. on the same basis as Vol. I. But again that's hoping too much, but I never give up hoping I guess.

Thanks again for that very enjoyable Vol. I. I have a feeling it may revolutionize Freshman Physics. It oughta. It beats Sears so hollow I hope we never have to mess much with that again.

Sorry if I sound off the deep end, but I am. Don't let those stuffed shirts discourage you; you've done a wonderful job. Don't stop now if at all possible.

Respectfully,

R. C. Fox

. . . .

## RICHARD P. FEYNMAN TO R. C. FOX, JANUARY 4, 1965

Mr. R. C. Fox
San Rafael, California

Dear Mr. Fox:

Thank you for your "fan" letter. I did have some trouble with "Editors." Volume I was edited by Leighton and he did a great job, leaving in all the little tidbits and sidelights. Volume II is out now. It was edited by an elaborate organization and I was embarrassed to have to say "put that back in" so often—who am I who thinks every word he utters is so wonderful? Also, the original lectures for Volume II weren't as good as for Volume I anyway. Have patience—Volume III is coming out in a few months—I hope. (Tensors are in Volume II, Quantum Mechanics in Volume III—sorry, no group theory—and no Volume IV,V, etc. I am done, for a while at least.)

Thanks again for the encouragement.

Sincerely,

Richard P. Feynman

## BETSY HOLLAND GEHMAN TO RICHARD P. FEYNMAN, DECEMBER 1, 1964

Dear Dick:

Many, many years ago, you were at Brookhaven and I was in a ridiculous show at the John Drew Theatre in Easthampton. If that doesn't jog your memory, let's try this: I'm a friend of your cousin Peggy Phillip's.

But, of course you remember.

Let's assume all the usual amenities have been exchanged, and I'll get down to business.

Your name has appeared in a number of places in recent years, with enough information to make your work sound interesting and important (which it always was, heaven knows). Has anyone done a long profile of you for a national magazine? If not, I would like to. It does not need to include anything about work you are doing which may be classified, but I do think a simple, clear, understandable explanation of just what you mysterious scientists are like would help to break down that barrier C. P. Snow is always ranting about. As I remember you, your language and imagery were fun, as well as instructive, and you were, if you'll pardon the expression, offbeat.

In recent years I have been working as a writer and editor for magazines, and have a book on twins and twinning about to be published by Lippincott (I have twins of my own). I have had some correspondence on multiple births with your confrere, Albert Tyler.

If I have your permission, I would like to submit you as a subject to either the *Saturday Evening Post* or *Fortune Magazine*. Please let me know if you are agreeable.

I heard from Peggy that you are happily married, and a father. I'm happy for you on both counts.

I look forward to hearing from you at your earliest convenience, as they say in business-like letters.

Best regards,

Betsy Holland Gehman

## RICHARD P. FEYNMAN TO BETSY HOLLAND GEHMAN, JANUARY 4, 1964

Mrs. Betsy Holland Gehman
Carmel, New York

Dear Betsy,

Of course, I remember you very well.

You certainly flatter me by wanting to do a profile for a great big national magazine. I was tempted, but I think I had better crawl deeper into my ivory tower and let C. P. Snow go on ranting. Perhaps for scientists, as for women, our charm is in our mystery. Presumptuous, no—surely women have charm, but scientists?

Anyway, thanks for thinking of me in such a flattering way.

Sincerely,

Richard P. Feynman

## LAWRENCE CRANBERG TO RICHARD P. FEYNMAN, JANUARY 6, 1965

Dear Professor Feynman:

I have just come across your article "The Relation of Science and Religion." This is a topic some of whose aspects we might try to include in a course presently being organized on science for undergraduates. I would

welcome an exchange of views with you on this challenging subject. At the moment let me take issue with you on your statement that "moral questions are outside of the scientific realm."

It was pointed out by Darwin (*Descent of Man*, Chap. IV) that ethical codes are not unique to homo sapiens, and represent a form of social adaptation favorable to survival. As such are they not properly subject to scientific study and improvement?

The distinction between "will" and "should" propositions may be more one of emphasis than of kind. I suggest that moral exhortation is an abbreviated form of conditional prediction, where the condition is often omitted because it is essentially the sine qua non of society: survival. What is the logical difference between statements such as: "if a particle goes from A to B in the minimum time it will follow a parabola", and the statement "if X is to survive he should learn to live with his neighbors"? Considering the very idealized circumstances under which the first is true, and the role of the uncertainty principle, the "will" is an overstatement about any physical system, and the two statements are logically identical. May not the golden rule be a sort of Fermat principle applied to society?

To insist that ethics and science are separate may confuse and weaken both. If Darwin is right, ethics require constant change to adapt to the changing requirements of survival. To exclude science—the quest of reason—from this process of adaptation seems to exclude it from its primary role. The effective functioning of science itself depends vitally on adherence to canons of conduct which are clearly "ethical"—a fact conspicuously clear in our times when we see those canons suffering significant erosion (while the leadership of science stands by).

Sincerely yours,

Lawrence Cranberg

Professor of Physics

## LAWRENCE CRANBERG TO RICHARD P. FEYNMAN, JANUARY 6, 1965

Dear Prof. Feynman,

Rereading my too hastily dispatched letter of this morning I wish I had omitted the phrase "in the minimum time" in the midst of the third paragraph, and the query at the end of that paragraph.

On the golden rule, let me refer instead to the final page of the chapter of Darwin, cited in my original letter.

The term "leadership" in the last paragraph is to be interpreted as referring to the leadership of several important organizations of scientists.

Sincerely yours,

Lawrence Cranberg

## RICHARD P. FEYNMAN TO LAWRENCE CRANBERG, MARCH 3, 1965

Professor Lawrence Cranberg

Department of Physics

University of Virginia

Charlottesville, Virginia

Dear Professor Cranberg:

Thank you for your letter discussing "The Relation of Science and Religion." I didn't really intend to insist that ethics and science are separate, but rather that the fundamental basis of ethics must be chosen in some non-scientific way. Then, when this is chosen, of course, science can help to decide whether we should or should not do certain things. Science can help us see what might happen if we do them, but the question as to whether we want something to happen depends on a choice of the ultimate ethical good. As you mentioned, such a choice then does not say there's a separation of the fields, and we cannot argue about the choice for the ultimate good of each. You have chosen survival as an ultimate value, then there is no longer any non-science ethic that serves the ultimate value. If we have two alternate ways in which we might survive, one in which the survivor is secure, but miserable, and the other that we are equally secure in the survival, but not happy in the living or our willingness to choose between them, from survival of what is the right race of individuals—how to balance the two— could the survival of the German Reich justify the actions of the tyrant and a religious martyr's values because they put individual survival below some greater good?

All I am trying to do is cast some doubt or confusion into the principle that survival permits ethics without question and that all people will agree that survival is the real determinate of good. If you can see that there may be some doubt about that, who would resolve the doubt for science?

So, there is no logical difference between the statements "if a particle

goes from A to B in the minimum time, it will follow a parabola," and "if x is to survive, he should learn to live with his neighbor."

Sincerely yours,

Richard P. Feynman


## RICHARD P. FEYNMAN TO ALAN SLEATH, APRIL 7, 1965

*The following letter refers to a transcript of Feynman's Messenger Lectures, later published in book form as* The Character of Physical Law.

Mr. Alan Sleath

The British Broadcasting Corp.

London, England


Dear Alan and/or Fiona:

I am returning Lectures 4 thru 7 to you. I think they are rather poor in their present forms as the English is horrible—the sentence structure atrocious! (I appreciate that these atrocities are of my own doing.) I do not have the time to edit these lectures into reasonably-readable English. I have made some minor corrections to make the physics clearer, but I do not have time to do more than this.

I understand that your intention is to publish these so that you have something to give to the people to refer to. I am willing to go along with that even if you publish it in its present form; however, to protect my reputation in this regard, could some statement be made or explanation given. Possibly in the preface you could say that these lectures were not given from a prepared manuscript—that this is a verbatim, direct report, presented before a live audience who could see me waving my arms, etc.

You may wish to decide after reading the results of these lectures in printed form that you don't want to publish them at all. If this is the case it will be perfectly satisfactory to me.

Kindest personal regards and many thanks.

Sincerely,

Richard P. Feynman


. . . .

## RICHARD P. FEYNMAN TO R. E. MARSHAK, MAY 18, 1965

Mr. R. E. Marshak
Department of Physics and Astronomy
The University of Rochester
Rochester, New York

Dear Marshak:
I am sorry, but I shall not have time to write the article you want for the Bethe volume. You make me feel uncomfortable—I like Hans so very much that I feel I "ought" to do what you want—but who invented this infernal idea of writing an article for a guy when he gets to be 60? Isn't there an easier way to show friendship and regard? I feel like I feel on "Mother's Day."
Sincerely,
Richard P. Feynman

## BARBARA KYLE TO RICHARD P. FEYNMAN, AUGUST 13, 1965

To Richard Feynman from an ignorant layman on first hearing (but not yet having read—it's on order) the Messenger Lectures.

What do I understand? That when you go to count which particles through which holes have made their way, the light you shine to let you see, changes the situation and makes them—as who would not—disappear.

I understand that when you want to see how fast they go or exactly what they look like, you alter their speed and change their character.

So we, under new carbon street lights, turn green and turn away, both from the ugly glare and your too curious inspection; and having escaped detection—reappear.

At a deeper level I can understand that you feel the need to question your assumptions—data—what you take for granted.

If you've invented them to play out roles in your equations, perhaps there are more of them than you've allowed for, and these slip past unrecognized, uncounted through the holes.

I can understand that you want all your not-yet-proved-wrong hypotheses to be together reconcilable, and that just now this is not so.

The maths give you, you say, infinity when you'd expected (what was it?

I didn't catch it) was it zero? Or another number preconceived? If zero—is that so different from infinity? Both of them circles?

I know, I know, to ask this question, or any others I am likely to thinking, is to offer a six-figure combination for a known five-figure lock and I apologize for this futility.

Then as a layman what from your lectures do I learn? What Message gets across to my rapt ignorance? I only know the Lecturer has an honest face.

All that I understand of what he says makes sense. All that I fail to understand feels like it's sense. I understand that as the explanations click into place the whole is beautiful, and then the tourists come and leave their litter.

But then you'll move on to biology, and I wish I might be there to see it.
Barbara Kyle

## RICHARD P. FEYNMAN TO BARBARA KYLE, OCTOBER 20, 1965

Miss Barbara Kyle
Dorking, Surrey
England

Dear Ms. Kyle,
May I thank you for your letter.

From the list of "What do I Understand" I am very happy to see so much really understood. You get a high grade from the professor—perhaps 90%. Not 100% because you didn't understand why getting infinity from a calculation is so annoying.

Thinking I understand geometry and wanting to cut a piece of wood to fit the diagonal of a five foot square, I try to figure out how long it must be. Not being very expert, I get infinity—useless—nor does it help to say it may be zero because they are both circles. It is not philosophy we are after, but the behavior of real things. So in despair, I measure it directly—lo, it is near to seven feet—neither infinity nor zero. So, we have measured these things for which our theory gives such absurd answers. We seek a better theory or understanding that will give us numbers close to what we measure. We are seeking the formula that gives the square root of fifty.

Will you think it is only politeness if I say sincerely, I rarely find from laymen such real understanding as I find in your letter?
Sincerely yours,
Richard P. Feynman

# 1965

---

∴∴∴
∴∴∴∴∴
∴∴∴∴∴
∴∴∴

---

## The Nobel Prize

A telegram arrived on October 21, 1965.

> Royal Academy of Sciences today awarded you and Tomonaga and Schwinger jointly the 1965 Nobel Prize for physics for your fundamental work in quantum electrodynamics with deep-ploughing consequences for the physics of elementary particles stop prize money each one third stop our warm congratulations stop letter will follow

A deluge of congratulatory letters and telegrams arrived soon thereafter. What follows is the best of the bunch—letters from colleagues and other scientific luminaries, long-lost friends and acquaintances, former teachers and students. One friend from Far Rockaway High School days wrote, "I only remember you, of course, as a skinny—bright boy who loves music—fun too—with a wicked twinkle in your eye. From your picture in the paper—you look very much the same." Another wrote with congratulations, remembering how Feynman contributed to their solid geometry class, the teacher "sitting back on his haunches letting you do his work for him. It's strange how that seems to be one of the few classes I can recall with any degree of vividness."

. . . .

## RICHARD P. FEYNMAN TO ERIK RUNDBERG, NOVEMBER 4, 1965

Professor Dr. Erik Rundberg
The Royal Swedish Academy of Science
Stockholm, Sweden

Dear Professor Rundberg:
Thank you for your letter confirming the telegram of my award of the Nobel Prize. You must know, of course, of the great excitement and pleasure such a telegram has meant to my family, my friends and acquaintances. I feel grateful and honored to have been considered for such a prize.

My wife and I are eagerly planning our trip to Stockholm. We expect to arrive late on December 7, and will surely attend, with great pleasure, the dinner on December 8 at your home. Thank you very much for the invitation.

I shall be glad to give my lecture on the afternoon of December 11, as you suggest. I shall send you the title as soon as I am sure of it.

We look forward to meeting you.
Sincerely yours,
Richard P. Feynman

## JULES GREENBAUM TO RICHARD P. FEYNMAN, OCTOBER 22, 1965
## TELEGRAM

Dr. Richard Feynman:
California Inst. of Technology Pasadena Calif. Congratulations on your well-earned achievement. Our best wishes to you.
Jules and Rosalee Greenbaum

## RICHARD P. FEYNMAN TO MR. AND MRS. JULES GREENBAUM,
## OCTOBER 23, 1965

Mr. and Mrs. Jules Greenbaum
New Rochelle, New York

Dear Rosalee and Jules,
I was surprised and happy to learn that I had won the prize, but to hear

from you stirred memories of long ago. You too, must remember someone who would have really been happy this day.

Always,

Richard P. Feynman

*Jules is Arline's brother.*

## LYNDON B. JOHNSON TO RICHARD P. FEYNMAN, OCTOBER 22, 1965 TELEGRAM

Good news is always good medicine. I was delighted to learn of your winning the Nobel Prize in Physics. The nation is grateful for your example and I am personally thrilled to share in your happiness. Well done.

Lyndon B. Johnson

The White House

Washington, DC

*President Johnson had recently undergone successful gallbladder surgery. Two days prior to sending this telegram he had displayed his twelve-inch scar to the entire nation during a televised address.*

## RICHARD P. FEYNMAN TO LYNDON B. JOHNSON, OCTOBER 27, 1965

President Lyndon B. Johnson

The White House

Washington, D.C.

Dear President Johnson:

One of the great pleasures of being awarded the Nobel Prize in Physics was receiving your telegram. I was surprised, honored, and delighted by it. May the news continue to be good and quickly restore your good health.

Sincerely yours,

Richard P. Feynman

.  .  .  .

## ROBERT R. WILSON TO RICHARD P. FEYNMAN, OCTOBER 22, 1965

Dear Dick:

What a pleasure and a gratification it has been to all of us in the Lab here that you have received the Nobel Prize you so much have deserved! We are looking forward, though, to your future work that we bet will match that already done. So get busy!

Warmest regards,

Robert R. Wilson

## RICHARD P. FEYNMAN TO ROBERT R. WILSON, NOVEMBER 23, 1965

Dr. Robert R. Wilson
Cornell University
Laboratory of Nuclear Studies
Ithaca, New York

Dear Bob,

You still think you are my boss telling me to get busy! Is there no rest for the weary? Now that I have received the prize, what more do you expect of me.

Thank you very much for your note.

Sincerely,

Richard P. Feynman

## HANS A. BETHE TO RICHARD P. FEYNMAN, OCTOBER 21, 1965

Dear Dick,

I just heard the wonderful news. You certainly deserved the Prize; I have thought so for at least ten years. I am very happy that the Nobel Committee showed such good taste.

Best wishes.

Yours sincerely,

Hans A. Bethe

. . . .

## RICHARD P. FEYNMAN TO HANS A. BETHE, NOVEMBER 30, 1965

Professor Hans A. Bethe
Laboratory of Nuclear Studies
Cornell University
Ithaca, New York

Dear Hans,
You know how much I owe to you so congratulations to you too.
Thank you very much for your letter. Best regards.
Sincerely,
Richard P. Feynman

*Bethe would receive his Nobel Prize in Physics in 1967.*

## EDWARD TELLER TO RICHARD P. FEYNMAN, OCTOBER 27, 1965

Dear Dick:
Congratulations! It is wonderful that you and Julian should share the
Nobel Prize, which I am sure you deserve in a like manner, although in no
other respect do you seem to resemble each other. I think you will illustrate to the Swedes in an excellent manner that not all Americans are alike.

I wish I could be there to observe you when you are on your best
behavior in your interaction with the King of Sweden. It will be quite a
phenomenon.

With best wishes,
Edward Teller

## RICHARD P. FEYNMAN TO EDWARD TELLER, NOVEMBER 30, 1965

Professor Edward Teller
University of California, Davis
Livermore, California

Dear Edward:
Thank you for your kind letter of congratulations. The phenomenon
you wish to observe with my interaction with the King of Sweden fright-

ens me as much as it interests you. Anything can happen, but I suppose nothing really will. I hope I get through it alive. It was good to hear from you.

Sincerely,

Richard P. Feynman

## LAURIE BROWN TO RICHARD P. FEYNMAN, OCTOBER 21, 1965 TELEGRAM

Dear Dick

Congratulations for a recognition long overdue. I have always been proud to know you, and grateful for what you have given me and for what you have added to the excitement, fun, and seriousness of our profession.

Laurie Brown

## RICHARD P. FEYNMAN TO LAURIE M. BROWN, NOVEMBER 2, 1965

Dr. Laurie M. Brown
Physics Department
Northwestern University
Evanston, Illinois

Dear Dr. Brown:

I am all excited by all the congratulations telegrams, but yours stood out and meant an especially great deal to me. It was like another little prize. Thank you very much.

Sincerely yours,

Richard P. Feynman

## RICHARD P. FEYNMAN TO HERBERT JEHLE, NOVEMBER 29, 1965

Dr. Herbert Jehle
Department of Physics
George Washington University
Washington, D.C.

Dear Dr. Jehle:

Thank you for your post card of congratulations. It is all because you showed me Dirac's Article in the library at Princeton. Thank you very much.

Sincerely yours,

Richard P. Feynman

## ISIDOR I. RABI TO RICHARD P. FEYNMAN, OCTOBER 27, 1965

Dear Dick:

I can't tell you how delighted I was to hear the good news, not only on your account, but also on Julian's. He was my first authentic, honest to goodness degree student to get the prize. Tomonaga is a good guy too so I expect you will have a good time together in Stockholm.

I do hope it will be possible for you to stop off in New York either going or coming so that I could convey my pleasure orally.

One word of advice from one whose great day was twenty-one years ago. Don't let it get you down. You will be even more on the shopping list of those who want to waste your time. Tell them to go to hell.

Cheers and best wishes.

Affectionately,

Rabi

## RICHARD P. FEYNMAN TO ISIDOR I. RABI, NOVEMBER 22, 1965

Dr. I. I. Rabi

Department of Physics

Columbia University

New York, New York

Dear Rabi,

Thank you very much for your note of congratulations and advice—I needed advice.

Right after I learned about winning the prize, I heard a refrain going around in my head "It ain't the money, it's the principle of the thing" and a vision of your playing paper and comb twenty-one years ago.

I am sorry, I shall not be able to stop off in New York, either coming or going as I have a letter from a professor suggesting that I take very good care of my time.

Kindest personal regards.

Sincerely yours,

Richard P. Feynman

## JULIUS ASHKIN TO RICHARD P. FEYNMAN, OCTOBER 29, 1965

Dear Dick,

I just want to tell you how everyone here, some who were at Los Alamos, others whom you haven't met, and especially the students, who are avidly reading your lectures, responded with spontaneous pleasure to the news of your Nobel prize. Beauty in physics is its own reward, but the formal recognition serves a general purpose.

With best wishes to you and Gweneth, and to Carl, from all of us,

Ash

## RICHARD P. FEYNMAN TO JULIUS ASHKIN, NOVEMBER 23, 1965

Professor Julius Ashkin

Carnegie Institute of Technology

Pittsburgh, Pennsylvania

Dear Ash:

Thank you very much for your kind note of congratulations. It was great to hear from you. I am sure that if the paper that I finally published was not so carefully corrected and studied by you before publication, it would not have been considered worthy by the Nobel Committee. So you see that I owe you not only the great pleasure of having had your personal acquaintance but also ultimately, no doubt, the fact that I received the Nobel Prize.

Best wishes to the entire family. I hope we meet together soon somewhere.

Sincerely yours,

Richard P. Feynman

## RICHARD P. FEYNMAN TO M. S. VALLARTA, NOVEMBER 22, 1965

Dr. M. S. Vallarta
Comision Nacional de Energia Nuclear
Insurgentes Sur 1079
Mexico 18, D. F. Mexico

Dear Manuel,
Thank you for your telegram of congratulations. May I thank you for your part in this. I appreciate what you taught me and the encouragement you gave me, including collaboration to publish my first paper.
Kindest personal regards.
Sincerely,
Richard P. Feynman

## FELIX BLOCH TO RICHARD P. FEYNMAN, OCTOBER 24, 1965

Dear Dick,
My warmest congratulations to you and to your wife for the Nobel Prize! I wish I could be at the ceremonies in Stockholm to hear what you are going to say.
Wishing you both a most enjoyable trip and with cordial regards
Your Felix.

## RICHARD P. FEYNMAN TO FELIX BLOCH, NOVEMBER 22, 1965

Professor Felix Bloch
Palo Alto, California

Dear Felix,
Thank you for your note of congratulations. I wish I knew what I was going to say at the ceremonies in Stockholm. Any advice on that and how to walk backward from the King without stumbling would be sincerely appreciated.
Kindest personal regards.
Sincerely,
Richard P. Feynman

## DONALD JONES TO RICHARD P. FEYNMAN, OCTOBER 21, 1965
## TELEGRAM

Congratulations on Nobel award. Delighted to see that good textbook writing finally is being recognized.
Donald Jones
Addison-Wesley Publishing Company, Inc.
Reading, Massachusetts

## RICHARD P. FEYNMAN TO DONALD JONES AND STAFF,
## OCTOBER 23, 1965

Mr. Donald Jones and Staff
Addison-Wesley Publishing Co.
Reading, Massachusetts

Dear Mr. Jones, et al:
Thank you for your telegram of congratulations. I was quite unaware of your enormous influence with the Nobel Committee and shall never underestimate the power of a publishing company.

Thanks to all of you who conceived and carried out this ingenious publicity maneuver.
Sincerely yours,
Richard P. Feynman

## TOM DAILEY TO RICHARD P. FEYNMAN, OCTOBER 21, 1965

CALIFORNIA INSTITUTE OF TECHNOLOGY
INTER-OFFICE MEMO
TO       *R. P. Feynman*
FROM     *Tom Dailey*
DATE     *10/21*
SUBJECT
I couldn't find an appropriate card to send you. Congratulations.

. . . .

*Tom Dailey's "card" on Throop Hall.*

## RICHARD P. FEYNMAN TO THOMAS M. DAILEY, OCTOBER 23, 1965

Mr. Thomas M. Dailey
Dabney House
Campus

Dear Tom,
You certainly did find an "appropriate card" to congratulate me. That great sign on Throop was, to me, one of the most exciting and meaningful signs of congratulation that I have received. Please accept my thanks and convey to all your cronies who risked their lives to put it up how much I appreciated and enjoyed it.
Sincerely yours,
Richard P. Feynman

## SANDRA CHESTER TO RICHARD P. FEYNMAN, OCTOBER 22, 1965
## TELEGRAM

Hail the Nobel Prize Committee for its recognition of your unsurpassed achievement in the field of bongo artistry.
Sandra Chester

## RICHARD P. FEYNMAN TO SANDRA CHESTER, DATE UNKNOWN

Dear Sandra,

I was delighted too when I heard about the Nobel Prize, thinking as you did that my bongo playing was at last recognized. Imagine my chagrin when I realized that there had been some mistake—they cited some marks I made on paper some 15 years ago—and not one word about percussion technique.

I know you share in my disappointment.

Thank you,

Richard P. Feynman

## RICHARD P. FEYNMAN TO DAVID KRAUS, OCTOBER 23, 1965

David Kraus
Science Department
Far Rockaway High School
Far Rockaway, New York

Dear Mr. Kraus, et al,

Thank you for your telegram of congratulations.

No one can do anything alone and all the influences of parents, teachers and friends add up to make a success. In my case, I have always been conscious of the very large contribution given me by my high school. I hope you are doing as well for all the other students as you did for me.

Sincerely yours,

Richard P. Feynman

## H. H. BARSCHALL TO RICHARD P. FEYNMAN, OCTOBER 21, 1965

Dear Dick:

When I was a graduate student, some of the other students were making nasty remarks about a young student who had recently arrived and who claimed to know so much that he did not even need to take any courses. A little later, after I had unsuccessfully approached several professors to obtain assistance with a calculation, I decided to try out the newcomer and was impressed and delighted by receiving promptly a complete solution.

Not long thereafter, the mood of the students changed to depression as

we began to wonder whether we were not in the wrong field of study because it was clear to all that we were not in the same class of intellectual ability with you. It was obvious that we had a potential Nobel Prize winner in our midst. My only surprise was how long it took the Swedish Academy to arrive at the same conclusion.

I was delighted to hear the announcement and should like to congratulate you on the well-deserved honor.

Sincerely,

H. H. Barschall

## RICHARD P. FEYNMAN TO H. H. BARSCHALL, NOVEMBER 30, 1965

Professor H. H. Barschall
Physics Department
University of Wisconsin
Madison, Wisconsin

Dear Heinz,

Thanks very much for your very kind letter of congratulations.

You have kept the secret a long time—I never realized that other students had made nasty remarks about me. Maybe it's too late now to become unhappy about it. Especially, after all the nice remarks you made about me in your letter.

I certainly enjoyed knowing you in our school days and also at Los Alamos. I appreciated your letter very much.

Sincerely,

Richard P. Feynman

## RICHARD P. FEYNMAN TO THE BROTHERS OF PI LAMBDA PHI, NOVEMBER 2, 1965

The Brothers of Pi Lambda Phi
Massachusetts Institute of Technology
Cambridge, Massachusetts 02139

Gentlemen:

Thank you very much for your telegram of congratulations. $\pi \lambda \Phi$ (as $\Phi \beta \Delta$) is the place where I was transformed from a one-sided boy scientist

to a more balanced man who could both integrate and dance (although not equally well).

I also remember I left the place with a credit in my account of $70, which I never collected. I hereby release you of that debt since I am feeling flush and happy these days.

Sincerely yours,

Richard P. Feynman

## WALTER BLEAKNEY TO RICHARD P. FEYNMAN, OCTOBER 22, 1965
## TELEGRAM

The entire Dept. of Physics at Princeton Univ. congratulates you on the Nobel award.

Walter Bleakney

## RICHARD P. FEYNMAN TO WALTER BLEAKNEY, OCTOBER 29, 1965

Dr. Walter Bleakney
Palmer Physical Laboratory
Princeton University
Princeton, New Jersey

Dear Dr. Bleakney:

Thank you for your telegram of congratulations in behalf of the Princeton Physics Department.

It seems to me that you ought to congratulate yourselves for having made it possible.

Sincerely yours,

Richard P. Feynman

## LARS SÖDERSTRÖM TO RICHARD P. FEYNMAN,
## OCTOBER 27, 1965

Dear Sir,

The Union of Student of Science at the Stockholm University congratulates you sincerely on your being awarded the Nobel Prize.

On December 13th we are having our Lucia celebration, a specifically Swedish tradition commemorating the return of light to our winter's darkness, to which solemn occasion we have made it a tradition to invite the Nobel Prize winners of science.

We should very much appreciate your attending our celebration during which you will be dubbed a knight of the Order of the ever Smiling and Jumping Frog like formerly most Nobel Prize winners of chemistry and physics.

On your arrival in Sweden we shall have the pleasure of sending you a formal invitation, and we hope that you will reserve this evening for us.

Yours sincerely,

Lars Söderström

President

## RICHARD P. FEYNMAN TO LARS SÖDERSTRÖM, NOVEMBER 19, 1965

President Lars Söderström

Naturvetenskapliga Föreningen

vid Stockholms universitet

Stockholm, Sweden

Dear President Söderström:

Thank you very much for your letter of congratulations. I shall be most happy to become a knight of the Order of the ever Smiling and Jumping Frogs. I have been that way ever since I was advised that I had won the Nobel Prize in Physics.

Sincerely yours,

Richard P. Feynman

## SONNY LUNDIN TO RICHARD P. FEYNMAN, NOVEMBER 13, 1965

Dear Sir,

This is to convey our warm congratulations for your outstanding contributions to the science of physics.

We have been informed, however, that you very likely will have the opportunity to personally visit our city and receive the Nobel Prize from the old, but well balanced and graceful hand of our supreme ruler HM the

King of Sweden, and, we pray and presume, to give a lecture at our institute, as the custom has always been.

We would also be very glad indeed, had you and your possible family the time and opportunity to be present at our "Lucia-festival" in the morning of December 13th. The Lucia morning is something very Swedish, a custom which probably seems quite exotic to the rest of the world. The party is very informal and beings at 7:30 A.M. in the localities of the Student Union, where professors and students of the department for Physics have coffee and glögg (a drink especially composed for this occasion) and enjoy the serene Lucia procession.

We fully understand that you as all academics highly appreciate a morning of sounds sleep, and we are quite aware of our intrusion, but still we innocently hope to see you the morning in question.

Hoping that you will have a pleasant stay in our country, we remain, Dear Sir,

Sincerely yours,

For the Section of Technical Physics

at the Student Union of the Royal Institute of Technology

Sonny Lundin

Chairman

## RICHARD P. FEYNMAN TO SONNY LUNDIN, NOVEMBER 22, 1965

Professor Sonny Lundin
Technical Physics Section
Student Union
Royal Institute of Technology
Stockholm, Sweden

Dear Professor Lundin:

Thank you very much for your letter of congratulations. I also appreciate your kind invitation to attend your "Lucia-festival" on December 13.

I want to reply in the affirmative; however, I have become somewhat confused by several invitations. I am supposed to become a frog sometime during the day. If there is no conflict in time, I shall be happy to be awakened early—thank you.

Sincerely yours,

Richard P. Feynman

*Red carpet entrance at Hughes Aircraft Company.*

## RICHARD P. FEYNMAN TO JOHN M. RICHARDSON, NOVEMBER 1, 1965

Dr. John M. Richardson
Hughes Research Laboratories
Malibu, California

Dear Dr. Richardson:

Boy, was I surprised by the reception I got when coming to Hughes last week. Please convey to all the people there how happy they made me by the reception and how immodestly I enjoyed the applause upon my arrival. I would suspect that in Stockholm the circumstances can be no more exciting than cutting that wonderful cake. I do not think they have Swedish cake makers who can design such an appropriate icing. My congratulatory card is admired by all who come into the office.

I have been avidly reading the book about the Nobel Prize so I shall have some idea of what I am in for. Please give my sincere thanks to everybody, it really made me feel good.

I promise this Wednesday to get back to work and give a lecture on superconductivity. Again my appreciation,

Sincerely yours,
Richard P. Feynman


## MARTIN B. EINHORN TO RICHARD P. FEYNMAN, OCTOBER 23, 1965

Dear Dr. Feynman,

As one of the guinea pigs of the original "Feynman Lectures" of three years ago, I find that I have been entrusted with an amusing responsibility. Now that the lectures have become popular reading among all physics students, a person who heard them is presumed to have understood them all. Joking aside, I enjoyed them very much, and, having now begun graduate study in theoretical physics, I appreciate even more the "head start" I got at Caltech because of the revision of the physics curriculum.

Congratulations on your Nobel award. In an age of "publish or perish" it has been my pleasure to have come into contact with a good physicist who was also a good teacher.

Sincerely,
Martin B. Einhorn

*Leaving Hughes Aircraft Company.*

## RICHARD P. FEYNMAN TO MARTIN B. EINHORN, NOVEMBER 22, 1965

Mr. Martin B. Einhorn
Palmer Physical Laboratory
Princeton University
Princeton, New Jersey

Dear Martin:

Congratulations on having survived the experiment three years ago. It's nice to know that some of the students still exist.

Thank you for your note of congratulations.

Sincerely yours,

Richard P. Feynman

. . . .

## RICHARD P. FEYNMAN TO THE *CALIFORNIA TECH*, NOVEMBER 2, 1965

Dear Sirs:

I prefer to believe that it was a lack of journalistic experience rather than of goodwill which accounts for the qualities of that "extra" which you put out in connection with my receiving the award of the Nobel Prize.*

First of all, that you have used a candid shot instead of a posed picture on such an important occasion is incredible. What will the Nobel Committee think after the months of deliberation and careful selection, to find out how their choice is considered by a newspaper of an institution so closely connected to the winner? And, the misspelling of "says" as "sez" requires no further comment for careless writing and careless proofreaders are all one can expect of the Tech.

There are two glaring faults which show that your reporters are most amateurish in the performance of their duties. The most amazing blunder was to apologize for taking up my time when they interviewed me and to introduce the request by saying "I know you must be very busy with more important matters!" Secondly that the entire article was clear, comprehensible, well-written and accurate can only be explained by a complete disregard of professional standards in these matters. For example, I was not able to find, attributed to me, a single quotation of anything that I did not say myself.

I wonder if you could send me a dozen copies of this article—I should like to use it when more professional organizations ask for an explanation of what I did to win the prize. I should like to be able to give them a copy of what not to write.

Sincerely grateful to you all for a good job.

Sincerely yours,

Richard P. Feynman

## HUBERT H. HUMPHREY TO RICHARD P. FEYNMAN, NOVEMBER 12, 1965

Dear Dr. Feynman:

May I add my heartfelt congratulations on the great honor you have received? The Nobel Prize is fitting recognition of your outstanding achievements in physics.

★ See Appendix IV.

I hope you will continue to make many more notable contributions to man's unending quest for knowledge.

You can be very proud in the honor you have brought to our nation and to international science.

Best wishes.

Sincerely,

Hubert H. Humphrey

## RICHARD P. FEYNMAN TO HUBERT H. HUMPHREY, NOVEMBER 22, 1965

Mr. Hubert H. Humphrey

The Vice-President

Washington

Mr. Vice-President:

Thank you for your very kind letter of congratulations on my winning the Nobel Prize. As you can well imagine receiving this letter from you has added greatly to the excitement and happiness of my family. It was a special pleasure to me to receive a letter from a man that I have admired for so long.

Sincerely yours,

Richard P. Feynman

## JOSEPH E. JOHNSTON TO RICHARD FEYNMAN, OCTOBER 22, 1965

Dear Richie,

CONGRATULATIONS! Too bad they took so long to get around to you. You had me worried for the past ten years even though you did win the Einstein Award a few years back. However the papers clarified things a bit this morning explaining the delay. For a time I thought the bongo drums or the hexaflexagons were diverting you from the main stream. But we all knew it had to happen—and it did.

It's a far cry—thirty six years to be exact—from the days when you used to pedal your bike from old P.S. 39 down to the new high school and ask me for some experiments to do. The school had just opened and all the equipment was brand new and I used to trot out the magnets, overflow cans, or whatever you wanted to play around with. Since it was after three

o'clock you had the place to yourself after I let you in the door near the laboratory.

Then a year or two later, you entered as a student and it wasn't long before you were at it again, this time as a chemistry student. You used to keep the pressure on me until I finally decided to let you do an experiment that was fairly difficult for a high school pupil, namely, the determination of the Avogadro Number. Once we assembled the rather crude apparatus, I left you on your own and a day or so later you came up with an answer of 5.6 x 1023, which I thought was quite amazing and worked with you on refining it. I repeated it several years later in another school where I became chairman of science and then decided to publish it. A copy is enclosed.

I've missed meeting up with you a couple of times along the way—at a Physics Teachers' meetings in N.Y. and once when I lectured at the Mellon Institute in Pittsburgh some years back. There they asked me if I ever came across any outstanding students and I mentioned your name only to be told that you had lectured there only a short time previously.

Whenever I see Abe Bader at Science Chairmen's meetings we usually exchange notes on you, and of course, last night we really went to town. Dr. Wierda and Mr. Barnes are retired and the latter is nearly 85 and living up state. When I write to him or see him on rare occasions he always inquires about you. I know he'd like to hear from you. He lives at 18 Congress St., Trumansburg, N.Y.

Incidentally, we would all like to felicitate you personally on one of your trips east. Your accomplishments make us all feel that teaching has been worthwhile even though we all realize that you would have accomplished what you did regardless of any teacher.

Again, my heartiest congratulations.

Sincerely yours,

Joseph E. Johnston

## RICHARD P. FEYNMAN TO J. E. JOHNSTON, NOVEMBER 24, 1965

Mr. J. E. Johnston
Eastern District High School
Brooklyn, New York

Dear Mr. Johnston:
Thank you for the congratulations! So exciting to hear from you. I

certainly remember all the times I came down on my bicycle from P.S. 39 to work in the High School Lab. I learned a lot from you and it was very inspiring. I was glad to hear all the news about Dr. Wierda and Mr. Barnes, but the story about the electrolysis experiment you had told me before. I certainly remember it very well, including the moment we thought of connecting an electric clock to the same switch that we used to turn on the current through the water. I have always felt though that we did not really measure Avogadro's number, but only what is called the Faraday. We had to use Milikan's value of the charge on the electron and that experiment is the one that deals with the very small quantities. That is the one where all the difficulty is. But I guess it wasn't bad for high school students. As a result of winning this prize I have received so many letters from old teachers and students from Far Rockaway High School that I have resolved to go back and visit the school for a day some time in the beginning of next year. The principal there invited me. I do not know yet which day.

You have apparently forgotten that we met near Columbia University one day in the street and went into a drug store to have sodas and had a long conversation.

Thank you very much again for your letter and more particularly thank you for all that you did for me before and during the time I was a student at Far Rockaway High School.

Sincerely yours,

Richard P. Feynman

## ABRAM BADER TO RICHARD P. FEYNMAN, OCTOBER 30, 1965

Dear Richard,

A belated letter of congratulation has a better chance of being read, since it isn't swamped by a host of others. In any case, you can apply time reversal to it and read it last week.

I was very happy to hear about the long-overdue awarding of the Nobel Prize. I can now bask in the reflected glory and claim that when the committee realized that you were once in my class, they had no choice.

The second volume of your textbook finally reached me and warmed the cockles of my heart when I read the kind reference to a discussion we once had. I have never forgotten another time when you asked Rube Miller and myself about the advisability of your taking courses in

Education so that you could become a high school teacher. I suspect that you would have been a good teacher, but what a waste of talent. We simply laughed at your thought.

A picture published in a Long Island newspaper shows a handsome young boy in your lap. My own son is five years old and insists he is going to win the Nobel Prize himself. He has already constructed a time machine with his Tinkertoy set and was somewhat disappointed when it did not send him back to World War II so that he could join me in the Royal Air Force (I was involved in radar work).

Again let me congratulate you and point out that there is no law or precedent against doing it again. By the way, the message from Mr. Seiden about your call to him was somewhat puzzling. What was it all about?

Best wishes to you and your family.

Yours sincerely,

Abram Bader

## RICHARD P. FEYNMAN TO ABRAM BADER, NOVEMBER 29, 1965

Mr. Abram Bader

John Jay High School

Brooklyn, New York

Dear Mr. Bader:

Thank you very much for your letter of congratulations. I am glad that you have seen the second volume of my textbook. The discussion we had about action was not the only one that I remember. I have a great deal to thank you for in aiming me and educating me in the direction of physics. I did not remember the time I asked you about the advisability of taking courses in education but I see that you supplied me with not only a good education but good advice. Another thing that I remember as being very important to me was the time when you called me down after class and said "You make too much noise in class." Then you went on to say that you understood the reason, that it was that the class was entirely too boring. Then you pulled out a book from behind you and said "Here, you read this, take it up to the back of the room, sit all alone, and study this; when you know everything that is in it, you can talk again." And so, in my physics class I paid no attention to what was going on, but only studied Woods' "Advanced Calculus" up in the back of the room. It was there that I

learned about gamma functions, elliptic functions, and differentiating under an integral sign. A trick at which I became an expert. Many years later when I was teaching mathematical methods of physics, to graduate students at Cornell, one of the students, wanting to object to the fact that I was dealing with such an advanced subject, asked me in what year I learned the subject. Before I could realize his purpose and how much my answer would bother him, I answered immediately, "in High School." It was very exciting for me to have you as a teacher in High School, and you certainly knew how to stretch a young boy's mind to the utmost of achievement. Thank you very much.

I am glad to hear about your son; mine is three and a half and does not have quite the control with his Tinkertoy set as your boy has, for he has not yet been able to construct a time machine, even one that does not work.

Perhaps by this time you know what the message from Mr. Seiden is all about. It is this. It had nothing to do with the Nobel Prize, and was initiated before that Prize was announced. The American Association of Physics Teachers representative, Mr. E. U. Condon, called me to say that they honor every year specially good teachers in physics, and that Professor Philip Morrison had been reading my textbook, and suggested that one possibility would be to honor you as an example of a teacher who knew how to handle (excuse my immodesty) an exceptional student. So, sometime in January, at a meeting in New York, they want to give you some sort of citation. I hope that they have been in touch with you about it. The reason Mr. Seiden was involved is that I had to call him to get your address.

It certainly is a pleasure to have been in touch with you again and to find out that you are well and happy. Best wishes to you and your family.

Sincerely yours,

Richard P. Feynman

. . . .

## RICHARD P. FEYNMAN TO HOWARD LEMBERG, DECEMBER 6, 1965

Mr. Howard Lemberg, Editor
The Chat
Far Rockaway High School
Far Rockaway, New York

Dear Sir:

Thank you very much for your congratulations on my receiving the Nobel Prize.

When I saw the envelope I suddenly remembered from long ago the name of a paper I had read often but had nearly forgotten. I was very pleased by your offering me a chance to write in the paper, and I put your letter aside hoping each day to send you an answer. But I have been far busier than I expected and have not had the opportunity to answer you.

In the meantime, however, another possibility has arisen. I have decided to come and visit my old school sometime in January. At that time I hope I will have a chance to talk to the students directly. If you wish, at that time I could be interviewed by your reporters or do whatever you would like.

I know that the newspaper business depends on timing of the news and articles, so I hope you will excuse my late answer to your special delivery letter. I shall try to make it up to you as well as I can this January, when I hope I may also meet you personally.

Sincerely yours,
Richard P. Feynman

*In an interview on January 10, 1966, Feynman told* The Chat *that when he was a student at Far Rockaway High School he was "no good in English, no good in languages, impossibly poor in drawing, and a goody-good boy in school. I've changed."*

## HAROLD I. LIEF TO RICHARD P. FEYNMAN, NOVEMBER 10, 1965

Dear Dick:

I have been meaning for a few weeks to add my congratulations to the accolades you have been receiving for the most honored recognition society can bestow.

If our teacher of solid geometry back at FRHS, Mr. Augsbury, is still

alive, he must be proud indeed. It's amusing to think how you used to teach that class with "Old Augsbury" sitting back on his haunches letting you do his work for him. It's strange how that seems to be one of the few classes I can recall with any degree of vividness.

I am so happy to see your work earn the world-wide recognition that it assuredly deserves.

With warmest regards,

Sincerely yours,

Harold I. Lief, M.D.

Professor of Psychiatry

## RICHARD P. FEYNMAN TO HAROLD I. LIEF, NOVEMBER 30, 1965

Harold I. Lief, M.D.

Tulane University

School of Medicine

New Orleans, Louisiana

Dear Harold:

Thank you very much for your note of congratulations. One of the big pleasures of getting this prize was to hear from very old friends, that I would not have ordinarily expected to hear from. I remember you whispering German words in my ear, when I was stuck trying to read in the German class. Many thanks for your assistance in getting through.

Sincerely yours,

Richard P. Feynman

## RICHARD P. FEYNMAN TO MORRIE JACOBS, NOVEMBER 24, 1965

Mr. Morrie Jacobs

Wantagh, Long Island

New York

Dear Morrie,

I was disappointed by your note of congratulations, you did not tell me what has happened to you, and how you are making out. Old friends should tell each other what they are doing. I have a boy now of three and a

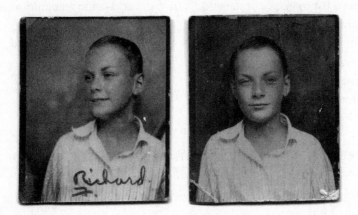

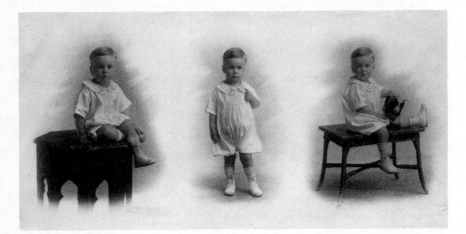

*TOP* Feynman, *1928.*

*BOTTOM* Feynman, *1920.*

half, and a very sweet English wife (guess which came first). I remember you my good friend very well, so many hours spent in the back of your father's luggage shop, discussing serious subjects, and watching you paint signs. Please write me a note telling me more about your present situation.

Sincerely Yours,

Richard P. Feynman


## RICHARD P. FEYNMAN TO BERTHA COHEN, NOVEMBER 15, 1965

Mrs. Bertha S. Cohen

New York, New York


Dear Mrs. Cohen,

Thank you very much for your note about the Nobel Prize. It was good to hear from such an old friend of the family.

I suspect it was something you must have taught me when holding me when I was a baby that enabled me to receive the award.

My best regards to Johanna and to you.

Sincerely yours,

Richard P. Feynman


## RICHARD P. FEYNMAN TO JESSE M. DAVIDSON, DECEMBER 6, 1965

Mrs. Jesse M. Davidson

New York, New York


Dear Aunt Bea:

Thank you so much for your note of congratulations. It is good to hear from someone who had known me for so long. You have gone through all the stages with mother, from ruined linen towels to mom's worrying about whether I would blow up the house with my laboratory.

It all came out pretty nicely, you see.

Love,

Richard P. Feynman


.  .  .  .

## RICHARD P. FEYNMAN TO MIMI PHILLIPS, NOVEMBER 15, 1965

Miss Mimi J. Phillips
Miami Beach, Florida

Dearest Mimi,

Thanks very much for your note of congratulations. It was fun to hear from you and to find out that you are doing so well as a nurse.

Thanks also for sending the newspaper article, I had not seen anything from U.P. before that came. I did appreciate the thought in your sending it along.

Your comment that I have gotten fat from Gweneth's cooking is mistaken. I was thin before when you last saw me because of worry about things and all the running around. . . now I am settled down. Well, maybe the food has something to do with it too!

Love,

Richard P. Feynman

## REVD. JOHN ALEX AND MRS. HOWARD TO RICHARD P. FEYNMAN, NOVEMBER 16, 1965

*Reverend John Howard christened Feynman's English nephew, Christopher Shaw.*

Dear Dr & Mrs Feynman,

This sounds awfully formal, but I don't wish to sound in any way cheeky. Actually, we think of you as Richard and Gweneth, here. But you will not think we do it to be rude, but rather that because of Eric and Jackie and Auntie Minnie, we feel you are part of the family here.

Marjorie and I want to say how tremendously thrilled we were to hear about your being awarded the Nobel Prize. This is simply wonderful, and really a great honour indeed, and everyone here is so very glad for your sake, and sharing in the joy and gladness that has come to you. As a matter of fact, I heard it over the radio, but I just didn't believe it, and thought— "No—it can't be, but the name is very *like* yours!" Actually, they pronounced it "Fineman." But Auntie Minnie said it really *was* you, and that this is how they pronounced your name in the States. You have our sincere congratulations on what must be a very wonderful event in your life, and

we hope that when you go to Sweden it will be a very exciting and uplifting time for you all. I wonder if we shall see you when you pass through? I dare say you won't get to Britain this time.

We saw your book advertised in the Radio Times, and I hope it isn't too difficult for amateurs, for I intend to buy it at Christmas time and read it. This work is all very, very important indeed, and the things that are happening, and those that *will* happen, will stagger the whole world, for we can make life very different altogether for the year 1980 and onwards. I wonder if we shall be able to use all our knowledge and technology to deal with the de-salination of the sea, and the re-forestation of the areas that need trees, and the reclaiming of the desert? It is wonderful to hear of these new petroleum products with which one can spray the deserts, and which allow moisture through, but don't allow it to evaporate out through the sun's heat.

But as well as dealing with the technology, and droughts etc, we have to deal with people, with the color problem, with nationalism, with Communism, and with the greatest of all problems—the materialism and selfishness in all our hearts. Yet I believe that could we rise to the challenge, a new world could be born. The question is, whether we will, because our aims and motives are so small and very often don't go beyond our own families. Yet I'm sure that you people in America think much more globally than we do, and the true American has a big and generous heart. I'm aware that there is sometimes a very different image of an American. So there is of British people!! But underneath, there are still millions of fine people, serving all over the world, and I feel that Vietnam is something that must be fought out, and I back what America is doing, for despite any mistakes made or wrongs done, it is still the Western World that has in it the seeds of Christian Faith, which is better than any alien ideology of hate or class. Our job—yours and mine—is to live this out, and I know that wherever you as a family go, your naturalness and love, and simplicity will give many people new hope and faith. So many thanks to you for your work. Our love from us both. God bless you all.

Yours sincerely,

Marjorie and John Howard

. . . .

## RICHARD P. FEYNMAN TO REVD. JOHN ALEX AND MRS. HOWARD, DECEMBER 6, 1965

Revd. John Alex and Mrs. Howard
Ripponden, Near Halifax,
Yorkshire, England

Dear Marjorie and John:
This sounds very informal, but I want you to continue thinking of us as Gweneth and Richard.

I was very pleased by your fine letter of congratulations. Everything that comes from Ripponden pleases me because the people took me in so nicely that I feel like I am part of it, for part of my family live there. Although we only met on a few occasions, one of my friends in Ripponden is you.

You write that true Americans have a big and generous heart, which shows only what a big and generous heart you have. For you must know that a great nation, at least one where the British ideas of freedom flourish, is very complex and side by side lie the great and the mean, the generous and the selfish, just as they lie side by side in each man.

To see generosity you must be generous enough not to see the meanness, and to see just meanness in a man you must be mean enough not to see the generosity.

I am sorry I will not be able to stop by on the way to Stockholm, but surely I will come back to Ripponden again soon—just to show off how Carl is doing. So we shall have another chance to talk.

Sincerely yours,
Richard P. Feynman

## RICHARD P. FEYNMAN TO ARNOLD H. PHILLIPS, NOVEMBER 18, 1965

Mr. Arnold H. Phillips
Phillips News Co., Inc.
Wheeling, West Virginia

Dear Arnold,
Thanks for your letter of congratulations. We have to do something to make news so you will have something to distribute, so I felt happy to help your business. I heard from Mimi Jean as you might have guessed and she

seemed pretty excited. I keep thinking of her as a little girl, but she isn't anymore, I guess.

Probably you have heard that Joan had a girl baby and Mom was visiting when the prize was announced, so she came running home. Mom seems to be happy in her new apartment which she recently moved into.

Do not be too mad at Mike for his C in physics. I got a C in English Literature. Maybe I never would have received a prize in physics if I had been better in English.

Anyway, thanks for your letter, I hope to see you all again very soon. Gweneth joins me in sending our love to you and yours.

Sincerely,

Richard P. Feynman


## RICHARD D. PETTIT TO DR. AND MRS. RICHARD FEYNMAN, OCTOBER 25, 1965

My dear folks:

Congratulations to you both on the announcement of the Nobel Prize Award to Richard in physics. This is a great achievement and I am very happy for you.

May I add that I was very pleased with the modesty which you met the press. I felt this was a true scientific reaction to such an award, and think the more of you for it.

With sincere personal regards,

Richard D. Pettit, M.D.


*Dr. Pettit was the obstetrician who delivered Carl.*


## RICHARD P. FEYNMAN TO RICHARD D. PETTIT, NOVEMBER 15, 1965

Richard D. Pettit, M.D.

Pasadena, California


Dear Dr. Pettit:

Thank you very much for your letter of congratulations. I was very pleased by it.

*Richard and Carl, 1965.*

I was surprised to read your comment about my meeting the press that you did not mention how cute and wonderful my little boy looks. Could that be modesty?

Sincerely,

Richard P. Feynman

## RICHARD P. FEYNMAN TO DR. AND MRS. HERMAN F. DAVIS, NOVEMBER 29, 1965

Dear Dr. and Mrs. Davis:

Thank you very much for your telegram of congratulations. It is fun to get a telegram from someone you can yell across the street to. So I guess you won't mind a letter even though I could have yelled back, "Thanks."

Sincerely,

Richard P. Feynman

## BETSY HOLLAND GEHMAN TO RICHARD P. FEYNMAN, OCTOBER 27, 1965

Dear Dick:

Your mysterious charm (a direct quote from your own writings) seems to be pushing you out of your ivory tower again.

I know how you hate notoriety, so my sympathies are with you on this sad occasion. Those Nobel people are no respecters of the scientists' rights to pretend they are not great copy.

This is the penalty you pay for thinking. At least for thinking in that extraordinary way.

Despite evidence to the contrary, I hold proof that your thought processes are not too far different than those of the igNobel masses: on the note you sent me on January 4 of this year, you wrote the dateline 1964. . . even as I do for nearly the entire month of January. It takes me that long to admit I'm a year older, too.

Congratulations. Now, back to the I.T.

Yours,

Betsy

## RICHARD P. FEYNMAN TO BETSY HOLLAND GEHMAN, NOVEMBER 23, 1965

Mrs. Betsy Holland Gehman
Carmel, New York

Dear Betsy:

Thank you very much for your note of congratulations. It is very hard to be pushed out of the ivory tower, the light is so strong that it hurts. What hurts worse is to think of myself in tails receiving something from the King of Sweden, while the television cameras are watching. It would be all right if they would just give you the prize, but to make you go through the ringer to get it is a dirty trick.

Thanks for your note, it was fun to hear from you.

Yours,

Richard P. Feynman

. . . .

## JACK LIBERMAN TO RICHARD P. FEYNMAN, OCTOBER 31, 1965

Dear Dick,

I hope after all these years, I may address you in this fashion.

First of all, let me offer my congratulations on your having shared this year's Nobel Physics Prize.

Before I go any further, let me explain who I am. When you were a Senior at M.I.T., I was a Freshman and a fellow member of Phi Beta Delta. I doubt if you remember me, but even then it was easy to see that you were destined to go places. However, it is the human traits I remember. One weekend, for some dance, I had my girl from New York come up and you did not. So when I took her up to the second floor meeting room to neck, you joined us to talk and then remark that we should forget you were there—you just wanted to watch—funny thing, as I recall it, we did forget you were there. I remember your spending a week-end with some deaf and dumb people you met learning the alphabet—I remember a discussion at meal time about how close did froglegs or shrimp (I forget) taste like chicken.

I'm sure I had a hero's crush on you. And since that time, I have followed your career with interest and pride.

Today, I teach Physics and Chemistry at Sharon High School in Sharon, Massachusetts. Actually it's a unified two year sequence based on Phillips Exeter Academy's Course. Then for the past two years, for my Advanced Physics class, I have used your first volume of "The Feynman Lectures on Physics" as the basis for the course. It has been very challenging.

I've been married since 1943, have a 20 year old son and 16 year old twin girls. . . We've lived in Boston, New Hampshire, Florida and now Massachusetts.

So this has just been a note to congratulate you, to wish you continued success, and just to enjoy a letter of your reflected fame as one having known you, years ago.

Sincerely yours,

Jack Liberman

. . . .

## RICHARD P. FEYNMAN TO JACK LIBERMAN, NOVEMBER 30, 1965

Jack Liberman
Bridgewater, Massachusetts

Dear Jack:

You were wrong in your letter of congratulations, I do remember you and was very glad to hear how you were making out. My God, you must have had a hero's crush. Any normal guy under the circumstances would have told me three is a crowd and told me to go to hell, please accept my apologies after all these years for bothering you.

The discussion you described all about how close froglegs taste like chicken seems to be typical of me, I have never tasted froglegs. I seem always to get into discussions of that kind, where I know nothing about the subject. That does not stop me from giving strong opinions however. I am married now, and have only one son of three and one half. If you ever get out this way, please let me know and we can talk over old times.

Sincerely yours,
Richard P. Feynman

## ALAN FARLEY TO RICHARD P. FEYNMAN, OCTOBER 29, 1965

Dear Prof. Feynman:

I'm sure that you won't remember me, but I was one of the original members of your "Physics X" discussions that began after dinner in Blacker House one night. I know that I speak for all the members of that group, as well as all Caltech students who have ever come in contact with you, when I congratulate you on winning the Nobel Prize; we always hoped and expected that you would someday receive it.

My best wishes to you on this happy occasion.

Sincerely,
Alan Farley
Acting Chairman
Dep't of Mathematics
Morehouse College

*"Physics X" was an unofficial, non-credit class Feynman held at Caltech where students prepared physics questions. He would entertain any question*

*about physics, whether it dealt with his specialty or not. Other faculty members were not allowed to attend.*

## RICHARD P. FEYNMAN TO ALAN FARLEY, NOVEMBER 29, 1965

Dr. Alan Farley
Department of Mathematics
Morehouse College

Dear Dr. Farley:
Thank you very much for your note of congratulations. I see that my "Physics X" discussions did not do any good, you ended up as Acting Chairman of the Department of Mathematics. Well, all we can do is try.
Sincerely yours,
Richard P. Feynman

## EDWIN J. WESELY TO RICHARD P. FEYNMAN, OCTOBER 22, 1965

Dear Dick:
My family and I, as well as millions of others, salute you on the great honor of being awarded the Nobel Prize. While I am in no position to judge, I understand from members of your particular "racket," who must be disciples of yours, that the honor is at least a bit over-due.

While your drumming was limited to file cabinets at Telluride House, is the rumor really true that you intend to use the proceeds of the prize for the purpose of developing an improved file cabinet-bongo drum?

Incidentally, while you may recall the double date we had with the Brazilian twins, I am not entirely sure that it has ever been reported to you that I married one of them and we now have two daughters who are ten and eight years old.

Congratulations again on your very great achievement.

With kindest best wishes and warmest regards, I am

Very sincerely yours,

Ed

.  .  .  .

## RICHARD P. FEYNMAN TO EDWIN J. WESELY, NOVEMBER 30, 1965

Mr. Edwin J. Wesely
Winthrop, Stimson, Putnam and Roberts, Law Offices
New York, New York

Dear Ed,

The greatest fun of getting this prize is hearing from old friends. I was very happy to hear from you and I vaguely remember that I knew you were married to one of the Brazilian twins. What's happened to the other one? Oops—I almost forgot that I am a married man and have a boy of three-and-a-half.

It was at Telluride that I did do all that stuff for which I got the prize, so I look back at those days with nostalgia.

Thank you very much for your letter.

Sincerely,

Richard P. Feynman

## DAVID L. ANDERSON TO RICHARD P. FEYNMAN, NOVEMBER 8, 1965

Dear Professor Feynman:

Early in September I suggested your name to the Oberlin Assembly Committee as a possible speaker.

With magnificently inept timing, President Carr, has now, I understand, gotten around to writing to ask you to come. Probably, as the result of recent events, you have had to hire a secretary to do nothing but write letters declining invitations to speak. But I simply want to add to whatever the president said, the hopes of the Physics Department that you might have a free Thursday sometime in the next era of geological time, and that you will be able to come. Articulate scientists are not easy to find.

You may—or may not—remember me as a fellow-tenant of Dorm T-101, or whatever it was, at Los Alamos, and as a fellow participant in the Mack-Brixner week among the Navaho.

In any event, let me be the 13,795th person to congratulate you on the well-deserved prize.

With best wishes,

David L. Anderson

## RICHARD P. FEYNMAN TO DAVID L. ANDERSON, NOVEMBER 22, 1965

Dr. David L. Anderson
Chairman, Department of Physics
Oberlin College

Dear Dr. Anderson:

I am sorry to have to tell you that you guessed right and I am too busy, on account of recent events, to be able to accept your invitation to speak at Oberlin. Would you please communicate this bad news to President Carr.

I certainly remember you very well and the trip with Mack and Brixner among the Navaho. I have pictures of that trip I still like to take out and look at. We went this summer to Monument Valley and that part of the country, with my wife and boy. It has, as you will not be surprised, changed to some extent. But still, the situation of the Navaho seems to be about the same, although, they do have cars and refrigerators and the toys of the children are made of metal and there are fewer horses.

Thank you very much for your congratulations, it was good to hear from you.

Sincerely yours,
Richard P. Feynman


## RICHARD P. FEYNMAN TO ARNOLD O. BECKMAN, DECEMBER 6, 1965

*Arnold O. Beckman, Chairman of the Board of Trustees, drafted a resolution offering sincere congratulations and commendations to Feynman, "with the knowledge that he will continue in the forefront of theoretical physics as a productive member of the faculty for many years to come."*

Mr. Arnold O. Beckman
Chairman of the Board of Trustees
California Institute of Technology
Pasadena, California

Dear Mr. Beckman:

Could you convey to the Board of Trustees my appreciation of their resolution of congratulations. I was flattered by their kind, but exaggerated

evaluations of my past services to Caltech but shall try, in the future, to dis-appoint them as little as possible.

Sincerely,

Richard P. Feynman

## RICHARD P. FEYNMAN TO ALBERT RUDDOCK, NOVEMBER 23, 1965

*After watching Feynman on television, Mr. Ruddock wrote, "My wife and I think if science should ever pall on you that you have a future in broadcasting."*

Mr. A. B. Ruddock
Pasadena, California

Dear Mr. Ruddock:

Thank you very much for your kind note of congratulations. It is good to know that I can fall back on TV broadcasting if the Board of Trustees ever begin to find me out.

I am sorry that I am unable to accede to your desire that I do not answer your note, as the machinery that I have set up for answering congratulatory let-ters does not permit that degree of flexibility. We suffer from the computer age.

Sincerely yours

Richard P. Feynman

## MAX RAFFERTY TO RICHARD P. FEYNMAN, NOVEMBER 22, 1965

Dear Dr. Feynman:

The State Board of Education at its meeting in Los Angeles on November 12, 1965, adopted the following resolution:

WHEREAS the California State Board of Education on March 14, 1963, appointed Professor Richard P. Feynman of the California Institute of Technology to the State Curriculum Commission,

WHEREAS Professor Feynman served for more than a year on this Commission during which time he took the leading role in the selection of mathematics textbooks for California school children,

WHEREAS Professor Feynman has been awarded a Nobel Prize in Physics,

NOW THEREFORE BE IT RESOLVED THAT this Board congratulates Professor Feynman on the greatness of his achievement and calls the attention of the people of California to the service performed for California children by one who in the midst of creative work gave freely of his time to help ensure the understanding of future citizens and the readiness of some one some time to again break ground.

Cordially,
Max Rafferty


## RICHARD P. FEYNMAN TO MAX RAFFERTY, NOVEMBER 29, 1965

Dear Dr. Rafferty:
Please convey to the State Board of Education, my gratitude and appreciation for their resolution of congratulations. I have always felt uncomfortably unhappy about the fact that I felt that I must sever my connection with them just at the time when the science books were coming up for evaluation. I feel it is a bit ironic to be cited for service performed for California children for one year, by a group of men who serve these same children so regularly and so well over so much longer period of time.

I am sorry that I cannot be there to thank you each personally for your good wishes. For in a short time, when I did help the Board, I got to know and like each one of you individually, and have missed seeing you again.

I was surprised, happy and honored to read the resolution. Thank you again.

Sincerely yours,
Richard P. Feynman


## JON A. JOHNSEN TO RICHARD P. FEYNMAN, OCTOBER 21, 1965

Dear Prof. Feynman,
Even though I failed your beautiful physics course last year, I named my cat after you. I now find that my cat not only was given me by Urey's son, but is also named after Feynman (Her name is Richard P.)

Thank you for giving my little Siamese such distinction.

Congratulations!

Sincerely,

Jon A. Johnsen

## RICHARD P. FEYNMAN TO JON A. JOHNSEN, DECEMBER 14, 1965

Jon A. Johnsen

Petaluma, California

Dear Mr. Johnsen:

Thank you very much for your letter of congratulations.

Some measure fame by just a Nobel Prize but I have had a cat named after me! Thank you for such a distinguished and subtle honor.

Sincerely yours,

Richard P. Feynman

## RICHARD P. FEYNMAN TO MADAN MOHAN PANT, NOVEMBER 24, 1965

*Mr. Pant had recently read* The Feynman Lectures on Physics *(Volume III) and commented on how much he enjoyed Feynman's direct approach. He started to think of Feynman as his "pen teacher." He said he danced with joy when he saw Feynman's name in the papers as a Nobel Laureate and wrote to congratulate him.*

*He was also concerned about the first lines of the epilogue in Volume III, in which Feynman said he was going to quit teaching elementary physics, and wanted to know what Feynman was presently working on.*

Mr. Madan Mohan Pant

Allahabad, India

Dear "Pen student":

Thank you for your very kind letter of congratulations. It is a pleasure to hear that as far away as India there are some students who are inspired by some things that I might have done. I appreciate very much your interest in my work. At the moment I am not working very well. Ideas do not come so easily any more. So I suppose that it will be young men like you who

will solve the problems, but I am working on the problem of understanding the strongly interacting particles, i.e., protons, neutrons and mesons. I think that we have almost enough information now from experiment that a very clever man would be able to guess the laws which relate these particles.

May I thank you again for your very kind and flattering letter. I am taking the liberty of sending you a short biography and a picture which I have signed. I wish you the greatest success in your studies and in your life.

Sincerely yours,

Richard P. Feynman


## RICHARD P. FEYNMAN TO P. PAELINCK, NOVEMBER 24, 1965

Prof. Dr. P. Paelinck
Secretariat de la Reine
Brussels
Belgium


Dear Professor Paelinck:

Thank you very much for your kind note of congratulation. I remember very well our meeting at the Solvay Congress, and especially my most pleasant day in your home, meeting your family.

Perhaps, there has been enough time passed now to tell you an amusing story, without embarrassing you. After I got back to the United States, I expressed to my wife the pleasure I had in meeting you and talking with you and your family, and we resolved to send you and the children some books. So we purchased a number of books, most of them for the children, about cats and other childish things, and sent them to you as Secretary of the Queen. After some time we received a note saying that the Queen appreciated very much the books that we sent her. She must have thought that I was some kind of a nut to send her a book containing cat stories for children, but I couldn't figure out a way to straighten it, so I did nothing.

I hope very much to meet you again some day.

Sincerely yours,

Richard P. Feynman


. . . .

## ROGER SPERRY TO RICHARD P. FEYNMAN, DATE UNKNOWN

*A Caltech colleague, Professor Roger W. Sperry, would be awarded the Nobel Prize in Physiology in 1981 for his work on the human brain.*

Hi Dick

I've been waiting for the congratulatory stream to taper off a bit to tell you that we're all tremendously pleased for you over here—(and especially that we hope you may now be *motivated* to devote *full* time to psychobiology!?)

Best,

Roger Sperry

## RICHARD P. FEYNMAN TO ROGER SPERRY, NOVEMBER 30, 1965

Dear Roger:

I have at last gotten around to answering your note of congratulations. Please be advised that I have no intention of donating full time to psychobiology. Now that I have won the prize, I expect to relax and enjoy life and not do science anymore. I will however, be available to give opinions on administrative matters in all fields of science and humanities, and how the freshman can make dates.

Yours,

Richard P. Feynman

## BOB WILLIAMS TO RICHARD P. FEYNMAN, OCTOBER 21, 1965

Dear Dick,

Today's news is gratifying indeed. While there is no dearth of Feynman Appreciators (the list, if I am not wrong, is headed by R.P.F. himself) I nevertheless have always felt that I had some inside information—that this guy is Really on the Track. Glad to see that the rest of the world has caught up.

Congratulations—I wait with interest to see what the next twenty years will bring.

Yours,

Bob Williams

## RICHARD P. FEYNMAN TO BOB WILLIAMS, NOVEMBER 29, 1965

Dear Bob,

It was good to hear from you and get your note of congratulations. You guessed right about who heads the list of Feynman Appreciators, but I must say it is pleasant to know that I can look down the list from time to time, and see that I am not the only one there.

All is going well here and hope all is well with you. Again, thank you for your kind note.

Sincerely,

Richard P. Feynman

## RICHARD P. FEYNMAN TO KOICHI MANO, FEBRUARY 3, 1966

*A former student, who was also once a student of Tomonaga's, wrote to extend his congratulations. Feynman responded, asking Mr. Mano what he was now doing. The response: "studying the Coherence theory with some application to the propagation of electromagnetic waves through turbulent atmosphere. . . a humble and down-to-the-earth type of problem."*

Dear Koichi,

I was very happy to hear from you, and that you have such a position in the Research Laboratories.

Unfortunately your letter made me unhappy for you seem to be truly sad. It seems that the influence of your teacher has been to give you a false idea of what are worthwhile problems. The worthwhile problems are the ones you can really solve or help solve, the ones you can really contribute something to. A problem is grand in science if it lies before us unsolved and we see some way for us to make a little headway into it. I would advise you to take even simpler, or as you say, humbler, problems until you find some you can really solve easily, no matter how trivial. You will get the pleasure of success, and of helping your fellow man, even if it is only to answer a question in the mind of a colleague less able than you. You must not take away from yourself these pleasures because you have some erroneous idea of what is worthwhile.

You met me at the peak of my career when I seemed to you to be concerned with problems close to the gods. But at the same time I had another Ph.D. student (Albert Hibbs) whose thesis was on how it is that

*Receiving the Nobel Prize from King Gustaf VI Adolf.*

*Dancing with Gweneth at the Nobel Ceremonies.*

the winds build up waves blowing over water in the sea. I accepted him as a student because he came to me with the problem he wanted to solve. With you I made a mistake, I gave you the problem instead of letting you find your own; and left you with a wrong idea of what is interesting or pleasant or important to work on (namely, those problems *you* see *you* may do something about). I am sorry, excuse me. I hope by this letter to help correct it a little.

I have worked on innumerable problems that you would call humble, but which I enjoyed and felt very good about because I sometimes could partially succeed. For example, experiments on the coefficient of friction on highly polished surfaces, to try to learn something about how friction worked (failure). Or, how elastic properties of crystals depend on the forces between the atoms in them, or how to make electroplated metal stick to plastic objects (like radio knobs). Or, how neutrons diffuse out of Uranium. Or, the reflection of electromagnetic waves from films coating glass. The development of shock waves in explosions. The design of a neutron counter. Why some elements capture electrons from L-orbits, but not from K-orbits. General theory of how to fold paper to make a certain kind of child's toy (called flexagons). The energy levels in the light nuclei. The theory of turbulence (I have spent several years on it without success). Plus all the "grander" problems of quantum theory.

No problem is too small or too trivial if we can really do something about it.

You say you are a nameless man. You are not to your wife and to your child. You will not long remain so to your immediate colleagues if you can answer their simple questions when they come into your office. You are not nameless to me. Do not remain nameless to yourself—it is too sad a way to be. Know your place in the world and evaluate yourself fairly, not in terms of the naïve ideals of your own youth, nor in terms of what you erroneously imagine your teacher's ideals are.

Best of luck and happiness.

Sincerely,

Richard P. Feynman

# 1966–1969

Although the Nobel Prize brought Feynman greater international recognition, as evidenced by the letters that follow, such acclaim was not without its pitfalls. As people wrote in from Switzerland to Australia, India to Hungary, he began saying "no" more often to the many requests for his time and expertise.

A notable exception was made, however, for continuing work on the California State Curriculum Committee, though he had officially resigned in 1964. In a lengthy memo sent in April 1966 to a Mrs. Whitehouse, a member of the Curriculum Committee, he commented on the relative merits of science textbooks written for elementary school students. Among his remarks was a theme that he revisited many times throughout his career:

> One gets the impression then that science is to be a set of pat formulas to standard questions. "What makes it move," quickly all hands are eagerly raised, the lesson is learned, they are to say "Energy makes it move," "Gravity makes it fall," "The soles of our shoes wear out because of friction." Just words, nothing is explained. It is like just saying "Because of God's will" and having nothing left to look into.

Judging from this memo and other letters he chose to answer, the teaching of physics, and science in general, was now his central concern. I particularly enjoyed reading that he and my brother, Carl (then 3 1/2), did some of the experiments in one of the books under review. It was a harbinger of

the strong relationship and intellectual partnership that would develop between them.

In 1967, Feynman was offered his first honorary degree—from the University of Chicago. He declined to accept it, as he would every other such offer.

In 1968, I was adopted.

. . . .

## VIRENDRA K. SINGH TO RICHARD P. FEYNMAN, OCTOBER 17, 1965

Dear Mr. Feynman—

Long ago I started studying your books which I have finished: My heartiest congratulations for your achievement, as the books are a landmark. I just do not find words how to appreciate them. To put a most difficult theory in a most easy way is a big game and not everybody can do it. I have recommended these books to the head of department and emphasized that every student must have these copies: They are to be read very thoroughly like Ramayana of Hindus.

I have been tremendously influenced by your work and please let me know your academic achievements. As a matter of fact I consider you to be "My Ideal Lecturer" and request you for a full size photograph with your signatures: I would like to have it in my University Dept.

I hope you shall write back. Please do send photograph.

Thanks:

V. K. Singh

## RICHARD P. FEYNMAN TO VIRENDRA K. SINGH, NOVEMBER 23, 1965

Dr. Virendra K. Singh
Rajasthan University
Jaipur, India

Dear Dr. Singh:

Thank you very much for your kind and flattering note of congratulations. I am enclosing a photograph for you.

Sincerely yours,

Richard P. Feynman

## VIRENDRA K. SINGH TO RICHARD P. FEYNMAN, FEBRUARY 7, 1966

Respected Dr. Feynman,

It really hurts me that you have misinterpreted my deep sense of regards and intrinsic faith to you. You have perhaps misfitted the word "flattering." It was partly due to me as I knowingly did not write you I am a physics previous student here. I was afraid you wouldn't send me the photograph. Look! Unpremeditated stream of thoughts for one's praise (which one richly deserves) is NOT flattery. People have different techniques of putting things, mine was simple and straight forward.

I understand one cannot expect such letter (No.1) from a Professor, and it was natural on your part to understand it the other way round.

Believe it, I have just installed your photograph right on my working table.

Thanks:

V.K. Singh

Physics

PS You need not reply, because busy people like you cannot spare much.

## RICHARD P. FEYNMAN TO VIRENDRA K. SINGH, FEBRUARY 14, 1966

Dear Dr. Singh:

I am so sorry to have hurt you by my careless misuse of language. I did not mean to imply that your letter was insincere in any way. The word "flattering" was wrongly used, as I have just looked it up in the dictionary and found it has negative connotations, which I certainly did not mean. I should have used some word like "complimentary" instead.

If you knew me well, you would know that I am immodest enough to take all compliments at their face value and not to think that they might be idle flattering. I just didn't know what the word really implied.

I was sincerely pleased, even nearly overwhelmed by your comparison of my lecture notes with the Ramayana.

I hope I have made no further errors in this letter, and that you no longer feel hurt or uncomfortable in any way as a result of our correspondence.

Sincerely,

Richard P. Feynman

## RICHARD P. FEYNMAN TO J. M. SZABADOS, NOVEMBER 30, 1965

J. M. Szabados
Victoria, Australia

Dear Miss Szabados:

Thank you very much for your kind note about my lectures. I am glad
you like them, and glad you took the time to write to me and tell me. It
seems to me that there is some chance that you may be successful since you
say you have not studied physics in a disciplined fashion. So much the better, but study hard what interests you the most in the most undisciplined,
irreverent and original manner possible.

Best of luck in your endeavor.

Sincerely yours,

Richard P. Feynman

## RICHARD P. FEYNMAN TO E. U. CONDON, DECEMBER 6, 1965

Dr. E. U. Condon
Joint Institute for Laboratory Astrophysics
University of Colorado
Boulder, Colorado

Dear Ed:

I had trouble writing the citation for Abram Bader, because I am personally involved, it is almost like trying to write my own. The best I could do
follows:

"Mr. Bader is cited for his superb teaching of an exceptional student.
Usually the truly outstanding student is left to go his own way. But when
Abram Bader found Richard Feynman (who this year shared the Nobel
Prize in physics) in his class he gave the boy great challenges, good advice
and fascinating new information about physics. He excused him from paying attention in class but challenged him to use the time to learn advanced
calculus from a book from his personal library. He fascinated him by
explaining to him the principle of least action, a central point of almost all
of Feynman's work. Finally the student's love and admiration for his teacher
resulted in the boy's wanting to become a teacher and scientist. Bader gave
good counsel."

The remark at the end is supposed to be subtle humor. He persuaded me, with considerable effort, to become a scientist of course. You may prefer to make it clearer, or avoid the "humor" as it may bother many of the teachers. I trust your judgment. Change the whole thing in any way you wish.

Sincerely yours,

Richard P. Feynman

## RAYMOND R. ROGERS TO RICHARD P. FEYNMAN, DECEMBER 17, 1965

Dear Sir:

I watched and listened to your discussion with members of K.N.X.T.'s news commentators tonight and was amazed at the colossal ignorance and smugness of your learning and achievements. Your use of the term "you guys" in your talk was sickning.

Your comments on smog was of a man intirely ignorant of the problem. You say there are many other problems more important than smog. With a civilization slowly dying in their own filth, what may these problems be? It will be solved just as soon as the financial hurt to the manufucturers and oil companys will be overcome.

I have never progressed beyond high school. My ambition was to attend Throop College which is now Cal Tech. My I.Q. was too low to get in. I served my apprenticeship as a machinist starting at ten cents an hour. My whole life has been to be the best machinist there was.

When I was retired from Technical Labortories (now T.R.W. Systems) on account of age I had worked up as far as I could go without a degree. Your smart young men from Cal. Tech. came over to tell me how things should be done. It sounded like the prattle of small children.

One part of the O.G.O. satalite was so poorly designed I told them so, and I could design one that would really work. They laughed at me, (a poor slob without an education). Two years later when O.G.O was was put into orbit that part was on O.G.O. exactly as I designed it.

Your discussion on atomic Energy was hardly of a man of letters. Your technical gobbledegook did not impress me one bit. Some times I think education is a handicap.

How did you get the Nobel Prize?

Yours Truly,

Raymond R. Rogers

## RICHARD P. FEYNMAN TO RAYMOND R. ROGERS, JANUARY 20, 1966

Mr. Raymond R. Rogers
Gardena, California

Dear Mr. Rogers:

Thank you for your letter about my KNXT interview. You are quite right that I am very ignorant about smog and many other things, including the use of the finest English.

I won the Nobel Prize for work I did in physics trying to uncover the laws of nature. The only thing I really know very much about are these laws. I was asked by the TV station to appear on their news interview program, but what happened was that they asked me all kinds of questions about which I didn't know anything. I had to answer them somehow or other, and I did my best, which you say is none too good.

But, we are both in the same boat, because although you have become a very good machinist and I a good scientist, neither of us really know about the smog problem. Just as my comments on it seem ignorant to you, so your comments on it in your letter do not seem so wise to me.

How would you like to receive a prize for being a great machinist, and then get on the TV to be interviewed by a group of men who don't care a bit about machining and its problems, but instead ask questions about smog? The thing that hurts is that they don't ask you about things that you love and have devoted your life to and received the prize for.

So, please excuse the fact that I wasn't happy and polite during my interview, and had to answer questions about which I had no particular special knowledge.

By the way, one of my ambitions had been to be at least good in the machine shop, but everything I made fit poorly, my bearings wobbled, etc. Good machining is essential to building good apparatus for the precise and careful measurements required in physics to discover Nature's laws. So, we physicists have always worked close to and depended on men like you and some of us (like Rowland, who made the first very precise ruling engines to make diffraction gratings) have been great machinists.

About using the words "you guys"—I am sorry it offended you, but it is because I never believed that people who used big words and very fancy speech were especially smart or good. I think it is important only to express

clearly what you want to say. I admit though, that "you guys" doesn't sound polite, so I guess that wasn't so good.

Yours sincerely,

R. P. Feynman

## RICHARD P. FEYNMAN TO GILBERTO BERNARDINI AND LUIGI A. RADICATI, FEBRUARY 9, 1966

Professors Gilberto Bernardini

Luigi A. Radicati

Scuola Normale Superiore

Pisa, Italy

Dear Colleagues:

Very many thanks for your invitation. I liked Pisa and Tuscany and would surely enjoy being with you. My real trouble is that I like it here too.

I have a nice cozy house with a good family in it, and I do not like to move everything temporarily to Italy for such a long period. I am a stick-in-the-pleasant-mud type of guy.

Thank you again, anyway. Someday shorter periods might be a possibility but a year is too much.

Yours sincerely,

Richard P. Feynman

## J. W. BUCHTA TO RICHARD P. FEYNMAN, FEBRUARY 18, 1966

Dear Professor Feynman:

The following "problem" has been given to us as the topic for a note in *The Physics Teacher*. A lawn sprinkler of the type sketched here rotates clockwise when water is ejected from the nozzles. It also rotates clockwise when placed underwater. How would it behave when placed underwater but the direction of flow reversed, that is water *enters* the nozzles and flows into the connected hose? The problem was labeled *Feynman's* problem, I assume because you may have proposed it. Would you be willing to contribute a note on it for publication in *The Physics Teacher*?

The question appears to afford an opportunity to discuss symmetry,

*Feynman at home, 1966.*

reversibility as well as fluid mechanics. I am sure our readers would appreciate a note from you regarding this problem. I hope I may have a favorable reply from you.

Sincerely,

J. W. Buchta

## RICHARD P. FEYNMAN TO J. W. BUCHTA, MARCH 3, 1966

Mr. J. W. Buchta

Editor

The Physics Teacher

Washington, D.C.

Dear Mr. Buchta:

The lawn sprinkler problem should *not,* please, be labeled "Feynman's problem." I first heard of it from somebody at Princeton when I was a graduate student. We had several discussions, and I did an experiment verifying expectations (which ended in a minor catastrophic explosion). The problem appears in E. Mach's "Science of Mechanics" published in 1883 (translated 1893 by McCormack, published by Open Court Publishing Co., see page 299, in connection with Figure 153a).

I have no other comments to make for the Physics Teacher.

Sincerely yours,

Richard P. Feynman

*The lawn sprinkler problem also makes an appearance in* Surely You're Joking, Mr. Feynman!*: "The answer is perfectly clear at first sight. The trouble was, some guy would think it was perfectly clear one way, and another guy would think it was perfectly clear the other way."*

## THOMAS J. RITZINGER TO RICHARD P. FEYNMAN, MARCH 2, 1966

Dear Dr. Feynman:

I am a physics teacher in Northwestern Wisconsin, teaching five classes of PSSC Physics per day. Because of our geographical location, it is extremely difficult for my students to be able to talk to or visit with scientists and research people very easily.

I thought that, with your cooperation, it might be possible to give them an experience which, I am sure, they will remember for a long time to come. I was wondering if you would consider sometime in the future talking to my students by a long distance telephone connection by way of an appointment call. I would be very happy to fit such a call into your schedule and would be able to have all my students gather into our auditorium so that they might listen to you and possibly ask a few questions of you. I would be able to make all the arrangements through our local telephone company. I felt that, if you will, you could talk to the students for 20 or 25 minutes and then give them an opportunity to ask you some questions. The total time involved would be 35 to 40 minutes.

This is purely experimental on our part, but I think that it would be thoroughly inspirational to my 130 physics students, who represent one half of our junior class.

If you feel that you could take time out of your busy schedule for such an experiment as this, I would be most happy to communicate with you in more detail at your earliest convenience. Thanking you for your kind consideration, I remain

Sincerely yours,

Thomas J. Ritzinger

## RICHARD P. FEYNMAN TO THOMAS J. RITZINGER, MARCH 15, 1966

Mr. Thomas J. Ritzinger
Rice Lake High School
Rice Lake, Wisconsin

Dear Mr. Ritzinger:

What a wonderful idea! It sounds terribly expensive, but if you say so, it is OK with me.

Anyway, let's try this grand telephone call. I think it will work best if I do nothing but answer questions for the whole thirty-five to forty minutes. I'll probably go crazy trying to explain things without a blackboard. But it sounds like fun and I would like to try.

Wednesday and Thursday afternoons and Tuesday mornings are bad for me. Other times are OK except April 2 and April 22–27, when I am going to New York.

A great and original idea that I have never heard of before! (How expensive is it?)

Sincerely,

Richard P. Feynman

*On April 25, 1966, Tom Ritzinger wrote to thank Feynman. He said the youngsters "were most excited, and their remarks since our conversation would indicate to me that they gained much from your answers and remarks."*

## RICHARD P. FEYNMAN TO MARY E. HAWKINS, MARCH 21, 1966

National Science Teachers Assoc.
Washington, D.C.

Dear Mrs. Hawkins:

Please do not set up any News Conference for me (as suggested by M. Ruth Broom in her letter of March 7). My only aim is to talk, as requested on "What is Science" on Saturday, April 2, 1966 at your conference. The purpose of the talk is just to discuss the meaning of science with the people attending the conference. I will have nothing of really great value or importance say, as far as I know, and I surely have nothing of interest for the general public. I do not expect any "coverage" of my talk outside of the audience who comes to listen to me.

Since it is an education conference reporters will want to know about educational matters. I know practically nothing about such things, and therefore, do not think it is worthwhile for me to have any news conference. There will be no news.

Sincerely yours,

Richard P. Feynman

.   .   .   .

## RICHARD P. FEYNMAN TO MRS. WHITEHOUSE, APRIL 13, 1966

*Feynman was no longer officially on the California State Curriculum Commission, but he nonetheless reviewed science textbooks for the commission in 1966.*

Mrs. Whitehouse—Here are some notes on Grades 1 & 5 of six series on science.

**Scott Foresman**—Fair. A spotty mixture of good and poor.

In 1st grade, simple clear experiments on condensation, etc., but most of the stuff on animals is how they differ in superficial ways (nothing on how they grow, eggs, babies, etc.)

In 5th grade, chemistry and sound is good and clear, but material on weather and electricity are not very good. In particular, in both these parts (weather and electricity) the teacher's manual doesn't realize the possibilities of correct answers different from the expected ones and the teacher instruction is not enough to enable her to deal with perfectly reasonable deviations from the beaten track. Also, in these sections, difficult experiments are suggested which may not work out easily as expected, but the teacher is not given clues that this might happen or what to do about it.

**MacMillan**—Good.

First grade is readable and directly understandable. Nice experiments, and a reasonable amount of content. The fifth grade is good science and very real, including many real pictures (not artists' conceptions) but in my opinion, there is *very* much too much material.

**Laidlaw**—Poor, in both grades.

In first grade, much ado about classifying animals by an unsound classification scheme (by fur, wings, hard covers, scales). In teacher's manual, many confusing or unclear statements. Many questions but no direction or motivation (like why classify the animal). Idea is to get them to *say* some preconceived thing about living things.

Very many of the "guiding the discovery" questions have answers in the teaching manual that will require a great deal of interpreting and rewording before they can be understood by children. Not enough for the less scientifically trained teacher to go on in the manual to be able to do it herself.

In fifth grade again there is a lack in teacher's manual. I can rarely read more than one paragraph in the student text without finding something a

little mixed up or imprecise (where to say it precisely and correctly is no harder). It appears as though made by a confused mind rather than a clear thinking one. There is not enough in the teacher's edition to enable the teacher to remedy this lack. Even in the teacher's edition, a very large part is a little bit wrong.

### Heath—Good.

In first grade, simple, correct and good. Not a great deal of stuff. Dealt with straight-forwardly. The teacher's edition is a good clear science lesson for teachers, putting them much in advance of pupils.

In fifth grade, there is a healthy emphasis on making and calibrating good instruments and doing significant experiments. There is a good, practical viewpoint. But there are so many experiments, so many devices to make, etc. Is it not too much?

### Harcourt, Brace & World—Fair.

In first grade an otherwise good text has a very serious weakness that a scientifically minded teacher could remedy, but that a poor teacher would develop happily. In fact, it is the very first lesson which is the most dangerous in this way and everyone can easily start off on the wrong track. I use this first lesson of an example to show what I am worrying about. It begins with a question, "Why does the toy dog move when it is wound up?" A good beginning if we were then to take it apart, see the gear and levers driven by the taunt spring, and see in detail how the thing goes. But the direct question has an obscure and meaningless (to children, and nearly meaningless to me) answer, "Energy makes it move." This same answer is to be given to the same question about real dogs, motorbikes, etc.

One gets the impression then that science is to be a set of pat formulas to standard questions. "What makes it move," quickly all hands are eagerly raised, the lesson is learned, they are to say "Energy makes it move," "Gravity makes it fall," "The soles of our shoes wear out because of friction." Just words, nothing is explained. It is like just saying "Because of God's will" and having nothing left to look into.

Energy in particular, is a very subtle concept, very difficult for first grade to understand (but easy, of course, to learn in a formal, parrot way without understanding). Force is much easier. It is unfortunate that this series starts so poorly. This tendency to try to elicit answers for a certain *form* from the students occurs in the other parts of the first grade book.

Another, less important weakness, I think, is to use artists' imaginary pic-

tures as "evidence" of what happened. (For example, on page 34, the teacher's edition says, "Let them verify their predictions by looking at the fourth picture" etc.). Again, there are no real pictures, for example, of a fossil, and no suggestion is made that a real fossil be brought in.

The fifth grade is good science, rather detailed without many experiments. It seems fine to me if it is not too much.

### Harper & Row—Good.

In first grade, seems very good and goes far. The fifth grade is also very good, in particular processes, as well as facts, are dealt with well, and the discussions of how we learn things is good. Possibly too much (see, for example, vocabulary p. 180).

I have noticed quite a few rather serious errors or misleading statements in both grades—but they are isolated, and can be remedied.

(I cannot help adding, unscientifically that my 3 1/2 year old boy has gotten his hands on this set which the company sent me when I was on the Commission. He often asks me to read to him from them. We have done some of the experiments. I considered them very poor books, and felt silly reading some of the things in them. In comparison to other books, and on looking at the teacher's edition, I realize now Grade 1 and Grade 5 are rather good.

I still feel other grades may have too much of this "we observe, for no good reason." But I cannot include this as I haven't seen the other grades for the other sets.

At any rate, be warned that reading only Grade 1 and Grade 5 may not suffice to judge a set of books from 1 to 6.)

### Conclusion

There are many good sets. But can I express an unscientific opinion? I believe that some of the best are trying to teach too much science. There are many things in many subjects for children to learn, and the detailed names of the parts of a nerve cell, etc., is really unnecessary in the 5th grade. Science should not overwhelm the other subjects. Too much of a good thing will give everyone indigestion. Also, are we already in danger of a general crisis and collapse from teacher overload?

You ask if I would advise co-basal adoption. If that means a school can select one set or another set, perhaps ok. But I do not think that two sets are necessary in the same class, (except possibly for one copy in a library as reference). My cursory examination did not show serious weaknesses in

any of the sets marked "Good" that *needs* to be supplemented by another point of view of another text. There is a tendency for too much science already; we need no more, except for references. (I assume you don't get tempted to adopt Laidlaw because of its apparent simplicity and low coverage.) The coverage level is fine, but the material is not covered clearly. Your teachers may think they like it because it is less to teach and seems simple enough, but rote learning from a teacher following that teaching edition is not a useful training in clear thinking, one of the purposes of teaching science.

Is there any way you can recommend *not* teaching all the stuff in some of those good 5th grade books?

Finally, all the texts assume the existence of a great deal of equipment from snakes, to chicken eggs, to bell wire. How are you going to supply it? I think it is a responsibility for anyone recommending these books to recommend a practical way to help in the supply of adequate experimental materials to go with them.

Wish I can point out detailed errors I happened to notice in the various books.

I am sorry to have not had time to look at Merrill's "Principles of Science."

Regards to all my old friends.

Dick Feynman (signed)

Wed. Morning 3:30 A.M.

## RICHARD P. FEYNMAN TO RICHARD GODSHALL, MARCH 2, 1966

*In a letter of February 19, 1966, Mr. Godshall thanked Feynman for sending his article on "Textbooks for the New Math."\* Mr. Godshall asked for Feynman's opinion of SRA's Greater Cleveland Math Program and said he wanted to buy ten copies of Feynman's article to present to the principal and others at a school meeting. He also wrote at some length on his concerns about "new math." One of the examples Mr. Godshall cited was that after three months of SRA and "many days on properties, I asked a top student to tell me about what she was doing. We were on the distributive property of multiplication over addition.*

---

\* See Appendix V.

$$2 \times 2 \quad = (1 + 1) \times 2$$
$$= (1 \times 2) + (1 \times 2)$$
$$= \underline{\quad} + \underline{\quad}$$
$$= \underline{\quad}$$

*Her comment still rings in my ear, 'I know what 2 × 2 is but I don't get all this jazz.' Yes, and how many other students are asking, 'Why all this jazz?'"*

Mr. Richard Godshall
San Jose Christian School
San Jose, California

Dear Mr. Godshall:

I should like to comment on your remarks on the SRA program. They point up the greatest and most serious criticism of much of the "new math" program. It affects teachers' and parents' confidence. Students have a resilience and a skill at recognizing "all that jazz" as just "jazz." It was a child who understood the emperor's clothes!

I believe that a book should be only an assistance to a good teacher, and not a dictator. Please have confidence in your common sense and protect the children from being intimidated by the unnecessary abstractions and pseudo-sophistications of the school books. Stay human, and on your pupil's side.

If you need a trick way to remember the symbols for "greater" or "less than," just note the symbol > has a wide side (here, on the left where the two lines diverge) and a narrow side (the point). The number on the wide side is bigger than that on the narrow side ($9 > 5$ and $5 < 9$).

I did have a hand in selecting the books and reluctantly settled on SRA despite many faults (the worst of which is the lack of word problems). We had to select something from what was offered and all the others were even worse (hard though that may be to believe!). We recommended some supplementary books to go with SRA to help offset its formal austerity, lack of word problems, etc. to help make a balanced program. At the last minute the state senate decided to save a little money and cut out all supplementary books.

Enclosed are the ten copies you requested.

Sincerely yours,

Richard P. Feynman

. . . .

## DRAFT OF LETTER WRITTEN BY FEYNMAN

*It is not known to whom it was written, the date it was written, or whether it was ever sent.*

Dear Sir,

Thank you for your criticism of my article on the new textbooks for the elementary grade mathematics.

You are right that I am not an expert on the matters in my essay. I have not tried to write a unit in any subject for these grades, nor do I have a firm grasp of the state of the students' experiences or the ability of the teachers, etc., etc. The only experimental material I had was the textbooks themselves—all of which I read carefully.

You are quite right when you say that it is impossible to make a critique of a popular article, for such an article is filled with examples which are easily misunderstood. Your letter proves your point; it criticizes one misunderstood point after another. I am not responsible for your misunderstandings —many seem to come from careless reading.

First you deny that mathematics which is used in engineering and science is largely developed before 1920 by giving examples all from before 1920. Next you read "a good deal of the mathematics which is used in physics was not developed by mathematicians alone, but to a large extent by theoretical physicists themselves," as saying "most applicable mathematics has been invented by non-mathematicians." "A great deal" is not "most" and certainly not "all." Your examples only show that not *all* applicable mathematics was invented by non-mathematicians.

I deny your assertion that I imply that mathematicians *are at fault* for not concerning themselves with applications. I merely said that pure mathematicians do not concern themselves with applications, and that users of mathematics must pay more attention to the connection between mathematics and the things to which it applies than pure mathematicians are likely to do. Is that not so?

I am not against the abstraction of mathematics—that is what makes it useful. Please read my article again. I don't know from very much experience (for example with many mathematically trained writers of articles on quantum mechanics) that abstraction alone is not enough for many practical purposes. It is also (read "also," not "only") important to understand the connection of the symbols to what they are applied if mathematics is to be truly useful. I have all these textbooks before me. It is easy to see that 17

amperes and 15 volts will be added together rarely if ever. But what do we do with a textbook which (unlike most of the others) says mathematics is useful in the sciences and then gives the example: "Red stars have a temperature of 8000 degrees, Blue stars, 12000 degrees. John sees three blue stars and one red star. What is the total temperature of the stars John sees? I was so pleased to find a text which tried to give any scientific example of the use of mathematics and so shocked at the example (not just one—*all* the examples in the section were adding or subtracting temperatures of various colored stars) that you must not blame me for thinking something is wrong with the textbooks offered for adoption of the State of California. Did you read some of these textbooks?

I do not condemn technical words—I object to learning only the words and nothing to say with the words. About "words" and "facts," I do not object to "words" in my article, only to "words" and *no* "facts." Do you not think some facts should come with the words so something about the mathematicians' words is learned other than *only* the words the mathematicians use? You may think I exaggerate—but read some of the books—for example on geometry I found hundreds of definitions with only two facts (the facts were a closed figure divides a plane into two regions, and the diagonals of a rectangle are equal). I just think that more semi-intuitive knowledge of geometry might be developed in the considerable time the students were studying it than that. If not, then save the time for another subject in mathematics.

You said, "You denounce the use of any base but 10."

Freshman—Oh excuse me, I didn't know you were reading.

Professor—I wasn't.

Sincerely Yours,

[unsigned]

## RICHARD P. FEYNMAN TO EDWIN H. LAND, MAY 19, 1966

Mr. Edwin H. Land

Polaroid Corporation

Cambridge, Massachusetts

Dear Mr. Land:

I had a great time visiting you and your colleagues (whose names I unfortunately do not remember). The experiments provoked many

thoughts, and I'm still thinking. It is too bad I am not at M.I.T., so I could come over and bother you more, for there are several effects which I would like to see again in more detail. There were so many new things that I did not realize until later the relative importance of some of them. I shall write you some of my ruminations, not that you have not already thought of them, and that I am telling you anything new, but rather as a continuing informal conversation to see how well I am learning my lessons.

On the plane I started to think of why the "positive in red, negative in white" gives no color, and I immediately realized there was something else far more fundamental that I did not understand (and this may just be a special case of it). The puzzle is, why, when the regular red and white pictures (giving good color sensation) are slightly offset the color disappears completely over the entire picture. In short, I think I understand (via the retinex view) how to determine the color to be seen, if you see it. But what principle determines the condition under which you do not see it, and why?

The right answer must be determined by experiment and pure speculation has no place—but not being able to experiment I could not keep my mind from wandering on the plane. So I should like to report on my pure speculations just for fun, not in the sense that I think these things are necessarily so, but only for the fun of thinking how exciting it might be.

Our first principle is that the thing is made to live. An animal must recognize a bug as the same bug no matter at what distance, what angle, what illumination, etc. and this is of first importance for the simplest seeing animal. It must sort the clues and *interpret* them. I mean by "interpret" that it must discover (in some way) some "object concept" which will explain what it sees. If you see (in the primary sense of light on the retina) an ellipse, you must perceive it as a circle inclined in space (if that is what it is) immediately and directly. The "circle inclined in space" or the "bug" etc. are what I mean by "object concepts." They are really like theories to explain what you are seeing (in the light-on-retina sense). So size constancy shape constancy (e.g., trapezoids or parallelograms seen are perceived as a rectangle in various positions), lightness constancy, color constancy, etc. are all examples of the action of the "interpretation" step in vision. What is illumination gradient, and what is color variation of an object is sorted by this "interpretation", etc.

I shall call it an "interpreter" although I have no idea of how it works. Our man-made machines are simple, but cannot do this very well. We tend to think that the first stages of vision must be simple and mechanical in this

way too. For example, we suppose gadget A measures the mean illumination, controls level of sensitivity of gadget B, which measures intensity at a point, etc., etc., to explain lightness constancy, and an entirely new mechanism to convert binocular discrepancy to distance, and still another to how we determine that a row of dots form a straight line, and another complex imaginary machine for recognizing ellipses as circles, etc., etc. I do not know how any one of them really works. But it is evident that the net result and function of all are qualitatively very similar—so for the time being, perhaps it is very wise just to recognize (1) that it is done, (2) that it is vital to survival of the simplest sensing animal—so it can use its senses to distinguish food from pebbles, (3) there must be many characteristics of this process that are common from one case to another. Therefore, let us study "interpretation" and its characteristics for itself—leaving detailed mechanisms for future study. (I guess what I call "interpretation" must be what psychologists call "forming the gestalt.")

Therefore your experiments on color vision are of vital psychological importance because they can study this "interpretation" process in an example which is easy to control experimentally and is as little as possible under the influence of conscious effort.

Let us suppose the "method of interpretation" is to supply a theory (or object concept) to explain the clues seen and, if the theory fits all clues, perceives the "vision" as the object, and sends his opinion to the next mental layer above it for a further level of interpretation. I imagine the hierarchy: interesting spots of light; straight lines; rectangles in space; a box lying on the table; Jack's coffin; tragedy and sadness as some kind of series of "interpretations" on one level after another working up through the levels of the brain. We are studying just one interpreter, that of "colored objects."

Thus the lights from the Mondrian are perceived as colored objects, for the theory fits perfectly. And if the illumination changes the theory of "colored objects in unusual illumination" still works perfectly and the colors don't change. Likewise, the red and white superposed pictures produce retinexes for which "colored objects in some illumination" is an adequate theory fitting nearly all the facts (binocular clues, other information changes theory) so one is really aware that there are no objects, and it is only a picture on a screen but they cannot countermand the color interpreter for that is at a lower level, out of voluntary control). But if the pictures are slightly offset, the theory does not work, the "objects" are too complicated, and there are regularities in the double image not explained by it. We no longer see "colored objects."

Is it unbelievable that a part of the brain can make one do much "think-ing", and interpreting and we would be unaware of it and not be able to control it? Maybe. But maybe not. The simplest animal must be able to think at this kind of level if he sees at all. Is it not conceivable that this mar-velous "process of interpretation" once invented by evolution is used over and over again in one layer on top of another as the complexity of the brain evolves? That higher "thinking" is much like lower thinking? That is, that the lower interpreting appears too much like thinking to us. Would it not be advisable to let the lower thinking processes in vision go on auto-matically out of conscious control for efficiency reasons? It is built in and unlearned in man, just as the whole behavior is built in and little learned by insects and other simple animals.

Physicists looking at psychology, through force of habit, try to suggest some simple element be found to study, rather than the whole human brain at once. Often they have suggested using simple animals but the difficulty is our brain is not hooked to them and we have experimental difficulties in determining "how the world looks to a caterpillar." But it is possible we have "simple animals" right in our own brain, left relatively intact in evolu-tion, which our higher centers are hooked to, but which are not under easy voluntary control of these higher centers. One of these "simple animals" may well be the "color interpreter."

Thus I would like to know all you can tell me about when we do and when we do not see the colors. My first guess about interpreters would be two laws: (1) If an "object concept" explains all dues adequately, the scene is interpreted as that "object." (2) If other strong regularities appear which are not explained by that "object concept," the interpretation will either not be made, or it will be made from time to time with other interpretations tried from time to time. If an interpretation is not made, the uninterpreted pic-ture is sent up to higher centers (maybe—whatever that means).

I thought of a few other things, less important. First, what physically comes from the screen from each point of one of your red and white pro-jections is various kinds of "pink light" (where by "pink light" I mean light which is the sum of some amount of the white projectors light plus some amount of the red filtered projectors light. "Pink light" can appear green or brown, etc.). Hence a single slide in one projector containing certain red dyes plus gray absorption could imitate it exactly and it would in prin-ciple be able to send to all your friends' quickie demonstrators: one slide that viewed outside the projector by looking through it appears only gray and red, etc. will produce a "full" color projection. ("Full" means as full as

*Caltech talent show, 1966.*

your red and white projections appear to be—greens, oranges, brown, reds, etc.)

Second, and this is harder, it should be possible to produce an object, a kind of painting in red and gray, which when placed against a dark velvet background and illuminated by a spot of light (which lights it only, and the rest of the room only indirectly) looks to be in "full" color.

I had other thoughts, but this letter, which I started as soon as I got back, has, through many interruptions, taken me two weeks to write.

The threat still holds, if you do not come here and give talks on your experiments, I shall give them.

Yours Sincerely,

Richard P. Feynman

## TOMAS E. FIRLE TO RICHARD P. FEYNMAN, AUGUST 7, 1966

Dear Dr. Feynman:

I need to thank you! In an essay a while ago you expressed your thoughts in some paragraphs starting "... I stand at the seashore..."

When I first heard these expressions I had an immediate deep response of a good feeling, faith, of beauty. I took the liberty to graphically re-arrange your words to make them more expressive to me. Incidentally there are other interesting permutations possible.

On receiving news of my father's death in Germany last July 4, I wanted to share some feelings of words of creation and life with my stepmother and I found myself attempting a "translation" of your thoughts into German. Since it has been a long time since I left Germany my "transla-tion" is undoubtedly poor, but it conveyed something of my thinking and feelings, as expressed by you, to my stepmother.

Why do I write you this letter? Partly to extend my thanks to you, to tell you that with these, to you maybe unimportant lines, you have filled another human being's need. Also, frankly, I enjoy the sensitivity of your "thoughts" especially since to me they indicate the closeness of artistic cre-ation with scientific greatness.

Yours truly,

Tomas E. Firle

*July 7, 1966*

I stand at the seashore,
alone, and start to think...
There are the rushing waves...
Mountains... of molecules,
each stupidly minding its own business...
trillions apart....
yet forming white surf in unison.
Ages on ages...
before any eyes could see...
year after year thunderously pounding the shore as now.
For whom, for what?...
on a dead planet,
with no life to entertain.
Never at rest...

tortured by energy
wasted prodigiously by the sun. . .
poured into space.
A mite makes the sea roar.
Deep in the sea, all molecules
repeat the patterns of one another
till complex new ones are formed.
They make others like themselves. . .
And a new dance starts.
Growing in size and complexity. . .
living things,
masses of atoms, DNA, protein. . .
Dancing a pattern ever more intricate.
Out of the cradle onto the dry land. . .
Here it is standing. . .
atoms with consciousness. . .
matter with curiosity.
Stands at the sea. . .
wonders at wondering
I. . .
a universe of atoms
an atom in the universe.

Richard P. Feynman, *Science and Ideas,* edited by A. B. Arons (Prentice-Hall, 1964), p. 5.

## RICHARD P. FEYNMAN TO THOMAS E. FIRLE, OCTOBER 4, 1966

Mr. Thomas E. Firle
Del Mar, California

Dear Mr. Firle:
I need to thank you. Thank you for noticing and enjoying what I tried to "hide" in my lecture. Really, I thought of it poetically also, but was afraid of ridicule if I tried to write a poem in a lecture. The rearrangement is almost exactly the way I would have done it. In fact, you should see my hand written lecture notes that I made to guide me as I gave the lecture for the first time—they are written out in lines just as you "deciphered."

Naturally, I am very pleased and flattered that you thought them worth translating to German. But, it means even more to me that you found the thought of some relevance to you in the face of your father's death.

Sincerely yours,

Richard P. Feynman

## RICHARD P. FEYNMAN TO R. HOBART ELLIS, JR., OCTOBER 3, 1966

*As a reply to a questionnaire from* Physics Today, *Feynman wrote, "I never read your magazine. I don't know why it is published, please take me off your mailing list. I don't want it." On August 25, 1966, the editor wrote to say that Feynman's reaction "poses some interesting questions for us." His main concern was where the magazine had failed. "Is it in the nature of the purpose or in the way we are serving it? . . . If* Physics Today *is not the magazine physicists want and need, we would like to supply what they do want and need" Here is Feynman's response.*

Mr. R. Hobart Ellis, Jr.

Physics Today

New York, New York

Dear Sir:

I'm not "physicists," I'm just me. I don't read your magazine so I don't know what's in it. Maybe it's good, I don't know. Just don't send it to me. Please remove my name from the mailing list as requested. What other physicists need or don't need, want or don't want has nothing to do with it.

Thank you for spending all the time to write such a long letter to me. It was not my intention to shake your confidence in your magazine—nor to suggest that you stop publication—only that you stop sending one copy here. Can you do that, please?

Sincerely yours,

Richard P. Feynman

## IRWIN L. SHAPIRO TO RICHARD P. FEYNMAN, OCTOBER 21, 1966

Dear Professor Feynman:

I thought you might be amused to know that after a film of your superb

lecture on the "Great Conservation Principles" was shown here last week several in the audience were overheard proposing that you run for governor. It was not clear, however, whether the intention was for you to be a candidate in California or here in Massachusetts.

Sincerely yours,

Irwin L. Shapiro

## RICHARD P. FEYNMAN TO IRWIN L. SHAPIRO, DECEMBER 6, 1966

Professor I. L. Shapiro
Massachusetts Institute of Technology
Lincoln Laboratory
Lexington, Massachusetts

Dear Professor Shapiro:

They were, of course, speaking of California. At this time I feel it is better to issue releases denying an interest but not completely dashing the hopes of those working at the grass roots. At the appropriate time, I shall announce my reluctance but pleasure in accepting the responsibilities urged upon me by my constituents. Until then, please make no public statement, although privately you may feel free to give a word of encouragement and hope to those who are working hardest on my behalf.

Thank you for your report on conditions in Massachusetts. Please be assured that I shall not forget my good friend, Irwin Shapiro, when I am in a position to help you more directly.

Sincerely,

Richard P. Feynman

## RICHARD P. FEYNMAN TO MIKE FLASAR, NOVEMBER 9, 1966

Mr. Mike Flasar
Boston, Massachusetts

Dear Sir:

The information you have about the number and need for theoretical and experimental physicists is quite correct. The B in mathematics is of no

consequence, as that kind of mathematics, although popular, is not really necessary in physics, theoretical or experimental.

Work hard to find something that fascinates you. When you find it you will know your lifework. A man may be digging a ditch for someone else, or because he is forced to, or is stupid—such a man is "toolish"—but another working even harder may not be recognized as different by the bystanders—but he may be digging for treasure. So dig for treasure and when you find it you will know what to do. In the meantime, you don't need to make the decision—steer your practical affairs so the alternatives remain open to you.

In any of the graduate schools you mentioned there is always the opportunity to change from theory to experimental or vice versa at any time.

While looking for what fascinates you, don't entirely neglect the possibility that it may be found outside of physics. The man happy in his work is not the narrow specialist, nor the well-rounded man, but the man who is doing what he loves to do. You must fall in love with some activity.

Yours sincerely,

Richard P. Feynman


## RICHARD P. FEYNMAN TO JEHIEL S. DAVIS, DECEMBER 6, 1966

Mr. Jehiel S. Davis
Van Nuys, California

Dear Sir:

I am sorry, but I have no information regarding the color television used at the Chicago Worlds Fair in 1922 nor do I know who to refer you to for this information.

Sincerely yours,

Richard P. Feynman

. . . .

## RICHARD P. FEYNMAN TO ASHOK ARORA, JANUARY 4, 1967

Master Ashok Arora
Debra Dun (U.P.) India

Dear Master Arora:

Your discussion of atomic forces shows that you have read entirely too much beyond your understanding. What we are talking about is real and at hand: Nature. Learn by trying to understand simple things in terms of other ideas—always honestly and directly. What keeps the clouds up, why can't I see stars in the daytime, why do colors appear on oily water, what makes the lines on the surface of water being poured from a pitcher, why does a hanging lamp swing back and forth—and all the innumerable little things you see all around you. Then when you have learned to explain simpler things, so you have learned what an explanation really is, you can then go on to more subtle questions.

Do not read so much, look about you and think of what you see there.

I have requested the proper office to send you information about Caltech and the availability of scholarships.

Sincerely yours,

Richard P. Feynman

## RICHARD P. FEYNMAN TO TORD PRAMBERG, JANUARY 4, 1967

Mr. Tord Pramberg
Stockholm, Sweden

Dear Sir:

The fact that I beat a drum has nothing to do with the fact that I do theoretical physics. Theoretical physics is a human endeavor, one of the higher developments of human beings—and this perpetual desire to prove that people who do it are human by showing that they do other things that a few other humans do (like playing bongo drums) is insulting to me.

I am human enough to tell you to go to hell.

Sincerely,

Richard P. Feynman

# A. R. HIBBS TO RICHARD P. FEYNMAN, JANUARY 10, 1967

*Dr. Albert Hibbs was a graduate student of Feynman's, a co-author (of* Quantum Mechanics and Path Integrals*), and a very close friend. He held many positions at the Jet Propulsion Laboratory over the years, but he was best known as the "voice of JPL," for his work as a broadcast spokesman for many of its planetary missions. (Incidentally, my husband and I were married at his home; Dr. Hibbs performed the ceremony, as he was an ordained minister—by mail—of the Church of the Mother Earth.)*

Jet Propulsion Laboratory
Interoffice Memo
Subject: Application for Astronaut

Here is the reference form which I have spoken to you about. My application as an astronaut is made with considerable optimism. I am both over-age and over-height. Nevertheless, I am counting on the phrase which the National Academy has used in describing these limitations: "Exceptions to these requirements will be allowed in outstanding cases." Obviously, I have to represent myself as outstanding in some particular area. I have a good background in both space science and particularly in space instrument systems. The broadness of this background may be unique, but this may not be too important for the Apollo Project. Nevertheless, I am counting on the desire on the part of the Academy and NASA to find what they term, "astute and imaginative observers," and I hope that my background qualifies me in this area.

But I think the one field where I might be called exceptional is my experience in communicating scientific results to others. So it is as a combined observer and communicator that I hope to present myself as an exceptional case.

Thank you in advance for your help.

—Al

.   .   .   .

## BETTE BRENT TO THE NATIONAL ACADEMY OF SCIENCES, JANUARY 25, 1967

Scientist as Astronaut
National Academy of Sciences
National Research Council
Washington, D.C.

Dear Sir:
I am attaching a Confidential Reference Report for Albert Roach
Hibbs as completed by Professor Richard P. Feynman of Caltech.
I have taken the liberty of typing Professor Feynman's comments for
easier reading for S–1, D–8, D–9 and D–10.
Sincerely yours,
(Miss) Bette Brent
Secretary to Richard P. Feynman

### Comments and Summary
Applicant: HIBBS, Albert Roach

### S–1
Hibbs' only weakness is that he is not a very highly-trained specialist in some particular scientific corner. For your needs this
should prove an advantage—for his scientific understanding and
spirit is ideal—just what is needed to study unexpected phenomena. Too much training, say in geology, could lead to expecting
features on the moon like those on the earth—whereas a more
open but extremely careful and observant mind (as Hibbs has)
would see more clearly what is really there. His general calmness,
yet deep and wide scientific attitude and interest make him an
ideal observer. Finally, do not forget the wide experience in
explaining science on TV and radio—how well he will be able to
tell the world what he has seen, what it means and what the entire
lunar program means.

### Personal Characteristics

### D–8
No opportunity to observe. Knowing the man I would guess that
he would remain unusually clear and observant—he always is—he

takes his scientific spirit of observation into all his experiences and he seeks out unusual experiences.

### D–9
He has done a great deal of exposition both very technical (we wrote a book together) and to non-technical audiences on TV and radio and is very good at it—always maintains his good sense and wisdom about what is important.

### D–10
He has had many positions of responsibility and leadership and I have not heard of any "personality difficulties" in the jobs he has had—although I do not have first-hand information from his colleagues and workers below him in more than two or three cases. I have worked with him intimately without anything but pleasure.

## RICHARD P. FEYNMAN TO GEORGE W. BEADLE, JANUARY 16, 1967

Dr. George W. Beadle
The University of Chicago
Office of the President
Chicago, Illinois
Dear George,
Yours is the first honorary degree that I have been offered, and I thank you for considering me for such an honor.

However, I remember the work I did to get a real degree at Princeton and the guys on the same platform receiving honorary degrees without work—and felt an "honorary degree" was a debasement of the idea of a "degree which confirms certain work has been accomplished." It is like giving an "honorary electricians license." I swore then that if by chance I was ever offered one I would not accept it.

Now at last (twenty-five years later) you have given me a chance to carry out my vow.

So thank you, but I do not wish to accept the honorary degree you offered.
Sincerely yours,
Richard P. Feynman

.    .    .    .

## RICHARD P. FEYNMAN TO TINA LEVITAN, JANUARY 18, 1967

*Tina Levitan wrote to Feynman requesting a biographical sketch and a black and white photograph for inclusion in her book-in-progress,* The Laureates: Jewish Winners of the Nobel Prize.

Miss Tina Levitan
New York, New York

Dear Miss Levitan:
It would not be appropriate to include me in "Jewish Winners of the Nobel Prize" for several reasons, one of which is that at the age of thirteen I was converted to non-Jewish religious views.
Sincerely yours,
Richard P. Feynman

## TINA LEVITAN TO RICHARD P. FEYNMAN, JANUARY 30, 1967

Dear Dr. Feynman,
Your letter of January 18, in which you state that it would not be appropriate for me to include you in my book of Jewish Winners of the Nobel Prize, has been received.

My listing of Jewish Nobel Prize winners includes not only professing Jews and those partly of Jewish ancestry but also those of Jewish origins for the simple reason that they usually have inherited their valuable heredity elements and talents from their people.

Under these circumstances may I include you? I will not represent the fact that at the age of thirteen you were converted to non-Jewish religious views.

If there is another reason why you do not wish to be included, please let me know.
Sincerely yours,
Tina Levitan
(Miss) Tina Levitan

. . . .

## RICHARD P. FEYNMAN TO TINA LEVITAN, FEBRUARY 7, 1967

Dear Miss Levitan:

In your letter you express the theory that people of Jewish origin have inherited their valuable hereditary elements from their people. It is quite certain that many things are inherited but it is evil and dangerous to maintain, in these days of little knowledge of these matters, that there is a true Jewish race or specific Jewish hereditary character. Many races as well as cultural influences of men of all kinds have mixed into any man. To select, for approbation the peculiar elements that come from some supposedly Jewish heredity is to open the door to all kinds of nonsense on racial theory.

Such theoretical views were used by Hitler. Surely you cannot maintain on the one hand that certain valuable elements can be inherited from the "Jewish people," and deny that other elements which other people may find annoying or worse are *not* inherited by these same "people." Nor could you then deny that elements that others would consider valuable could be the main virtue of an "Aryan" inheritance.

It is the lesson of the last war not to think of people as having special inherited attributes simply because they are born from particular parents, but to try to teach these "valuable" elements to all men because all men can learn, no matter what their race.

It is the combination of characteristics of the culture of any father and his father plus the learning and ideas and influences of people of all races and backgrounds which make me what I am, good or bad. I appreciate the valuable (and the negative) elements of my background but I feel it to be bad taste and an insult to other peoples to call attention in any direct way to that one element in my composition.

At almost thirteen I dropped out of Sunday school just before confirmation because of differences in religious views but mainly because I suddenly saw that the picture of Jewish history that we were learning, of a marvelous and talented people surrounded by dull and evil strangers was far from the truth. The error of anti-Semitism is not that the Jews are not really bad after all, but that evil, stupidity and grossness is not a monopoly of the Jewish people but a universal characteristic of mankind in general. Most non-Jewish people in America today have understood that. The error of pro-Semitism is not that the Jewish people or Jewish heritage is not really good, but rather the error is that intelligence, good will, and kindness is not, thank

God, a monopoly of the Jewish people but a universal characteristic of mankind in general.

Therefore you see at thirteen I was not only converted to other religious views but I also stopped believing that the Jewish people are in any way "the chosen people." This is my other reason for requesting not to be included in your work.

I am expecting that you will respect my wishes.

Sincerely yours,

Richard P. Feynman

## RICHARD P. FEYNMAN TO TINA LEVITAN, FEBRUARY 16, 1968

*Miss Levitan wrote on February 16, 1967, agreeing to respect his wishes and not include him in her book. One year later she informed Feynman he was again under consideration, this time for an article on "The Scientist and Religion"— "a portrait gallery of Jewish scientists of great intelligence and creative accomplishment." She enclosed a questionnaire and asked for a photograph.*

Dear Miss Levitan:

I have your form letter and questionnaire of February 14. Please see my previous correspondence, in particular, my letter of February 7, 1967, to understand why I do not wish to cooperate with you, in your new adventure in prejudice.

Sincerely yours,

Richard P. Feynman

## RICHARD P. FEYNMAN TO JAMES D. WATSON, FEBRUARY 10, 1967

*Early in 1967, while both visiting the University of Chicago, James Watson gave Feynman a copy of the manuscript for the book that would later be published as* The Double Helix. *They had met when Watson visited Caltech to give lectures on the coding systems of DNA. The following letter was Feynman's reaction.*

Don't let anybody criticize that book who hasn't read it thru to the end. Its apparent minor faults and petty gossipy incidents fall into place as deeply meaningful and vitally necessary to your work (the book—the literary

work I mean) as one comes to the end. From the irregular trivia of ordinary life mixed with a bit of scientific doodling and failure, to the intense dramatic concentration as one closes in on the truth and the final elation (plus with gradually decreasing frequency, the sudden sharp pangs of doubt)—that *is* how science is done. I recognize my own experiences with discovery beautifully (and perhaps for the first time!) described as the book nears its close. There it is utterly accurate.

And the entire 'novel' has a master plot and a deep unanswered human question at the end: Is the sudden transformation of all the relevant scientific characters from petty people to great and selfless men because they see together a beautiful corner of nature unveiled and forget themselves in the presence of the wonder? Or is it because our writer suddenly sees all his characters in a new and generous light because he has achieved success and confidence in his work, and himself?

Don't try to resolve it. Leave it that way. Publish with as little change as possible. The people who say "that is not how science is done" are wrong. In the early parts you describe the impression by one nervous young man imputing motives (possibly entirely erroneous) on how the science is done by the men around him. (I myself have not had the kind of experiences with my colleagues to lead me to think their motives were often like those you describe—I think you may be wrong—but I don't know the individuals you knew—but no matter, you describe *your impressions* as a young man.) But when you describe what went on in *your* head as the truth haltingly staggers upon you and passes on, finally fully recognized, you are describing how science *is* done. I know, for I have had the same beautiful and frightening experience.

If you were really serious about wanting something on the flyleaf, tell me and we can work something out.

*The hard cover edition of Watson's book does indeed have a quote from Feynman on the dust jacket: "He has described admirably how it feels to have that frightening and beautiful experience of making a great scientific discovery."*

## DANNY ROBINSON TO RICHARD P. FEYNMAN, FEBRUARY 13, 1967

Dear Sir,

I am in the sixth grade and my name is Danny Robinson. In class my

teacher was reading to us about microminiaturization. The book was from *Life* and it was called *The Scientist*. In the book it said that you offered $1,000 of your own money to anyone who could build you an electronic motor that is no more than a quarter millionth of a cubic inch in volume. In the book it said a man made one. Is it true? If it is true could you tell us what kind of tools the man used? What is the purpose for it? How did it work out? How long did it take him to build it? Where do you keep it?

I thank you very much for reading my letter and will you please write back to —.

Thank you,

Danny Robinson

## RICHARD P. FEYNMAN TO DANNY ROBINSON, FEBRUARY 24, 1967

Mr. Danny Robinson
Petaluma, California

Dear Mr. Robinson:

You are right—there is such a motor as you described. It was made by a Mr. McLellan in response to a challenge I made in a lecture. I am enclosing a copy of my original lecture, and a description and photographs of the motor and its parts made by Mr. McLellan.

Several motors were made—I have one, there is one in an exhibit case here at Caltech and Mr. McLellan has several. They all still work very well.

There is no use to it—it was all done just for fun. And if you read the lecture carefully you will see that there is another prize offered for small writing, which has not yet been won.

Sincerely yours,

Richard P. Feynman

## RICHARD P. FEYNMAN TO ARON M. MATHIEU, FEBRUARY 17, 1967

Mr. Aron M. Mathieu, Publisher
Research and Development
Cincinnati, Ohio

Dear Sir:

My doctor forbids me to serve as a consulting editor because it is bad for my blood pressure. I end up wanting to write the book and am frustrated by the actual author.

Enclosed find $25.00 draft sent to me by some nut wanting to throw money away. Buy your wife some flowers with it.

Sincerely yours,

Richard P. Feynman

P.S. The outline you sent me shows no imagination, it is hard to believe that anything very good can come from it.

P.P.S. I am returning the two chapters which you sent in subsequent letters. I have had no opportunity to look at them.

## R.B.S. MANIAN TO RICHARD P. FEYNMAN, MARCH 6, 1967

To,

Dr. Feynman,

Blazer of new trails,

Sir,

It would seem to you most inappropriate and ludicrous to receive a missive (or a missile!) from a nobody like me. I am post-graduate student in physics and came across your "Lectures in Physics" series. When I went through the forward I was staggered. Physics is an elixir no doubt but to ram it in one measure down the throat is entirely another thing. All these diverse facets translate past (as the armoury does on *Red Square*) your brain swiftly. After the vistas melt it is only a daze felt in a maze. You don't quite get what I am up to in one stroke: *It is too much for the freshman or the sophomore.* A great hullabaloo and tom tomming by Physics Revision Committee has not made things all right. I hear that most of the students take it on the lam after some months of turmoil. If things go on like this Caltech cannot produce scholars. One cannot or should not bite more than he can chew. Physics should be stagger-tuned, i.e. given in successive slabs so that student can lap it up steadily. Even if a dog drinks water in a river, it can force it in minute quantities. Your vol. II, III, IV we do in our post graduate curriculum.

Even electro dynamic books by such well-known persons like "Reitz

and Milford" and Tralli men are for graduate course as mentioned in their forward. Matrix formulations, tensions, group theory, operational calculus are not for sophomores. You may now well be in the process of retrospection and a sentiment of guilt would have been left after thoughts evaporate. Not even a beaver like you should be eager to introduce reforms that are short-sighted. Your books are no doubt superb but it is a square block in a round hole. Really, inclined planes may be obsolete but they are stepping stones to more refined topics. A climb to the rotunda needs stairs or a lift: one cannot be at the top of Empire State Building pronto. How did you become like this. Not by high-faluting modern way of teaching physics. You did your climb in orderly way. If my views are myopic or I am derived from Conservative forces, please enlighten me. Take me also in the bandwagon and let me get inured to the kaleidoscope of ever changing patterns.

Please Reply.

I believe in your rectitude and bonhomie.

Yours Sincerely,

R.B.S. Manian

## RICHARD P. FEYNMAN TO R.B.S. MANIAN, MARCH 14, 1967

R.B.S. Manian

Bombay, India

Dear Sir:

I think you may be quite right in your criticism. On the other hand, it is equally wrong to insist that all students go up by slow and easy steps along the same route. All students are different, to some one way applies, to others, another. If you do not like my book because it is too advanced for one reason or another, there are many more elementary books than mine. Possibly in your case there is one much more useful to you.

If the school at which you are learning has chosen my book for the Freshman and Sophomore courses, criticize them, not me. The thing I was trying to do when we started these lectures was to teach the students that I had when I gave the lectures. The decision to edit and publish these lectures and to use them in the ensuing years was not made by me. I am proud of these books as Physics books, but on the problem of when to use them, by whom, and where, I have no opinion.

Thank you very much for your remarks. If you need this letter to help you influence the school not to use the books by the elementary students, please do so. Good luck in your venture.

Sincerely,

Richard P. Feynman


## RICHARD P. FEYNMAN TO BERYL S. COCHRAN, APRIL 27, 1967

Mrs. Beryl S. Cochran
Madison Project
Weston, Connecticut

Dear Mrs. Cochran:

As I get more experience I realize that I know nothing whatsoever as to how to teach children arithmetic. I did write some things before I reached my present state of wisdom. Perhaps the references you heard came from the article which I enclose.

At present, however, I do not know whether I agree with my past self or not.

Wishy-washy.

Richard P. Feynman


## RICHARD P. FEYNMAN TO HUGH DEGARIS, APRIL 27, 1967

*Mr. DeGaris, a second-year physics, math, and philosophy student, said he would like to take over Feynman's work when Feynman was "too old." Mr. DeGaris was concerned about losing half his productive years to learning instead of creation and thought perhaps Caltech would be a more conducive atmosphere than the one he was in. He agreed with Murray Gell-Mann's position on unified field theory—"the greatest adventure of our time"—and said he would like to be "in it" as well.*

Mr. Hugh DeGaris
c/o Queens College
Victoria, Australia

Dear Mr. DeGaris:

If you want to be "in it" it doesn't make any difference where you are.

You must learn to develop and evaluate *your own* ideas. To start with, why not choose your ideas of fractional dimensions, as a purely mathematical idea, and develop it. You are bound to learn something. If it is no good or gets wound up in uninteresting things (which is nearly impossible) you must find another of *your own* ideas and work that out.

At the same time study physics in the conventional way in schools or through books or encyclopedias. That may give you other ideas—but you need to know something about the problems in physics you are trying to solve (so that you can judge how likely it is that "perhaps he ties up with this idea somehow" etc.).

There is no quick shortcut that I know.

Sincerely yours,

Richard P. Feynman

## MARGARET GARDINER TO RICHARD P. FEYNMAN, MAY 6, 1967

Dear Professor Feynman,

I very much hope that you will agree to sign the enclosed statement which it is proposed to publish as an advertisement in the (London) Times, and if funds reach that far, in one of the more popular, high-circulation papers as well. I am looking for the signatures of a smallish number— around 50—of eminent Americans whose names are also well known in this country (not always overlapping categories). Among those who have already signed are Naum Gabo, Joseph Heller, Professor Stuart Hughs, Thomas Merton, Professor Anatol Rapoport, Professor Meyer Schapiro, William Schirer and Professor Victor Weisskopf. I am hoping to get a number of signatures of those whose names are not yet publicly linked with this protest—or at least, not in this country.

We believe that this advertisement could have a significant effect on public opinion here, where protests about our Government's support of the war are constantly met with taunts—both official and unofficial—of anti-Americanism and 'it's all Hanoi's Fault'. We also believe that since Britain is the only major power that explicitly supports the war, even a partial disassociation on the part of our Government might be of importance.

I propose to publish the advertisement on June 1st or 2nd, when

Parliament is sitting again and while Mr. Wilson is visiting Mr. Johnson. So if you agree to sign, as I hope you will, I'd be grateful for an early reply.

Yours Sincerely,

Margaret Gardiner

London, England

Should you wish to contribute towards the cost of this advertisement please will you make your cheque out to "Gardiner and Kustow—Vietnam Account."

STATEMENT TO BE PUBLISHED AS AN ADVERTISEMENT IN *THE (LONDON) TIMES* and to be signed by eminent Americans who are also well known in Great Britain.

We, citizens of the United States, who are deeply concerned over the war in Vietnam, wish to put it on record that we do not subscribe to the official view of our country and of yours, that Hanoi alone blocks the path to negotiations. On the contrary, there is considerable evidence which has been presented to our Government but which has never been answered by them, to show that escalation of the war by the United States has repeatedly destroyed the possibilities for negotiations.

We assure you that any expression of your horror of this shameful war—a war which is destroying those very values it claims to uphold—ought not to be regarded as anti-American but, rather, as support for that American we love and of which we are proud.

## RICHARD P. FEYNMAN TO MARGARET GARDINER, MAY 15, 1967

Miss Margaret Gardiner

London, England

Dear Miss Gardiner:

I would like to sign your letter for I am completely in sympathy with its spirit and its last paragraph. However, I am, unfortunately not familiar with the evidence that escalation has destroyed the possibilities for nego-tiation. Certainly escalation has failed in its attempt to "force" Hanoi to negotiate—but I have not been following things closely enough to know that there would have been any real possibility of negotiation without escalation. It has seemed to me that Hanoi's policy has never included

such an alternative—but that that is no justification for our being there and destroying what we claim to wish to save.

I feel unhappy that I am not sure enough of my position to be able to sign your letter. As next best alternative I am enclosing a small check to help to make sure your advertisement is published.

Sincerely yours,

Richard P. Feynman


## RICHARD P. FEYNMAN TO DONALD H. DENECK, JUNE 27, 1967

*Whether in response to a letter or phone call is unclear, but on June 13, 1967, a physics editor at a New York publisher wrote Feynman to apologize for any misunderstanding concerning a promotional letter sent his way. "Although the letter may have looked unique, it is just another example of the advanced state of a part of the graphics industry. Briefly, the letter was a precise duplicate of other letters in a mass mailing to something over 3,000 physics professors in the country." He hoped it had not caused Feynman any embarrassment or offense.*

Donald H. Deneck
John Wiley and Sons, Inc.
New York, New York

Dear Sir,

You have me all wrong—I am just trying to cut down on the amount of mail I get. It has nothing to do with being an author, or being embarrassed. I am trying to get off mailing lists. Won't you please take my name out of your mailing lists? I ask this of all publishers, including Addison-Wesley. Thank you. No offense meant.

Yours sincerely,

Richard P. Feynman


## RICHARD P. FEYNMAN TO MARK KAC, OCTOBER 3, 1967

Professor Mark Kac
The Rockefeller University
New York, New York

Dear Kac,

I'm sorry, but I don't want to go anywhere to give lectures. I like it here and I want to work in peace and quiet—preparing, going, lecturing, and returning is too much of a disturbance to my tranquil life.

Thanks for the invitation, though.

Yours,

Richard P. Feynman

## RICHARD P. FEYNMAN TO *THE NEW YORK TIMES MAGAZINE,* OCTOBER 1967

To the Editor:

It was fun to see my name and my dog's picture in *The New York Times Magazine* ("Two Men in Search of the Quark," Oct. 8).* Although I did do many of the things described in your article, I am really not one of the men responsible for starting scientists thinking about quarks. It was the result of one of the great ideas that Gell-Mann gets when he is working separately.

Richard P. Feynman

California Institute of Technology

Pasadena, Calif.

## RICHARD P. FEYNMAN TO MRS. ROBERT WEINER, OCTOBER 24, 1967

*As a response to some remarks he made in the* Los Angeles Times *on the subject of modern poetry, Mrs. Robert Weiner wrote to Feynman. She felt that Feynman's remarks "added up to a complaint that modern poets show no interest in modern physics. . . The truth of the matter is that modern poets write about practically everything, including interstellar spaces, the red shift, and quasars." Concluding that Feynman had a reputation "for liking forbidding difficult stuff," she enclosed a copy of W. H. Auden's "After Reading a Child's Guide to Modern Physics":*

If all a top physicist knows

About the Truth be true,

Then, for all the so-and-so's,

* See Appendix VI.

Futility and grime,
Our common world contains,
We have a better time
Than the Greater Nebulae do,
Or the atoms in our brains.

Marriage is rarely bliss
But, surely it would be worse
As particles to pelt
At thousands of miles per sec
About a universe
In which a lover's kiss
Would either not be felt
Or break the loved one's neck.

Though the face at which I stare
While shaving it be cruel
For, year after year, it repels
An aging suitor, it has,
Thank God, sufficient mass
To be altogether there,
Not an indeterminate gruel
Which is partly somewhere else.

Our eyes prefer the suppose
That a habitable place
Has a geocentric view,
That architects enclose
A quiet Euclidean space:
Exploded myths—but who
Would feel at home astraddle
An ever expanding saddle?

This passion of our kind
For the process of finding out
Is a fact one can hardly doubt,
But I would rejoice in it more
If I knew more clearly what
We wanted the knowledge for,

*Carl and Kiwi in the desert, 1968.*

Felt certain still that the mind
Is free to know or not.

It has chosen once, it seems,
And whether our concern
For magnitude's extremes
Really become a creature
Who comes in a median size,
Or politicizing Nature
Be altogether wise,
Is something we shall learn.
                    —W. H. Auden, from *About the House*

Mrs. Robert Weiner
Del Mar, California

Dear Mrs. Weiner:

I haven't answered your July 7th letter before now as I was off on vacation and am just now getting down to answering the mail that accumulated this summer.

The point of my remarks about poets was not meant to be a complaint that modern poets show no interest in modern physics—but that they show no emotional appreciation for those aspects of Nature that have been revealed in the last four hundred years.

Mr. Auden's poem only confirms his lack of response to Nature's wonders for he himself says that he would like to know more clearly what we "want the knowledge for." We want it so we can love Nature more. Would you not turn a beautiful flower around in your hand to see it from other directions as well?

Of course, men want knowledge for many other purposes also, to make war, to make a commercial success, to help the sick or the poor, etc., motives of various values. These obvious motives and their consequences the poets do understand and do write about. But the emotions of awe, wonder, delight and love which are evoked upon learning Nature's ways in the animate and inanimate world, together (for they are one) is rarely expressed in modern poetry where the aspect of Nature being appreciated is one which could have been known to men in the Renaissance.

And the crassness of our time, so much lamented is a crassness that can be alleviated only by art, and surely not by science without art. Art and poetry can remind the mind of beauty and gradually make life more beautiful.

My lament was that a kind of intense beauty that I see given to me by science, is seen by so few others; by few poets and therefore, by even fewer more ordinary people.

On the other hand, you might be right, for I have read very little. But at least the example you enclosed only confirms my view, for a modern poet is directly confessing not understanding the emotional value of knowledge of nature.

Sincerely yours,

Richard P. Feynman

P.S. You can find my more complete remarks first-hand in *The Feynman Lectures on Physics*, Volume I, pgs. 3–6 footnote.

.   .   .   .

## RICHARD P. FEYNMAN TO ROBERT F. BOHEEN, FEBRUARY 16, 1968

Dr. Robert F. Boheen
President's Room
Princeton University
Princeton, New Jersey

Dear Sir:
I am sorry, but I do not wish to accept the honorary degree you have offered me. I already have one degree, honestly earned, from Princeton. At the commencement when I received it, I remember watching the honorary degrees being conferred and feeling that an "honorary" degree was a debasement of the idea of a degree that confirms that certain work has been accomplished.
Sincerely yours,
Richard P. Feynman

## RICHARD P. FEYNMAN TO BRUCE JOWERS, APRIL 25, 1969

*Mr. Bruce Jowers, a high school student who described himself as a future scientist, wrote Feynman two letters—though unfortunately, only one remains—about fusion devices, describing what he saw as the most basic faults and then offering suggestions.*

*1. He thought the attempt to heat gas until it has enough energy to fuse was too much. "Why don't you keep the heat of the gases down to, say, 3/4 of the needed heat. Then take the resulting plasma and run it through a plasma-accelerator. You then take this plasma beam and pass it through another head on. This should give the gas particles enough energy to fuse."*

*2. Instead of attempting to control fusion by containing it, Mr. Jowers suggested letting it move through a long magnetic pipe, and noted that this would let the reaction happen at a high velocity, and the inertia of the high speed would help to control the reaction.*

*3. He thought carbon or another catalyst should be added to help reaction along. "All the astronomy books I have read have informed me that the sun's fusion reactions use carbon to catalyze them. I have not read of you people trying to apply this."*

*A diagram of the recommended device was enclosed. In closing, Mr. Jowers*

*said, "I hope this letter has been of some value to you. If at all possible I would like you to rip my ideas to shreds and send me the remains. I need the experience, for I am still in high school. Thank you."*

Mr. Bruce Jowers
Sepulveda, California

Dear Sir:

You sent me two letters asking me to tear your ideas to shreds. You will be happy to learn that I won't do that for they are, on the whole, mainly correct. On the other hand, don't be too disappointed to hear that many of them have been thought of before.

About the reactor—catalysts have been suggested, but not carbon. We use atoms easier to make fuse than $H^1$. We use $H^2$ (deuterium) or $H^3$ (tritium) which react easier, and sometimes add catalysts like lithium if that would speed the reaction. The sun has very little deuterium or tritium and must start with nearly pure $H^1$, so carbon does help there as a catalyst. Also solar rates of reaction are too slow for our use, so we must try other reactions. Moving beams of plasma into each other, etc., is a good idea, and people try things like that. The confinement problem means helping the beams or plasmas from spreading out of the apparatus—for example, sideways.

On your nuclear glue, you are mainly right. The nuclei with large numbers of neutrons are unstable, however, not because the electrons pull them out but because there is a rule that many particles of the same kind cannot be put into the same space without a lot of energy. (That is why, for example, all the electrons in an atom don't all settle close to the nucleus—they stay out of each other's way in "shells" or motions of wider special extent.) The electrons have only a tiny effect on nuclear processes, as we know from experiments of the kind you suggest. Different chemical forms of an element in different substances (for which the motions of the electrons would be altered near the nucleus) have only very tiny effects on nuclear energies and rates of decay. The nuclei are held together by strong forces between neutron and proton (and also between proton and proton, and also between neutron and neutron). They are modified by the smaller electrostatic repulsion between protons. The problems I personally work on now are related to what makes this nuclear force.

Sincerely yours,
Richard P. Feynman

## RICHARD P. FEYNMAN TO SANDOR SOLT, APRIL 25, 1969

Sandor Solt
Director, Muszaki Konyvkiado,
Budapest, Hungary

Dear Mr. Solt:

I was very pleased to hear that you have thought so well of the *Feynman Lectures* that you are translating them into Hungarian. It is good to know that you want to make them available to your Hungarian students.

You have suggested leaving out several sections (for example, pages 46–8) and have asked my permission to do so. I cannot give you this permission because it may prove embarrassing to you if your translation is compared to other editions in many languages. The reason for your leaving out the sections (that they are repetitious or that modern research show them to be incorrect) may be misunderstood and trust in the fidelity of your translation of other parts might be lost.

Instead, I suggest that you translate everything but, in the case of paragraph 46–8 (and any others that you may choose) and a "translator's footnote" in which you point out the parts that you think are erroneous and why. It would be very good for the students to learn that science is an ongoing subject with new research always modifying old ideas, and I myself would be happy to see the careful criticism of my ideas that your physicists have made (for I am not yet aware that those ideas had been found to be invalid). These criticisms would enhance the book and I urge you to include them, if you wish, in such footnotes, but the main text must be left intact.

Again, thank you for your patient work of translation and criticism.

Sincerely yours,

Richard P. Feynman

.   .   .   .

## RICHARD P. FEYNMAN TO H. DUDLEY WRIGHT, MAY 1, 1969

*The Feynman Fellowship was established by Dudley Wright, a longtime friend of Feynman's. The recipient for the 1969–1970 year was Mr. Kenneth Young, a Caltech student majoring in physics.*

Mr. H. Dudley Wright
Geneva, Switzerland

Dear Dudley:

Dean Bohnenblust wrote you about the choice for your Feynman Fellowship next year, and I've been aiming to write a note to go with it but these deans are more efficient than absent-minded professors and beat me to it.

It is a very good guy we have as recipient—the best we have. Sometimes it is easy because one guy stands out clearly among the others and this is it. We're off to a good start.

There has been talk that our grad students are getting poorer year by year (maybe from the competition from other schools). We had this great undergraduate and thought he ought to be rewarded. We found out later that he had been talking to his friends about going to Princeton, but the fellowship changed his mind, so we thank you.

What's new? Will we see you again soon?

Here things are as usual. Everybody in the family is OK. News is we have adopted a baby girl. We've had her since she was two months old and she is eight months old now, and of course cute as can be. Also I had a little art exhibit, with an opening, etc., of my drawings here at Caltech in the basement of the Athenaeum. So you can never tell, you had better hold on to those behinds I sent you.

Love,
RPF

·   ·   ·   ·

# RICHARD P. FEYNMAN IN A LETTER TO THE FACULTY, MAY 12, 1969

Letter to the Faculty
c/o The California Tech
Winnett Student Center
Campus

It is generally believed that the Caltech student is unhappy, and that all he needs is social recognition by the faculty. For that reason, there have been many studies and committees working on the problem of student-faculty relations. Up to now, we of the faculty have so often made every conceivable effort with no response by the students at all. I wish to report here, however, some glimmering of a beginning of a small student effort in that direction. An effort, which I believe should be answered by the faculty.

Having been invited to Page House last Saturday evening, I was watching a moving picture, when I was captured by a number of men and women in togas and wooden swords, and carried off to Fleming where I was dressed (by lovely handmaidens) in a toga, given a crown and declared emperor (of what the more straitlaced among you might call a Roman orgy).* I was served breads with four kinds of cheeses and wines while watching gladiators battle for my amusement. An oracle declared that the emperor needed a consort, so four maidens were selected for me to choose from. After watching them dance for me, I found myself with my head in the lap of one of the most beautiful and gracious ladies of Claremont, being fed grapes, of course, while we watched a play. (I had difficulty keeping track of the play.) Whatever I wished was granted—my back was rubbed with special lotions prepared for the occasion (by the Zoltan Tokes I understand); a pan of hot water was brought, my lady removed my shoes and socks and I soaked my feet, while a group of musicians on recorders and drums played for me. A dancing Bacchus proffered wine and more wine; and a most delicious hot buttered rum was prepared (more than once) for the emperor.

As in all such situations, jealousies arose, and there was an attempt to depose the emperor, but having been previously warned by a spy among my coterie of handmaidens I found it easy to put down the attempted revolution.

* It should be noted that the first women were admitted to Caltech in 1970, so any women on campus were from nearby colleges.

Finally it was decided (after some tests) that I need not drive my own chariot home, and a sober charioteer was found to escort me.

Taking account of their inexperience in entertaining faculty (they were unable to produce a kumquat on demand), I think this represents a sincere first step in an attempt to improve relations. I have every reason to believe that this represents a new policy. I should suggest therefore that next time you are invited to student houses, you accept for you will surely be treated better than I, as they gain experience and develop competition among the houses.

Incidentally, in the closeness of the relationship developed during the evening via the hot buttered rum, I found out what their real demands are; what will bring peace and happiness to this campus while other colleges have chaos. It is not a demand for more girls, or more student say in administration or faculty decisions, or any of those complex psychological things. These are just covers for their true underlying uncertainty. They asked only one thing and, under the circumstances that I found myself in, it seemed very easy to grant. I needed only to announce that all Fleming men were guaranteed an A in physics to make them deliriously happy.

I am sure my colleagues will find it easy to arrange to make my promise into reality—in the name of student faculty relations, so this campus may never fear the troubles that beset others.

And to the men of Fleming, and their ladies, thank you. I had the time of my life!

Ricardus Feynmanius, Emperor

## RICHARD P. FEYNMAN TO ANDY WANG, SEPTEMBER 30, 1969

*Mr. Andy Wang, a foreign student from Hong Kong, wanted to study physics in college but had had previous difficulties with the subject. In a letter to Feynman, he enclosed his academic report from the previous year for reference.*

Mr. Andy Wang
Haverford College
Haverford, Pennsylvania

Dear Mr. Wang:
I am sorry, but of course I cannot advise you without knowing you better. Sometimes a situation like yours arises from a little block of misunder-

*Carl, Michelle, and Richard on Michelle's first birthday, 1969.*

*Richard and his children at the beach.*

standing that can be found and cleared away. At other times it may be harder to straighten out and really not worth it. Your 93 in Electricity and Magnetism looks good. But it is not good to hit yourself on a stone wall, either, so what can I say?

I say this. Try to find some friends who are also somewhat interested in physics and try to discuss physics things with them. If you find yourself able to explain things in your own words, so that *they* are led to understand things from what you say, you are OK. Soon you will find yourself able to explain things to yourself. Otherwise, give up and plan for a different career. If you can't find such friends, try to tutor elementary physics, and see how it goes.

Sincerely yours,

Richard P. Feynman

# 1970–1975

Feynman's ongoing emphasis on education brought him the American Association of Physics Teachers Oersted Medal in 1972. It was a distinction that he shared with many of his correspondents over the years: Hans Bethe, J. W. Buchta, Freeman Dyson, David L. Goodstein, Philip Morrison, Frank Oppenheimer, Isidor Rabi, Victor Weisskopf, John Wheeler, and Jerrold R. Zacharias. The following year he received the Niels Bohr International Gold Medal as well.

His physics work at this time consisted of the development of an important concept of partons (that is, asymptotically free quarks), which is still used today. More lectures were turned into highly specialized textbooks (*Statistical Mechanics: A Set of Lectures* and *Photon Hadron Interactions*), and Feynman continued work with relativistic quarks. He became increasingly intrigued by computers, and in 1973, he began discussions with Edward Fredkin at MIT about the possibilities of "artificial intelligence."

Interestingly, and perhaps not surprising given the tumultuous times, many of the people who wrote to him during this period wanted to engage him in political issues. Among the defenses of his decision to work on the atomic bomb and his comments about women's capacity for science are a number of profoundly affectionate and candid letters.

. . . .

## MEMO TO DELEGATES, MAY 1, 1970

TO:            Delegates to the XVth International Conference on High
               Energy Physics
FROM:          Henry Abarbanel, Princeton University
SUBJECT:       Political Exclusion of Delegates from International
               Conferences in the Soviet Union

As many of you are aware, since June, 1967 the Soviet Union has made it a practice to exclude from International Conferences held within its borders delegates from Israel. The method of exclusion has been to invite the delegates then to refuse to grant them visas. Since the exclusion of any group from a scholarly conference for clearly political reasons is intolerable, I propose we act in the following manner in an attempt to avoid such an occurrence at our meeting.

Send a petition (see below) to the Soviet scientists organizing the conference, to the Soviet Academy of Sciences and to the Soviet Foreign Ministry saying, in essence, that the undersigned will refuse to attend the Kiev meeting if any group is excluded for political reasons. Since the Israeli delegation has requested that the visas be delivered by 1 June 1970, the petition will be sent to the Soviets after that date, if it is necessary. If you agree with the basic idea, then please return the signed copy of the petition attached, along with an address where I can contact you during the month of August, 1970, especially during the week before the conference. I will see to it that those interested receive, at least 72 hours before the start of the meeting, notification at the given address whether or not the Israelis or any other group have been refused visas. At that point any action you take with that knowledge is up to you. Timing is an important problem since it usually takes some time for anyone to receive a visa from the Soviets, and it may arrive only weeks or days before departure even when they want you.

Let me say that I would be pleased to accept suggestions on the implementation of this idea and sincerely hope nothing like this will need to be done. It is clearly better to be prepared, however, thus I write you this note.

A copy of this petition along with the names of all delegates to the Kiev Conference who signed it will be forwarded to the Organizing Committee, the Soviet Academy of Sciences, and the Soviet Foreign Ministry. Please return this by May 25, 1970. If you have passed your invita-

tion on to another physicist, could you please pass this note along too. Thank you.

The undersigned invited delegate to the XVth International Conference of High Energy Physics support the idea of a meeting open to all physicists. We will refuse to attend this Conference if it becomes known that any group has been selectively excluded from attendance for clearly political reasons.

## RICHARD P. FEYNMAN TO HENRY ABARBANEL, MAY 14, 1970

Dr. Henry Abarbanel
Physics Department
Princeton University
Princeton, New Jersey

Dear Dr. Abarbanel:
I already refused to go to many conferences in Russia, including the Kiev conference. I do not visit Russia for scientific purposes because of its government's policies maintaining the right to control which of its own scientists go where and do what, scientifically. I agree with your petition but cannot sign it implying I would go if this Israeli thing is fixed up, for I won't go, feeling the disease is worse than this one symptom.
Sincerely yours,
Richard P. Feynman

. . . .

## RICHARD P. FEYNMAN TO K. A. CARDY, AUGUST 27, 1970

*A member of the Royal Society UNESCO (United Nations Educational Scientific and Cultural Organization) Committee proposed to nominate Feynman for the Kalinga Prize for the Popularization of Science.*

Mr. K. A. Cardy
National Commission for the United Kingdom
London, England

Dear Sir:

I was honored to hear that you would consider me for nomination for this year's Kalinga Prize for the popularization of science. I am sorry that I don't consider myself fit for the Prize. Much as I might be tempted by the proffered trip to India, I don't feel I could discharge the responsibility to interpret Indian science upon my return.

I am sorry to be so late in answering you, but I was on vacation when your letter arrived and have only just become aware of it.

Sincerely yours,

Richard P. Feynman

## JOHN SIMON GUGGENHEIM MEMORIAL FOUNDATION CONFIDENTIAL REPORT ON CANDIDATE FOR FELLOWSHIP, MURRAY GELL-MANN, DECEMBER 9, 1970

Requested of:
Mr. Richard P. Feynman

Report:

There is almost no theoretical discovery in the field of high energy physics that does not carry Gell-Mann's name. Virtually all that we know of the symmetry of these particles is a direct result of his work. You could not contribute to the development of physics in a more important and more certainly fruitful way than to help this candidate to do whatever he needs to do. It would be a credit to your foundation to accept his request.

Signed Richard Feynman

*Whether Feynman's letter had anything to do with it or not, Dr. Gell-Mann received a Guggenheim Fellowship in 1971.*

*Richard and Michelle, 1970.*

*Feynman giving a talk at Argonne National Laboratory, 1970.*

# RICHARD P. FEYNMAN TO STANISTOW KRUK, JANUARY 18, 1972

*Mr. Stanistow Kruk, an eighteen-year-old from Poland, wrote to Feynman, asking, "What is your attitude with respect to the world of science? How did you make your great scientific career? What are your ideas on the still unknown laws of physics? . . . When you were 18 as I am now, did you know what a great future you had before you?"*

Mr. Stanistow Kruk
Kielce, Poland
Dear Mr. Kruk:

I am sorry but your questions are too big to be given brief answers. I can only refer you to my lectures published in English called *The Character of Physical Law*. Lectures more detailed in physics published in Polish are "Feynman Wykxady Z Fizyki, Warszawa 1970, Panstwowe Wydawnictwo Naukowe," but these you may have already seen.

When I was 18 I did not know what the future might bring, but I did know that I must be a scientist, that it was exciting, interesting and important.

Sincerely yours,
Richard P. Feynman

# DR. VERA KISTIAKOWSKY TO RICHARD P. FEYNMAN, FEBRUARY 11, 1972

Dear Professor Feynman:

I would like to comment briefly on the remarks concerning women that you made at the recent A.P.S. meeting and the criticism of the anecdotes in your book. I have just finished writing a report as chairwoman of the A.P.S. Committee on Women in Physics and, therefore, these are very live issues for me.

Your statement to the effect that women have been discriminated against and that this is ridiculous, was enormously helpful, especially since it was made to an audience containing so many physics teachers. It was also unusual. Although most physicists pay lip service to equal treatment for equal ability and performance, they dismiss existing inequalities as the results of the problems of matrimony and motherhood. Even statistics showing that married women are somewhat more successful than single women are dismissed, because "there are so many factors involved."

Therefore, if your prestige gains acceptance for your statement, you definitely are in line for the Elizabeth Cady Stanton medal.*

However, I must disagree that concern about the anecdotes in your book is energy wasted over trifles. I say this although I have used the book as a text with real pleasure and no twitch of the feminist subconscious. This was several years ago before I started reading and thinking in any coherent way about women's problems. Since then, my observations of my own reactions and those of my daughter support the results appearing in many psychological and sociological studies—that all the media and most socialization push the female toward a lowered opinion of her capabilities and toward a very ambivalent feeling concerning success. Very briefly, then, let me suggest what might be the subconscious effects of anecdotes such as yours: Feynman (the great physicist) portrays women as less intelligent and not as members of the set containing physicists; I am a girl; maybe I really am excluded from being a physicist. This is possibly an over-simplification, but I think it has a lot of truth in it. Try replacing "woman" by "negro" in those anecdotes to get the flavor. The fact that you are famous greatly heightens the effect. Therefore, I don't think this is trivial, it is more at the heart of the problem than equal treatment for Ph. D. women physicists.

Very sincerely yours,

Vera Kistiakowsky

Massachusetts Institute of Technology

Cambridge, Massachusetts

## RICHARD P. FEYNMAN TO VERA KISTIAKOWSKY, FEBRUARY 14, 1972

Dear Dr. Kistiakowsky:

Thank you very much for your letter. It is true that the quotations in my book are unfortunate and with the increasing sensitivities of today, I probably would have been more careful and not made them. Actually, however, I agree with my sister that in a way they are trivial unless they are missed. "Feynman portrays women as less intelligent and notes members of the set containing physicists" may be a subconscious conclusion but not a fair one (in particular the latter half). Although there was a lady driver

---

* Elizabeth Cady Stanton was a contemporary of Susan B. Anthony and a pioneer in the women's rights movement.

and "the changing of a woman's mind" as two phrases in the vast book there is also "a very ingenious mother" and the "experiments of Miss Wu of Columbia" to offset the impressions (the true tale about the girl friend of the nuclear physicist is about one particular individual, not about women in general).

My sister's point is that physics is a difficult subject requiring objective thinking and staying on the subject without distraction by minor subsidiary things. She thinks that someone who is so easily put off when considering the subtleties of definition of velocity, or of universality of ability to measure, as to be seriously concerned with the illustrations incidentally used will have a hard time with physics anyway.

She may be wrong. But I am inclined to agree that we should all try to be more rational and attempt to see what is really there and control our subconscious if it is drawing exaggerated conclusions. There is enough problem in what is real.

As you say, you, yourself, felt no subconscious twitch. That is better evidence than your theory of what the effect might conceivably be on somebody else.

Sincerely,

Richard P. Feynman

## RICHARD P. FEYNMAN TO A. V. SESHAGIRI, OCTOBER 4, 1972

*Nineteen-year-old A. V. Seshagiri wrote an eight-page letter from India. He had a severe stuttering problem. He said he was being tortured and tormented as a result of his speech impediment, and he also had problems with teachers. Mr. Seshagiri felt that they "discourage the students. They kill the enthusiasm of the students. . . . They do not want to give away their knowledge to the students." He thought Caltech might instead be a place where he could study "calmly and peacefully without any anxiety."*

Mr. A. V. Seshagiri
Bombay, India

Dear Mr. Seshagiri:

Thank you for your letter.

It is fortunate that you are interested in physics because such a study is not seriously impeded by a speech difficulty. In fact physics must be studied

alone—you must teach yourself. Do not worry so much about your instructors. There are very many books of varying degrees of difficulty and written in many styles, and on different aspects of physics. You must find the ones which suit you—which you enjoy reading and from which you can learn most rapidly and easily. In case you find my book interesting—although at present it might be too difficult for you—I'm sending a little book of problems to go with it. Do not be alarmed if you cannot do any of them now, take it easy, reading things you honestly understand first and your knowledge will grow.

Please be assured that the difficulties you are now having with your teachers is not an inevitable thing in India. I have spoken about what you wrote to a colleague of mine who lived in India several years and knows something of the situation. As you proceed to more advanced work you will encounter, ever more frequently, professors who are kindly and understanding of your problem.

It is extremely difficult to get into Caltech, as an undergraduate, from far away—but it is much easier as a graduate student (working for a Ph. D). In fact this year of the twenty new physics students, two are from India. I would suggest that you continue your studies there, in India, until you acquire a B. Sc. and also an M. Sc. Degree and then you can apply to graduate schools in the United States if you still wish to.

In the meantime, study calmly and quietly those things which interest you most and which you honestly can understand. I will not recommend any special books because it depends on your interest and level, you must find them yourself in the libraries of Bombay.

Sincerely yours,

Richard P. Feynman

## RICHARD P. FEYNMAN TO BART HIBBS, OCTOBER 13, 1972

*Bart Hibbs is the son of Dr. Albert Hibbs.*

Bart Hibbs
Pasadena, California

Dear Bart:

You asked me to write you why I thought the sun looks red at sunset. Air molecules scatter blue light more than red. The color of the sky (in

directions away from the sun) ordinarily is light scattered by the air and we all know the sky is blue. The light that is not scattered—that passes from the sun to the eye directly—has less blue in it—and even less blue the more air it goes through. Thus as it sets, and we look at it through a very long column of air it looks very red indeed.

The colors of clouds and sky at sunset are very beautiful but equally complicated—some are clouds directly illuminated by a setting sun—others are partly illuminated by light scattered a few times high in the sky, etc.

I hope that answers your question. If it only makes you think of more questions write me and I'll try to answer those too.

Yours,

Dick Feynman

## RICHARD P. FEYNMAN TO K. CHAND, M.D., NOVEMBER 6, 1972

*Feynman was on his way to the Conference on High-Energy Physics in Chicago when he tripped on a sidewalk partially concealed by long grass and broke his knee cap.*

Dr. K. Chand, M.D.
Chicago, Illinois

Dear Dr. Chand:

I am one of your patients who had a knee cap removed September 7. You asked me to write about my progress a month or two after I left the hospital. All is well and progressing exactly as you predicted—I can bend it 90 degrees today (that's why I remembered to write) and it has been increasing at about 1 degree per day. Thank you and Dr. Kuhlkani for doing such a good job. Please send my special greetings to Dr. Kuhlkani.

The trip in the airplane was very easy. They had wheel chairs ready at departure and arrival. They put me in a special seat near the entrance so there was leg room, but—better than that—since the plane was not full they arranged for the seat on each side of me to be unoccupied—so I had three seats together and rode the whole way with my leg comfortably on pillows up on the three seats (American Airlines).

I rather enjoyed my stay in the hospital and remember with pleasure many good people—even besides the doctors!—for example Henry

Pinicke the nurse, Miss Chan, the diet girl, Mr. Blas and Joyce in the physical therapy department, etc. (except I was astounded at how much the insurance company had to pay for each day I stayed there!)

When I break my other knee cap I hope it will be in Chicago so I can see you all again. I am making a trip there (to the National Accelerator Laboratory) in early December and if experience is any teacher I will break it there, so be prepared.

Yours sincerely,
Richard P. Feynman

## RICHARD P. FEYNMAN TO MALCOLM GIBSON, DECEMBER 29, 1972

*Mr. Malcolm Gibson, age fifteen, wrote from England to ask what Feynman's personal reasons were for working on the bomb, "knowing the consequences of your work." He could appreciate that Feynman's time was valuable, and "above all you must feel that you do not have to justify your actions to me." He ended by saying he was not sure whether Feynman had worked on the atomic bomb project and apologized for disturbing him if that was not the case.*

Malcolm Gibson
Yorkshire, England

Dear Malcolm:
I did work on the atomic bomb. My major reason was concern that the Nazi's would make it first and conquer the world.

Sincerely,
Richard P. Feynman

## RICHARD P. FEYNMAN TO NIELS FOSDAL, MAY 9, 1973

Mr. Niels Fosdal
Copenhagen, Denmark

Dear Mr. Fosdal:
I, of course, accept the great honor of your offer of the Niels Bohr International Gold Medal, what a pleasant surprise! My wife and I will be

delighted to be in Copenhagen from Sept. 30 to Oct 7 as you suggest. (We are contemplating bringing our 11-year-old son.) I will be glad to give a lecture, but I have still to decide on what subject.

Thank you for considering me worthy of the medal. We shall take full advantage of the chance you give us to visit Copenhagen. I know how wonderful it is, having already visited a few times, but my wife never has, and I shall be pleased to show it to her.

Sincerely,

Richard P. Feynman

## RICHARD P. FEYNMAN TO BEN HASTY, JUNE 1, 1973

*"Happy Birthday Richard Feynman! From a grateful General Physics class, which used your lecture book. Thank you?" The hand lettering on the accompanying card is decorated with mathematical symbols (arrows, symbol for sum, increment signs, infinity signs, carats, exclamations, dot product signs, integral sign for the "F" in his last name, etc.) and has sixteen signatures.*

Ben Hasty

Springfield, Missouri

Dear Ben:

I wouldn't believe it if I didn't see it myself. A grateful physics class that sends a birthday greeting to the author of their textbook! In my time it was conventional to hate the author of the textbook which brought such pain upon us all. Maybe times have changed—but I see you haven't fallen absolutely, there was, at least, a question mark after "thank you."

It was very nice of all of you and a very great surprise. One more unbelievable story to tell my grandchildren.

Thank you very much, and good luck.

Sincerely,

Richard P. Feynman

.  .  .  .  .

## RICHARD P. FEYNMAN TO AAGE BOHR, SEPTEMBER 6, 1973

Professor Aage Bohr
Copenhagen

Dear Aage:
I should be glad to talk to physicists or physics students at any time—but prefer not to give a formal talk—could we just try getting together and discussing things through questions and answers? If you don't think that is a very good idea let me know and I'll try to think of some prepared subject—but I hope I do not have to.
Sincerely,
Richard P. Feynman

## RICHARD P. FEYNMAN TO H. B. HINGERT, SEPTEMBER 12, 1973

*Professor Hingert wondered if he had misunderstood that Feynman had learned Spanish before lecturing in Brazil, and thought he must have learned Portuguese instead.*

Professor H. B. Hingert
Milan, Italy

Dear Professor Hingert:
Thank you for your note. *The Feynman Lectures* are published by Addison Wesley and are, I believe, available in English book stores.
I did learn Spanish before I went to South America. I had expected to travel somewhere in South American but I hadn't decided where and was learning Spanish in preparation. But then I received an invitation to visit Brazil to do research there for three months. The invitation came six weeks before I was to go and I spent the six weeks "converting" my newly learned Spanish to Portuguese. People like to make up stories about how I have been foolish on occasion. So I let this story stand, as if I didn't know they speak Portuguese in Brazil. I hope they will be satisfied with that story and not probe deeper to find out how much more of a fool I have been on other occasions and start stories about those.
Sincerely,
Richard P. Feynman

*Receiving Niels Bohr Medal in October 1973 from*
*Queen of Denmark, Margrethe II.*

## RICHARD P. FEYNMAN TO HUBERT SPETH, OCTOBER 10, 1973

Hubert Speth
San Gabriel, California

Dear Mr. Speth:
Thank you for the kind note of good wishes. We went to Copenhagen and have just come back. Carl was with us, and you were right, he was beaming with pride to see his daddy get a medal from the queen.
Sincerely,
Richard P. Feynman

.   .   .   .

## RICHARD P. FEYNMAN TO EDWARD FREDKIN, OCTOBER 18, 1973

Professor E. Fredkin
MIT

Dear Fredkin:
I have to thank you for a number of things.

There is a fellow here at Caltech, *M. Weinstein,* who is interested in "artificial intelligence," whatever that is. He suggested to me that I come to work regularly on a computer terminal that connects to all the advanced systems in the county, so we could see what happens. I guess I am to get familiar with present computer abilities and think about what we need next, etc., or something. I will get a lot of fun out of doing that. He said you suggested he try me, so I thank you.

Maybe, if you would like to, you can give me advice on what you would like to see me doing, or thinking about. I haven't yet made head or tail of what we are doing. I am starting out with MACSYMA. A man named Don Brabstone is teaching me. Do you have suggestions as to directions of thought? My very first impression is that MACSYMA is not smart enough to be very helpful in the little algebraic manipulations I need to do in my work from time-to-time, but I don't know if that is a relevant observation (if, indeed, it ultimately turns out to be true).

On another subject: You once gave Caltech money for me to use as I liked. I never could figure out a good use. But last year a good research man (has Ph. D.), named Finn Ravndal, felt he had to go back to Norway (his home) to get a job. I had worked closely with him and we had written a paper together, and I was rather unhappy about (a) that we had to separate and (b) he would get stuck in Norway in a job much below his talents and interest. Suddenly I remembered your gift— proposed a year appointment for him here using it—and saved the day. Our field (high energy physics) has suddenly become pregnant with ideas and excitement and our working together this year will be great.

And, in the meantime, he has been offered a job for next year at the Bohr Institute at Copenhagen, which is a very respectable and excellent place. So we saved a good man. Thank you.

I donated your Muse to the boys in one of the rooming houses after thoroughly enjoying it myself.

What new things are you up to these days?

Sincerely,

Dick Feynman

*The "Muse" Feynman spoke of was a digital music box of sorts. To Professor Fredkin's knowledge, this box was the first device made for ordinary people with digital circuitry in it.*

## RICHARD P. FEYNMAN TO R. B. LEIGHTON, APRIL 18, 1974

Subject: Memo re: Professor E. Fredkin

Prof. E. Fredkin, present director of project MAC, the computer science laboratory at MIT, has shown an interest in visiting Caltech for a year, next year. There are many branches of computer science and the interests of our people here in that field are varied, so that although the kind of work Fredkin does is of direct interest to some, that is not true of everyone in that division. Therefore, it would help if others would also support a recommendation that he come as a Fairchild fellow.

There is one direction of computer science that has particular personal appeal to me. (It is called by the unfortunate name of "artificial intelligence," unfortunate because in the past several obviously naïve ideas went under that name.) Virtually all computer programs today simply follow step by step instructions—they do exactly what you tell them. They can make no use of some sort of explanation of what you are about or what want done, and figure out how to do it themselves. As an example there is a big industry of programming. Given a problem, such as to make a computerized inventory system for a business, one must hand the problem over, with explanations of what you want, to a programmer. The programmer does a great deal of work to write a program of instructions for a computer to do what you wish. To what degree can we use machines to help in this programmers work—ultimately to make machines which program themselves with little more information than we now give the programmer who at present makes the program?

In psychology there is a profound question of what type of facility is it that permits a child to learn a language from simply hearing it spoken and seeing it used. We are far from knowing how it is done. It is even very hard to see how it can be done. But it is done by every child. We cannot expect

*Mexico, 1974.*

to solve such problems by studying machines. Nevertheless, it is an intriguing academic investigation to see, at least in principle, some way that it might be done by a machine. In that manner we could at least start to guess at what kind of facilities the brain might have that enables it to learn language. In addition, if such a machine (or one with less but analogous abilities) could actually be designed or implemented by software on present machines, it would be of just the kind capable of automatic programming.

It is understood at MIT that such problems are exceedingly interesting. They are doing very promising work on them. I intend to spend some part of my time next year to thinking about these thoughts. Discussions with Fredkin would be very valuable. That is why I personally would like to support this recommendation.

But Prof. Fredkin's visit would have much wider interest than just this. Prof. Fredkin has had experience in the practical and business sides of computers as well as having visions of how the science and the industry should develop. For example, he was one of the early investigators of the development of on-line and time-sharing computer terminals done largely at MIT. He is well-known for his invention and development of special software and also hardware for image-processing. He also was a prime mover in the interesting development of computer software to handle symbols, algebra, equations and calculus. (The system developed at MIT is called Macsyma.)

Having Prof. Fredkin here could help us still further broaden our range of computer interests. We could widen the opportunities of an ever greater number of our students interested in computers.

I think that this visit would be a good opportunity for Caltech, and I should like to add my name to those of the computer science department which have recommended him. I hope others of the physics department would agree, so that we can show a sufficiently campus-wide interest in his visit to be able to get him here.

Professor Feynman

c.c. Prof. F. H. Clauser

Prof. M. Weinstein

*Professor Fredkin came for the year, and incidentally, so did Stephen Hawking, another Fairchild Fellow.*

. . . .

## RICHARD P. FEYNMAN TO PAUL OLUM, OCTOBER 31, 1974

*Paul Olum, a colleague and friend of Feynman's since the days of their gradu-
ate studies at Princeton University together, wrote to ask Feynman to speak at
the University of Texas. Dr. Olum had become the Dean of the College of
Natural Sciences there, after a brief stint in administration at Cornell
University. Dr. Olum remarked that his decision to leave Cornell University
for the University of Texas had a lot to do with the fact that he felt it was "a lot
more exciting and creative to be part of building something than to try just to
keep the quality that is already there which would have been the case largely at
Cornell." The letter ended with Dr. Olum uncertain if he would stay at the
University of Texas, as the President had recently been fired.*

> Professor Paul Olum
> College of Natural Sciences
> The University of Texas
> Austin, Texas

> Dear Paul:
> Great to hear from you. I am sorry to hear that you have been unable to
> resist getting a case of our occupational disease. Fortunately I myself have
> not succumbed and am still doing physics with as much pleasure, although
> not as cleverly, as ever.
> I am happily married to an English wife and have two kids (boy 12, girl
> 6) all of whom are a delight and make me very content. I feel like at last all
> of life's problems are solved!
> But being so content at home and at work here makes me very reluctant
> to go anywhere else, even for two weeks. So although sorry not to see you
> again, I shall decline your invitation to visit Austin.
> Best regards,
> Richard P. Feynman

## RICHARD P. FEYNMAN TO DAVID PATERSON, NOVEMBER 19, 1974

> Mr. David Paterson
> BBC TV
> London, England

Dear David:

I saw your quark-hunting program the other night and want to congratulate you on a first-rate job. I know how difficult the subject must have been for you, as an outsider, but you must have conquered it completely to be able to put all those pieces of interviews and apparatus together into such a coherent entirety. It told the story of what we physicists like to think of as an abstruse subject in a very clear and simple way. I was surprised it could be done, but you proved it to me.

I also must admit being a bit proud of my colleagues when I saw how well they explained themselves. Physicists didn't seem to be such a bad lot, after all.

The importance of such communication goes, of course, without saying; we both understand that.

So thank you.

Sincerely,

Richard P. Feynman

## RICHARD P. FEYNMAN TO ROBERTA BERRY, DECEMBER 18, 1974

*Mrs. Berry wrote to obtain permission to reprint "What is Science?" for a college course called "The Citizen and Science."*

Mrs. Roberta Berry
Editorial Assistant
Indiana University
Bloomington, Indiana

Dear Mrs. Berry:

OK Mrs. Berry—but times have changed since I gave that speech in 1966. So some of the remarks about the female mind might not be taken in the light spirit they were meant. Perhaps there is nothing we can do after all these years—but if after you read those remarks you still want to print it as is, it is OK by me.

Sincerely yours,

R. P. Feynman

. . . .

## RICHARD P. FEYNMAN TO B. E. BUSHMAN, JANUARY 7, 1975

*B. E. Bushman wrote, wanting a reference for the idea of "partons."*

B. E. Bushman
Laguna Beach, CA

Dear Mr. Bushman:

I can't suggest a simple reference. The idea is discussed at length but very technically in my book "Photon Hadron Interactions" published by W. A. Benjamin.

However, the word means something simple—if we can suppose that protons, neutrons, pions etc., are not "elementary" but themselves made up of more fundamental parts, that is they are made of simpler particles (just like atoms are made of other particles, electrons and nuclei—or electrons, protons and neutrons) we needed a name for these unknown particles—that name is "partons." The next question is "what are the partons like?" that is, what charges do they carry, etc. (if they exist).

At present it does look like this idea is good and most of the partons are quarks (but there may be other uncharged kinds too.)

Sincerely,

R. P. Feynman

## DAVID A. MARCUS TO RICHARD P. FEYNMAN, JANUARY 13, 1975

Dear Mr. Feynman,

At the stage of atomic research and control would you consider nuclear energy the curse of humanity or the potential salvation of mankind?

As one who contributed so significantly to the means of man's utilization of this awesome force, what are your thoughts as you look back at this scientific development?

Do you look to the future with fear or with hope?

I am an amateur historian and sociologist and I would be most grateful for your comment.

Respectfully,

David Marcus, D.D.S.

## RICHARD P. FEYNMAN TO DAVID MARCUS, FEBRUARY 18, 1975

Dr. David Marcus
Palm Springs, California

Dear Dr. Marcus:
I am sorry to have to answer your question (as to whether I consider nuclear energy a curse or a salvation of mankind) that I really don't know. I look to the future neither with hope nor fear but with uncertainty as to what will be.
Sincerely,
Richard P. Feynman

## RICHARD P. FEYNMAN TO DAVID HAMILTON, MARCH 26, 1975

*On February 22, 1975, a retailer of office equipment wrote to Feynman seeking comment on an idea that he had conceived. Mr. Hamilton began by noting that he had already written to Feynman's colleague Murray Gell-Mann but as yet had received no reply. He hoped Feynman would respond, since he had so enjoyed Feynman's appearances on public television.*

*Mr. Hamilton's idea was to construct an accelerator for fundamental particles in the shape of a figure 8. Such a shape would have the advantage, he said, that the particles could collide with each other as they raced in opposite directions through the crossing point of their 8-shaped path. He asked whether new particles produced in such a collision might move faster than light, since the relative speed of the colliding particles would be almost twice that of light, and he suggested using a variant of a particle detector called a spark chamber to study the particles.*

Mr. David Hamilton
Venice, Florida

Dear Mr. Hamilton:
Your idea is a very good one. So good that it is in full use now, in a machine we call a colliding beam machine at the European Organization for Nuclear Research which is in Geneva, Switzerland. They use protons whose energy is so high their mass is thirty times enhanced (by relativity

effects). Their velocity is less than the speed of light by only one part in two thousand. They go around in two rings that intersect at one place where experiments are done to see what happens. The protons from an accelerator are "stored" in the rings where they go around and around, colliding from time to time in the intersection at A. To get the same kind of collision in the conventional way (by hitting one fast proton against another at rest in the laboratory), one would need an energy sixty times higher than one has in these rings (that is protons whose mass is enhanced 1800 times).

Although in the usual way of figuring, since each proton is going nearly at the speed of light, it seems their relative velocity is nearly twice the speed of light. But Einstein showed the regular way of thinking is wrong and relativity effects make the apparent relative speed of one as seen by the other less than the speed of light (by only one part in 7,000,000 in this case!).

Spark chambers (and other devices) are indeed used to study the particles produced in the collisions. Very interesting results are obtained which we are all puzzling over, trying to fully understand.

There is a similar colliding beam device at Stanford University (SLAC), but here there is only one ring with electrons going around one way, and positrons the other way. A few months ago an entirely unexpected new particle (called a ø, about three times as heavy as a proton) was found. It is destined to change, drastically, our ideas about what matter is made of.

So your idea is at the experimental forefront of high energy physics today. I hope you are not too disappointed that it had already been thought of.

Yours sincerely,
Richard P. Feynman

.  .  .  .

*One of "Ofey's" sketches.*

## RICHARD P. FEYNMAN TO WILLIAM L. MCCONNELL, MARCH 5, 1975

*William L. McConnell, who had been working pairing talented high school students with researchers in the St. Louis area, was intrigued by the notion that those with higher levels of intelligence could work in more disciplines than most people. He had learned that Feynman had taken up drawing and wanted a sketch for his office wall.*

Dr. William McConnell
Director, Science MAT
St Louis, Missouri

Dear Dr. McConnell:
I don't know much about the "general theory of intelligence," but I do remember when I was young I was very one-sided. It was science and math and no humanities. (Except for falling in love with a wonderful intelligent lover of piano, poetry writing, etc.) It is only as I became older that I tried drawing (starting 1964). Bongo playing has never been "music" to me—I

don't read notes or know anything about conventional music—it has just been fun making noise to rhythm—not much "intelligence" in the intellectual sense is involved.

I am sorry not to want to send you a drawing, because it has been my policy not to sell them to people who want them because it is a physicist who made them.

Sincerely yours,

Richard P. Feynman

## RICHARD P. FEYNMAN TO KENNETH R. WARNER, JR., APRIL 1, 1975

*On March 6, 1975, a man who had seen Feynman on a PBS* Nova *program and struggled to understand* The Feynman Lectures on Physics *wrote to him seeking help. Mr. Warner prefaced his question by expressing dismay at Feynman's apparent dismissal of dilettantes in the* Nova *program, since Mr. Warner's own life circumstances forced him to be a dilettante. He then described the great pleasure he got in figuring things out, such as why, as described in Feynman's book, a particle, bouncing off a wall, transfers a momentum 2mv to the wall (where m is its mass and v its speed). The factor 2 results from the particle first having to stop its motion and then start it again.*

*Mr. Warner then posed his question: In Chapter 15 of the* Lectures, *Feynman had postulated that it is impossible to determine the absolute speed of a moving ship by means of experiments performed inside it (for example, experiments that compare the ticking rates of two different kinds of clocks). Feynman then used this postulate to deduce behaviors of some laws of physics, such as the increase of a body's mass as it moves with higher and higher speed. Mr. Warner was puzzled about Feynman's premise that it is impossible to determine absolute speed. How could that premise be justified? "I fail to see it and even that failure to see fascinates me."*

*Feynman responded as follows.*

Mr. Kenneth R. Warner, Jr.

Pittsburgh, Pennsylvania

Dear Mr. Warner:

You certainly seem to understand much more than many dilettantes I have known. Anyone who stops to make sure of the 2 in $Dp = 2mv$ is either not a true dilettante, or one in grave danger of becoming not one.

Your vision of hints of a reductio ad absurdum with the clocks is because you didn't appreciate the source of something we put in. *We have assumed it to be impossible* to determine the absolute speed of our moving ship, not for any logical or necessary reason, but for the sake of argument, as a possible principle of nature. This principle was suggested by many experiments (e.g., one by Michelson and Morley) designed to measure the absolute velocity of the earth. They all failed, until it dawned on our dull human minds (in fact only on a few, like Poincaré's or Einstein's) that maybe it was impossible—and they set up to see what the consequences would be if we assumed this as a matter of principle. In the book I was following their type of argument. It was successful, the necessary consequential phenomena (like mass changing with velocity) were ultimately observed experimentally.

Sincerely,

Richard P. Feynman

## RICHARD P. FEYNMAN TO MICHAEL STANLEY, MARCH 31, 1975

*Mr. Stanley was a graduate student in pharmacology seeking advice on remaining fresh in methods of reasoning. "How is it possible to reach that high level of preparedness without stifling the creative process that permits the examination of problems in novel ways?"*

Michael Stanley
Department of Pharmacology
Mount Sinai School of Medicine
New York, New York

Dear Mr. Stanley,

I don't know how to answer your question—I see no contradiction. All you have to do is, from time to time—in spite of everything, just try to examine a problem in a novel way. You won't "stifle the creative process" if you remember to think from time to time. Don't you have time to think?

Sincerely,

Richard P. Feynman

. . . .

*Caltech Commencement, 1975.*

# RICHARD P. FEYNMAN TO EDMUND G. BROWN, JR.,
## JUNE 23, 1975

The Honorable Edmund G. Brown, Jr.
Governor of the State of California
The State Capitol
Sacramento, California

Dear Sir:

I should like to urge you to support the school programs for the mentally gifted by signing Senate Bill 480.

I am a theoretical physicist (Nobel Prize 1965) whose son (13) and
daughter (7) are very intelligent and intellectually active. I am often asked
where I send them to school so that they can best develop their active
minds. I am proud to answer, "the public schools of the State of California."
For in the regular public school they have both been happy students, where
I would have expected the tedium and boredom I experienced in my day
(in New York public schools). But schools have come a long way in
improving the curriculum, and especially in recognizing the problem of
providing materials in a variety to try to meet the variety in children's
minds. The biggest contributor to, in particular, my son's happiness and
development has been the special classes designed for his special type of
variation—the programs for the mentally gifted minors. To an extent it is
an oasis in a desert for both of them.

Keeping the better students intellectually alive and active is obviously
useful to them, and to society, for their special talents are developed
instead of being "turned off" by "education." In addition, the competition from private schools is such that it would be a shame if parents of
clever kids thought that to get them the "best" education they should
take them out of the public schools (when on the contrary, the public
schools are better!). For that would leave our schools relatively dull
places, with fewer bright flashes from students to light up a teacher's day
and to suggest ideas to other students. Ideas suggested by fellow students
shows the others what can be done and keeps standards and perform-
ances high.

My interest in state education is not just that for my own children, but as
a teacher (of university physics) I have a general interest. I served on the
State Curriculum Committee choosing textbooks for two years, 1964–65.

So I urge you to continue the program to make our schools interesting to our best students.

Sincerely,

Richard P. Feynman

Richard Chase Tolman Professor of Physics


## ILENE UNGERLEIDER TO RICHARD P. FEYNMAN, DATE UNKNOWN

Dear Richard
I've fallen in love with you
From seeing you on "Nova"
I'm so glad you're alive
I appreciate your: wit
                    wisdom
                    brilliance
                    looks
    You are a feyn-man
Are there lots of physicists with fans?
                    You have one!

Ilene Ungerleider


## RICHARD P. FEYNMAN TO ILENE UNGERLEIDER, AUGUST 11, 1975

Ms. Ilene Ungerleider
Seattle, Washington

Dear Ilene:

I am now unique—a physicist with a fan who has fallen in love with him from seeing him on TV.

Thank you, oh fan! Now I have everything anyone could desire. I need no longer be jealous of movie stars.

Your fan-nee, (or whatever you call it—the whole business is new to me).

Richard P. Feynman

. . . .

*On a camping trip, c. 1975.*

## RICHARD P. FEYNMAN TO WILLIAM NEVA, AUGUST 14, 1975

*"Being a layman, curious about creation and all its after effects, one question plagues me (as I'm sure it has you and thousands of others) and that is— CAN INVISIBILITY BE INDUCED, CREATED so as to render an object unseen by the human optical setup?"*

   *—Mr. William Neva, in a letter to Richard P. Feynman, July 23, 1975*


Mr. William Neva
Henrietta, New York


Dear Mr. Neva:
Thank you for your letter and for your question about invisibility. I would suggest that the best way to get a good answer to your question is to ask a first-rate professional magician. I do not mean this answer to be facetious or humorous, I am serious. What a magician is very good at is making things appear in an unusual way without violating any physical laws, but by arranging matter in a suitable way. I know of no physical phenomenon such as X-rays, etc., which will create invisibility as you want. Therefore if it is possible at all it will be in accordance with familiar physical phenomenon. That is what a first-rate magician is good for, to create apparently impossible effects from "ordinary" causes.
   Sincerely,
   Richard P. Feynman


## RICHARD P. FEYNMAN TO DAVID RUTHERFORD, AUGUST 14, 1975

*David Rutherford wondered if it would be possible to record dreams on tape, much like television shows are recorded.*


Mr. David Rutherford
Davis, CA


Dear Mr. Rutherford:
The difficulty in seeing what goes on in the brain by just measuring brain waves is that the impulses are not truly transferable into video pictures that are being seen in the brain. If they were transferable it would have to be via some sort of code but we do not know what code. However, I am

certain that the amount of information or detail that is in a dream image is enormously greater than that which is carried into the gross variations measured in the EEG. It is like trying to describe or "see" what a painting looks like just from some overall facts like the weight of the painting, or the amount of paint of each color used, etc. More detail, like where the paint is located, is needed. I am sure it is impossible to try to decipher the details of the impulses in the millions (?) of nerve fibers that constitute a dream image by just looking at the EEG effect.

Sincerely,

Richard P. Feynman


## BEULAH ELIZABETH COX TO RICHARD P. FEYNMAN, AUGUST 22, 1975

*Physicists think of electrical forces as produced by electric field lines that stick out of any electrical charge. Gauss's Law states that the total number of electric field lines that cross through any closed surface (such as a sphere or a cube) is proportional to the total electrical charge contained inside that surface. The following exchange concerns Gauss's Law.*

Dear Dr. Feynman:

I recently took a course in Elementary Physics at the College of William and Mary in Virginia. An exam question concerned Gauss's Law and conductors, namely, does a hollow conductor shield the region outside the conductor from the effects of a charge placed within the hollow but not touching it?

I read Chapter 5, Volume II of *The Feynman Lectures on Physics*, and understood all except the next to last paragraph, in which you say "... no static distribution of charges *inside* a closed conductor can produce any fields *outside*." This was confusing, as it seemed to contradict all your previous statements. My instructor showed me how a simple application of Gauss's Law, with the surface of integration enclosing the entire conductor, shows that the E vector outside the conductor is not zero.

Could you perhaps explain what the paragraph in question means? I would greatly appreciate a reply as I am now very confused. My address is above.

Sincerely,

Miss Beulah Elizabeth Cox

P.S. I must admit I have a devious motive in writing to you because on the exam I answered with the explanation that your book gave. However, my instructor did not give me any points, even after I found your book to validate my answer. If you could clarify this question for me, I would be very appreciative. Thanking you in advance.

## RICHARD P. FEYNMAN TO BEULAH E. COX, SEPTEMBER 12, 1975

Miss Beulah E. Cox
Williamsburg, Virginia

Dear Miss Cox:
Your instructor was right not to give you any points for your answer was wrong, as he demonstrated using Gauss' law. You should, in science, believe logic and arguments, carefully drawn, and not authorities.

You also read the book correctly and understood it. I made a mistake, so the book is wrong. I probably was thinking of a *grounded* conducting sphere, or else of the fact that moving the charges around in different places inside does not affect things outside. I am not sure how I did it, but I goofed. And you goofed too, for believing me.

We both had bad luck.

For the future I wish you good luck in your physics studies.

Sincerely,

Richard P. Feynman

## RICHARD P. FEYNMAN TO ALEXANDER GEORGE, SEPTEMBER 26, 1975

*The question of whether "an independent breakthrough" was still possible in the present scientific world was one that Mr. Alexander George posed on September 20, 1975. At the time Mr. George was doing independent research and had been repeatedly told that research teams were needed for discoveries.*

Mr. Alexander George
New York, New York

Dear Mr. George:
To answer your question, it depends on what branch of physics you are

working on. In high energy physics the experiments are so complicated and elaborate and require such expensive machines that almost all experiments are done by large teams. But when it comes to a realization of what an experiment might mean, or to inventing and producing a new clever way of doing something—that might be done by one fellow independently. Finally, good theoretical work seems to me to be much as it always has been—good ideas appear in individual brains, not in committee meetings. Of course, as always, reading others' work on conversations and discussions with colleagues helps a lot in preliminary stages of thinking.

Sincerely,

Richard P. Feynman

## RICHARD P. FEYNMAN TO R. H. HILDEBRAND, OCTOBER 28, 1975

Professor R. H. Hildebrand
Chairman, Appointments Committee, EFI
The University of Chicago
Chicago, Illinois

Dear Professor Hildebrand:

This is in response to your letter of October 1. I am sorry, but I have a general policy never to write evaluations of people for institutions where that person has recently spent time, or is still located. My reason is that the people at the institution have had ample opportunity to observe him themselves (more recently, and more closely, than I) and should be capable of making their own evaluation.

This is a general rule I follow and has, of course, nothing to do with the particular individual about which you asked me.

Also for that reason, I can make no exceptions, for otherwise others to whom I have applied my rule will misinterpret my refusal to write, if they were to find out that I sometimes do write such evaluations.

I am sorry, therefore, to have to refuse your request.

Sincerely,

Richard P. Feynman

*After the letter to D. S. Saxon on October 20, 1964, Feynman wrote this type of letter whenever he was pressed by an institution for an evaluation of someone who was still at that institution.*

## STATEMENT BY MAX DELBRÜCK AT THE PRESS CONFERENCE ON BEHALF OF ACADEMICIAN ANDREI SAKHAROV, ORGANIZED BY THE CALIFORNIA COMMITTEE FOR HUMAN RIGHTS, DECEMBER 9, 1975

*In 1968, decorated Soviet physicist Andrei Sakharov wrote an article that was smuggled out of the Soviet Union and published in* The New York Times. *Titled "Reflections on Progress, Peaceful Coexistence, and Intellectual Freedom," it was highly critical of the Soviet political system and highlighted the risks of radioactive fallout from nuclear tests to hundreds of thousands of innocent people. Sakharov's efforts to ban nuclear testing, as well as foster democracy, won him the Nobel Peace Prize in 1975. The Soviet government refused to let him attend the ceremonies.*

*The following statement was drawn up by Max Delbrück, winner of the 1969 Nobel Prize in Medicine and a friend of our family (we went on many camping trips with the Delbrücks). Feynman associated himself with the statement, as did fellow Laureates Harold Urey, Carl Anderson, and Julian Schwinger.*

Sakharov's plans for greater openness became widely known in the West in 1968 through his magnificent 10,000 word essay, "Progress, Coexistence and Intellectual Freedom," which was reprinted at the time in the *N.Y. Times.*

He argued that the capitalist and the socialist systems have great merits and demerits. There should be a freer flow of information, of visits, and of open critical discussion, to accelerate the natural process of convergence between the two systems, and thus lead to a more humane way of life for all of us, including the Third World.

The Nobel Committee of the Norwegian Parliament in Oslo wished to applaud Sakharov's ideas and to give them wide public attention. They also wished to give recognition to his unremitting efforts in organizing campaigns for human rights, especially the right to public dissent, and to his forceful and courageous stand in many individual cases of human rights.

The authorities of the USSR have, perhaps unwittingly, supported the Nobel Committee's intentions: by denying Sakharov permission to travel to Oslo they have enormously increased the world-wide attention given to his ideas. Thus they may not have done *him* a favor, personally, because he would have probably enjoyed the trip. But they have done *his cause* a favor, by enhancing public attention to it. Surely we would not be here today if Sakharov's presence in Oslo tomorrow were not very much in doubt.

Sakharov is a scientist of the highest caliber, and a great citizen of the world. He is, perhaps most of all, a great Russian patriot, belonging to what we would call his government's loyal opposition, a concept unfortunately not entirely accepted in the USSR.

We wish to join his many friends in the USSR and abroad in saluting him on the day of the award to him of the Nobel Prize for Peace.

# 1976–1981

In June of 1978, at the age of sixty, Feynman had his first operation for abdominal cancer. He had noticed a bulge in his side and went to the doctor to get it checked. By the time it was removed, the tumor weighed six pounds and was the size of a football. Both Richard and Gweneth were battling cancer at this time, and both bore their surgeries and treatments with courage. His illness did curtail both his teaching and his correspondence, and he became even more selective about the letters he answered.

By this stage in his career, however, Feynman stories were well-established Caltech lore. Around the time of his operation, some students at a Caltech dorm wrote to him about their attempts to honor him. They had painted a mural depicting an old western saloon—and an anatomically detailed female nude. The students signed the painting with "R.P.F" (though Feynman signed his art work "Ofey"), and soon the nude was cut from the canvas and passed around campus. It finally ended up in a lecture hall, where it offended department secretaries and was promptly removed from the premises. The students' letter concludes, "The masterpiece was never recovered but, as far as we know, it's still extant. We thought you might be interested in it as it is probably the first documented Feynman art forgery."

Feynman's reply: "I thought I could get in enough trouble myself but you guys are doing it for me without me having to do anything."

In the world of physics, he did substantial phenomenological work with quark jets (collaborator Rick Field doing the actual computer calculations). There was also a long arduous attempt to prove quark confinement in

*Caltech, 1976.*

QCD (the sole published evidence for this seems to be a 1981 paper, "The Qualitative Behavior of Yang-Mills Theory in 2+1 Dimensions").

. . . .

## RICHARD P. FEYNMAN TO VIKTOR WEISSKOPF, JANUARY 6, 1976

*A wager was made immediately after Feynman won the Nobel Prize. Professor Weisskopf was convinced that Feynman would succumb to a career in administration, or what Feynman once referred to, in a letter to Paul Olum, as an "occupational disease."*

Prof. V. Weisskopf
Physics Department
M.I.T.
Cambridge, Massachusetts

Dear Professor:
I have found the document describing our wager and find that you gave me too much money so here's $15 back. For your records, may I state in

writing that as of this date, January 6, 1976, I am not holding, nor during the last ten years have I held, a responsible position as defined in the contract of the wager. Therefore I consider that the wager has been paid by Professor Weisskopf and that's that!

Sincerely,

Richard P. Feynman

—On this the FIFTEENTH DAY of DECEMBER of the YEAR ONE THOUSAND NINE HUNDRED AND SIXTY FIVE, at a Luncheon given at the Laboratories of the European Organization for Nuclear Research (CERN), Meyrin, Geneva, the following WAGER was made between Professor Viktor F. WEISSKOPF and Professor Richard P. FEYNMAN.

The terms of the said WAGER are as follows:

—Mr. FEYNMAN will pay the sum of TEN DOLLARS to Mr. WEISSKOPF if at any time during the next TEN YEARS (i.e., before the THIRTY FIRST DAY of DECEMBER of the YEAR ONE THOUSAND NINE HUNDRED AND SEVENTY FIVE), the said Mr. FEYNMAN has held a "responsible position."

—Conversely, if on the THIRTY FIRST DAY of DECEMBER of the YEAR ONE THOUSAND NINE HUNDRED AND SEVENTY FIVE, the said Mr. FEYNMAN shall have held or be holding no such position, Mr. WEISSKOPF will be deemed to have forfeited his WAGER and will be in duty bound to pay the sum of TEN DOLLARS to Mr. FEYNMAN.

—For the purpose of the aforementioned WAGER, the term "responsible position" shall be taken to signify a position which, by reason of its nature, compels the holder to issue instructions to other persons to carry out certain acts, notwithstanding the fact that the holder has no understanding whatsoever of that which he is instructing the aforesaid persons to accomplish.

—In case of contention or of non-fulfillment of the aforementioned conditions, the sole arbiter shall be Mr. Giuseppe COCCONI.

Signed at Meyrin on this the FIFTEENTH DAY of DECEMBER of the YEAR ONE THOUSAND NINE HUNDRED AND SIXTY FIVE.

Richard P. Feynman

Viktor F. Weisskopf

Signed and witnessed: G. Cocconi

## A. M. HUGHES TO RICHARD P. FEYNMAN, JANUARY 19, 1976

Dear Sir,

Volume three of *A Supplement to the Oxford English Dictionary* is in course of preparation here in Oxford, and I wonder if you, as the coiner of the word *parton,* could give me some information as to its origins.

The *Supplement* is a historical dictionary that prints the earliest known occurrence of each word or phrase, and in the case of *parton* the first occurrence has proven elusive. I have found it used in the issue of *Physical Review* for September 25, 1969 (p. 1975), and also what appears to be a coinage by you in the proceedings of the conference on high-energy collisions held on 5 and 6 September 1969 (edited by C. N. Yang et al.). This would be straightforward did not the word also occur *earlier* in a report of R. H. Dalintz's Bakerian Lecture of the Royal Society (*New Scientist* for 26 June 1969, p. 679).

Can you tell me whether there is any printed use of the word *parton* before the *New Scientist* article? Dalitz refers to "the so-called 'parton' theory of Feynman" in his lecture, but I have failed to locate the word in earlier publications of yours. Perhaps at this stage you had not committed it to print? If you had, I should be very grateful if you could send me the precise reference.

At the same time you might care to comment on the definition provisionally proposed for the *Supplement*:

Each of the hypothetical point-like constituents of the nucleon that were invoked by R. P. Feynman to explain the way the nucleon inelastically scatters electrons of very high energy.

I hope this request will not be too much trouble, and look forward with interest to hearing from you.

Yours faithfully,

A. M. Hughes, Assistant Editor (Science)

. . . .

## RICHARD P. FEYNMAN TO A. M. HUGHES, FEBRUARY 4, 1976

A. M. Hughes
Assistant Editor
Oxford English Dictionary Supplement
Oxford, England

Dear Mr. Hughes:

In reply to your letter of 19 January, Ref. O.C./RWB, your research and definition of the word "parton" is admirable. I don't know of any earlier publication. I used the word in verbal discussions earlier, but as far as I know you have the earliest published reference.

Sincerely,

Richard P. Feynman

## RICHARD P. FEYNMAN TO ANDRZEJ TRAUTMAN, FEBRUARY 4, 1976

Professor Andrzej Trautman
Directory of the Institute for Theoretical Physics
University of Warsaw
Warsaw, Poland

Dear Professor Trautman:

We have a young Polish student here, Isabel Kierkowska, who is doing a remarkable job. The first thing she did is take a "transfer examination," to see if she qualified for our institution. This is a very difficult examination because we have an excellent reputation and many more applicants than we can accept, therefore we make an unusually high level of ability a criterion for admittance. Miss Kierkowska passed this exam with an exceptionally high grade and was immediately accepted to study here. And it was no mistake! She rose rapidly to the top of her class and is doing wonderfully. She is the kind of student this place is designed for, and for such a student we feel we really make a distinct and unique contribution. Their full scientific potential is developed and stretched to the limit.

To interrupt such a good start and to break so natural a connection between institution and student, as we have found here, would mean a tragic loss of human potential.

It has come to my attention that she is having some difficulty in obtaining the necessary exit visa to continue her studies here in the United States.

I am writing to you in the hope that you might be able, and wish to assist, one of your countrymen to develop her full capability to become as useful as possible to society. Is there anything you could do to clear up the difficulty with her papers?

Thank you very much.

Yours sincerely,

Richard P. Feynman

Professor of Physics

*Professor Trautman agreed that Miss Kierkowska should be given an opportunity to finish her studies at Caltech and wrote back with advice for filing a new application for an extension of Miss Kierkowska's passport.*

## RICHARD P. FEYNMAN TO DAVID PATERSON, FEBRUARY 11, 1976

Mr. David Paterson

Chief Producer Science Programs

British Broadcasting Corporation

London, England

Dear David:

I was glad to hear from you. I looked at your enclosure "Travelling in Time," but didn't read beyond the second sentence because I, also, believe that time travel cannot be done, and I thought my colleagues agreed with me. The science fiction writers who have interpreted my view of the positron as an electron going backward in time have not realized that that theory is completely consistent with causality principles, and in no way implies that we can travel backward in time.

Sincerely,

Richard P. Feynman

## RICHARD P. FEYNMAN TO THE EDITOR OF THE *CALIFORNIA TECH*, FEBRUARY 27, 1976

*In 1974, Caltech's literature faculty unanimously recommended Dr. La Belle, an English instructor, for tenure. Humanities Chairman Robert Huttenback*

*rejected it, changing the criteria for tenure (which she in turn met) and reject-*
*ing arbitration (which found in her favor). The Equal Employment*
*Opportunity Commission also found in Dr. La Belle's favor, and a settlement*
*with Caltech ensued. She became the first female tenured professor at Caltech*
*in 1979.*

I should like to commend the *Tech* for giving its entire first page to
Professor La Belle to outline her complex view of a sorry situation, and to
complement a perfectly clear and transparent account of the matter in the
previous edition. Miss La Belle need not have answered the other account.
A colleague of mine showed your reporter's account to me and said, "what
a whitewash." I had followed the situation as it developed and knew he was
right, but he had not; how did he know? He smiled and reminded me he
was an expert on judging evidence in difficult physics experiments. In
physics the truth is rarely perfectly clear, and that is certainly universally the
case in human affairs. Hence, what is not surrounded by uncertainty cannot
be the truth (unless it be written in Russian "pravda").

I have known Jenijoy ever since the first decisions on her tenure, for she
introduced me to literary research and the Huntington Library and the
wonders of holding in one's hand an old book written by Newton. I could
appreciate directly how much he knew and how much he didn't know, and
what expressions he used that we still use. I will be very sorry to see her
leave. Caltech is the loser.

I know nothing of English literature and cannot judge that, but from the
first I was surprised to learn that in Humanities, here at Caltech, the criterion
was "publish or perish"—or rather, "publish in the most prodigious journals
or perish"—or now, having published in the finest, it is "perish anyway." Why?

Most shocking of all, for me, was to see how the tenured English faculty
were treated during this entire process. I am glad Jenijoy, in her personal
anguish, could make this situation public. Nobody paid attention to them,
their opinions were not respected. They have done a great deal for us and
our students to make this place livable and human, as the Department of
Humanities is meant to. One purpose of this note is to get a chance to tell
them that, though I ridicule their subject through ignorance (for I know,
regretfully, less about their subject than they know about mine), I respect
them and their contribution to our institution and to you students, and
have felt sorrow and shame at how they have been treated.

Writing letters is dangerous—if you show an interest in a subject, those

cornered mongeese (Sec'ty; leave as is, I want to prove I know no English) making decisions are likely to put you on a committee. It is very hard to make decisions.

Sincerely,

Richard P. Feynman

## RICHARD P. FEYNMAN TO V. L. GINSBURG, MARCH 16, 1976

Professor V. L. Ginsburg
Lebedev Physical Institute of the Academy of Science
Moscow, U.S.S.R.

Dear Professor Ginsburg:

I have heard that there is a possibility that you may be able to visit us here at Caltech next year.

I was so glad to hear the news! We can look forward to exchanging many ideas again and learning things from each other just as we did in our brief encounter in Poland (1961?). But this time it could be much longer and we could go into the problems of physics so much more deeply. Here many people think the same way—they go in packs like wolves. It seems to me from what journals I read, and physicists I have met (in Poland and Hungary), that in Russia there is some more independent thought and different ideas or, at least, rather different emphasis. Therefore I expect an exchange of ideas will be very fruitful.

Please accept and come, I don't want to be disappointed.

Personal regards,

Richard P. Feynman

## RICHARD P. FEYNMAN TO JOHN ELLIOT, APRIL 7, 1976

*Fred Hoyle was a British astrophysicist and professor at the University of Cambridge. He generated a large number of creative ideas, some of which were spectacularly successful (for example, his ideas about the generation of heavy elements inside stars by burning of nuclear fuel, as described in the following letter), and many of which turned out to be wrong. He was most well known for arguing that there was no big bang origin of the universe, but instead the universe has always existed, with matter being continually created in the vast*

*reaches of interstellar and intergalactic space, to fill the void left by the expan-*
*sion of the universe. This steady state theory was later disproved by astronomi-*
*cal observations in the 1960s, and the big bang theory against which it*
*competed was confirmed.*

*Feynman knew Hoyle somewhat through Hoyle's frequent visits to Caltech,*
*and Feynman spoke to his friends rather positively about Hoyle. In 1976, the*
*BBC asked Feynman to participate in a film about Hoyle. Feynman did, after*
*some initial hesitation.*

Mr. John Elliot
British Broadcasting Corporation
London, England

Dear Mr. Elliot:

I am uncomfortable. I don't think I know or knew Hoyle so well that I
should be talking on camera in a programme on his life. I try to imagine
things I could say and I am unsure of myself about what he did or others
did. For example, you want me to talk "about the work on particles" but I
really don't know he did any work on particles and what it was.

A possibly appropriate thing I could say is that when he first came here
he gave a series of seminars on how the heavier elements might be formed
in stars if it all started as hydrogen (as his steady state theory proposed). It
was all very carefully analyzed and impressive in its attention to detail. He
concluded that it wouldn't work unless there was a nuclear level in carbon
close to 7.62 MeV. Since he believed his hydrogen theory he said there
*must* be such a level.

We were all very impressed. To find the position of a nuclear level (as yet
unknown) by looking not at nuclei in the laboratory, but at the stars in the
sky seemed to us most remarkable and courageous.

He was right, the level was soon found.

I had a few occasions to discuss other astrophysical problems with him
and William Fowler at other times. He also made a number of field theories
related to cosmology or gravity (is that what you mean by "particle
work"?) which I never thought were likely to be right. He has since aban-
doned them—but I don't think a technical evaluation like that is what you
would want.

Of course, I like the guy very much, personally, and would like to help.
There are so many others so much more closely entwined in his life, however, that I am surprised you would think a day of shooting me would be

appropriate. You may be under a false impression about how close Hoyle and I have been personally.

Sincerely,

Richard P. Feynman

## MARK MINGUILLON TO RICHARD P. FEYNMAN, APRIL 14, 1976

Dear Dr. Feynman:

I am a 19 year old student of physics and I have studied much about you and your famous contributions to science.

From what I can understand, you were on the Atomic Bomb Project staff and along with other scientists you paved the way towards the nuclear age.

In my chemistry class, we began a discussion about the Manhattan Project and about how many of the scientists who were once very involved and "pro atomic bomb" now shifted their thoughts and are now against the whole idea of atomic or nuclear power.

If I remember correctly, I read an article awhile back that made mention to the fact that, among other things, you were still pro nuclear power, and yet my teacher said that you weren't.

Dr. Feynman, I would be very grateful to you if you could please take a moment out of your very busy schedule to help clarify this mini debate between my teacher and I and possibly enlighten the whole class as to your views about nuclear power.

Thank you very much for your most precious time.

Respectfully yours,

Mark Minguillon

## RICHARD P. FEYNMAN TO MARK MINGUILLON, APRIL 23, 1976

Mr. Mark Minguillon

Poway, California

Dear Mark:

Of course it takes more than a "moment of my time" to discuss something as complex and uncertain as the atomic bomb and its relation to nuclear power. For a very quick summary I see nothing wrong with

nuclear power except questions of the possibility of explosions, sabotage, stealing fuel to make bombs, leaking stored radioactive spent rods, etc. But all these are technical or engineering questions, about which we can do a great deal. So I think the risks can be controlled and that nuclear power, if economical, should be developed. Problems about the atomic bomb and the future are much more complicated and I cannot make any short statement to summarize my beliefs here.

Well, I guess that means you win your debate—but that doesn't mean we know what's true. Just because Feynman says he is pro-nuclear power, isn't any argument at all worth paying attention to because I can tell you (for I know) that Feynman really doesn't know what he is talking about when he speaks of such things. He knows about other things (maybe).

Don't pay attention to "authorities," think for yourself.

Yours sincerely,
Richard P. Feynman


## MARK MINGUILLON TO RICHARD P. FEYNMAN, JULY 31, 1976

*A few months later, Mr. Minguillon had further reason to write Feynman.*

Dear Dr. Feynman:

I am a 19 year old student of physics at a local college here in Southern California and I have run into a little trouble as far as choosing the particular field of physics I would like to major in.

My interest in physics stems back to 1972, and it was during that year that my interest really solidified (after meeting a great professor in this field at a school in Spain that I attended) to the point of making my decision of what my future career would be.

I became interested in reading about the early scientific discoveries concerning the atom and its nucleus. Bohr, Thomson, Heisenberg and Chadwick became my "heroes." I soon became extremely interested in nuclear and partial physics (not to any real knowledgeable degree yet of course, that's why I'm studying physics).

My problem, Dr. Feynman, is that no one I have been able to talk to is sufficiently informed in this area to be able to tell me whether the field is opening up, or becoming more and more saturated. No one can tell me if

all the research that can be done, has been done. This is the field I would like to study, but for all I know, maybe there isn't anything left to do research on!

I have written to other scientists in the past but I guess a 19 year old student isn't important enough to bother with, for I never received an answer.

The reason I write to you now is because I know you have done research into nuclear physics (who can forget your great contributions to the Manhattan Project during your stay at Los Alamos) and particle physics and you are probably one of today's most respected physicists. If any one should know, you would. I am sorry to have to ask you for some of your time, but no one seems to want to, or be able to, help me. I would be very grateful if you would.

Thank you very much Dr. Feynman.

Most respectfully yours,

Mark Minguillon

## RICHARD P. FEYNMAN TO MARK MINGUILLON, AUGUST 16, 1976

Mr. Mark Minguillon
Escondido, California

Dear Mr. Minguillon:

Relax. In no field is all the research done. Research leads to new discoveries and new questions to answer by more research.

Particle physics is the frontier of unknown physical laws to be discovered. It is very active and completely open-ended now, but it is hard to get a job in it because so many people want to.

But have no fear—you are just starting out and should not pick a subject of physics so soon. Just learn more and see what interests you most and what you like to do best as you go along and you will not have any trouble choosing when you know more.

By the way, at about your age, I didn't even know what field I wanted. I entered M.I.T. in mathematics, changed to Electrical Engineering for a while and then settled in physics. What field of physics? Aside from deciding I liked theoretical work best I have wandered around from stresses in molecules to quantum theory electrodynamics, theory of liquid helium, nuclear physics, turbulence in the flow of water (I didn't succeed in those

The reason I write to you now is because I know you have done research into nuclear physics (who can forget your great contributions to the Manhattan Project during your stay at Los Alamos) and particle physics and you are probably one of today's most respected physicists. If any one should know, you would.

I am sorry to have to ask you for some of your time, but no one seems to want to, or be able to, help me. I would be very grateful if you would.

Thank you very much Dr. Feynman.

Most respectfully yours,

Mark Minguillon

---

Dear Mr. Minguillon.

Relax. No one will fill in all the research done. Research leads to new discoveries and new questions to answer by more research.

Particle physics is the frontier of unknown physical laws to be discovered. It is very active now and will probably keep on — but it is hard to get a job in it because so many people want to.

But have no fear — you are just starting out and

should not pick a subfield of physics so soon. Just learn more and see what interests you most and what you like to do best as you go along and ~~you~~ you will not have any ~~diffic~~ trouble choosing when you know more.

By the way at about your age I didn't know even what field I wanted. I entered M.I.T. in mathematics, changed to Electrical Engineering for a while and then settled in physics. What field of physics? ~~I have worked in~~ aside from deciding I liked theoretical work but I have wandered around from stresses in molecules to quantum ~~electro~~ theory electrodynamics, theory of liquid helium, and recently particle physics. You do any problem that you can, regardless of field.

Yours

R.

last two problems so nothing is published), and recently particle physics. You do any problem that you can, regardless of field.

Yours sincerely,

Richard P. Feynman

## RICHARD P. FEYNMAN TO SIDNEY COLEMAN, AUGUST 13, 1976

*The est foundation was interested in sponsoring physics conferences. Feynman and two other physicists, Geoff Chew and David Finkelstein, had suggested Professor Sidney Coleman as the consultant for the conference.*

*From Professor Coleman's letter, "(1) The participants in the conference will be as many of the following as are willing and are able to come: C. Callan, G. Chew, S. Coleman, R. Dashen, L. Faddeev, R. Feynman, D. Finkelstein, F. Goldhaber, J. Goldstone, D. Gross, R. Jackiw, T. D. Lee, S. Mandelstam, Y. Nambu, A. M. Polyakov, C. Rebbi, L. Susskind, G. 't Hooft, and K. Wilson (Roman Jackiw helped me construct this list)." A possible title was suggested of "Novel Configurations in Quantum Field Theory," and though Feynman and Geoff Chew were not working on the subject, they were encouraged to come, as they "would be a joy to have with us."*

Professor Sidney Coleman
Department of Physics
Harvard University
Cambridge, Massachusetts

Dear Coleman:

About the est conference. It seems to me that we should keep the conference as small as possible and have only guys that are really working actively in the subject attend. On item 1, what the hell is Feynman invited for? He is not up to the other guys and is doing nothing as far as I know. If you clean up the invitation list, to just the hard-core workers, I might begin to think about attending.

Sincerely,

Richard P. Feynman

*Professor Coleman wrote back to say that Feynman was off the list and urged him to come. Feynman attended the conference.*

## RICHARD P. FEYNMAN TO E. FRED CARLISLE, JANUARY 24, 1977

*Michigan State University was developing a special scientific writing course for science undergraduates and wanted to use "The Theory of Gravitation" from* The Feynman Lectures on Physics *in an anthology of scientific writings.*

Professor E. Fred Carlisle
Department of English
East Lansing, Michigan

Dear Professor Carlisle:
You certainly have my permission to use "The Theory of Gravitation" from *The Feynman Lectures on Physics* in the way you proposed in your letter. I hope you find it useful.

However, your course being a writing course you might be interested in how the material was written. It was given directly by me as a verbal lecture (from a page of very short "notes" to guide me) and recorded as given. Then some people, probably mainly Robert Leighton, edited it into proper English, in parts making omissions, inclusions or reorganizations of ideas, and in parts sticking closely to the original words. It was a big job for I don't speak writable English. I then went over it and made a few further modifications etc., we discussed it a bit and then I was off preparing the ideas for the next lecture.

Caltech owns the copyright, so I am asking them to send you a note confirming my permission.

Sincerely,
Richard P. Feynman

## RICHARD P. FEYNMAN TO RAMON CORTINES, APRIL 25, 1977

Superintendent Ramon Cortines
Pasadena Unified School District
Administration Offices
Pasadena, California

Dear Mr. Cortines:
My friend Ralph Leighton* has already talked to you about this problem

---

* Ralph Leighton is the son of Robert Leighton, who was a Caltech professor and an editor of *The Feynman Lectures.*

and suggested I write you this letter. In the coming school year my son Carl will be a junior at John Muir High School. He is required by state law to take Health and Driver Education, and required by the Pasadena Unified School District to take Consumer Education. He would have taken these courses this year had he been 16, but since he skipped kindergarten, Carl has been one year younger than most of his classmates.

Having to take Health and Driver Education as well as Consumer Education in eleventh grade will mean that Carl will have to drop one of his electives, i.e., physics, calculus, or Latin. He could of course take one of these courses in summer school, but he works hard in school and I think a young man should have a full three month vacation each year to refresh himself and enjoy some time away from school work.

I believe that Carl is capable of learning the material in Health and Driver Education, and in Consumer Education in less time than the full year required for these courses. I have found out, through Polytechnic School, that the University of California at Berkeley offers Health and Driver Education by correspondence, in which the student completes a number of assignments and takes a final test administered by a teacher at his school (or a public librarian). In addition Ralph Leighton, who has been a teacher in the Pasadena Unified School District, tells me that the material covered in every course taught in the district is on file at the Board of Education.

I therefore ask if my son could be permitted to take Health and Driver Education by correspondence, and Consumer Education by examination, with access to the material in the course. If so, I will see that Carl makes the necessary arrangements with some appropriate teachers at Muir, and with Mr. Snyder, if necessary, unless you advise otherwise.

Sincerely,

Richard P. Feynman

Professor of Physics

## RICHARD P. FEYNMAN TO J. T. DEVREESE, OCTOBER 26, 1977

*Professor Devreese wrote to say how much they missed Feynman at the Advanced Study Institute on Path Integrals. The meeting proceedings would appear in a volume in early 1978, and they proposed to dedicate the volume to Feynman in honor of his sixtieth birthday. As Feynman was the originator of*

*Family trip, 1977.*

the method, they thought it would be a great honor if he would provide an epilogue to the volume.

Prof. Dr. J. T. Devreese
Antwerp, Belgium

Dear Prof. Devreese:
I was very sorry that I could not attend your conference. My wife and I had planned an extended trip with the kids through western U.S. and Canada for that time. Many times lying out in my sleeping bag I wondered what was going on at your conference, and thinking about how I would enjoy being there.

I was delighted when the proceedings arrived and had a very great pleasure in reading them. Thank you very much for sending them.

Your thought, to honor me on my birthday year by dedicating the book is very nice. You take much of the pleasure out of it suggesting that I work for it (by writing something—an epilogue?). That is not fair, and I am too lazy to write anything.

It is nice of you but completely unnecessary to do something formal in my "honor." You, and Thornber, and so many other of the conferees have already honored me directly and very much by paying attention to my work and developing it further. It has given me great pleasure to see these works of others and to be seeing it all collected together in the conference proceedings. To me it represents a demonstration that what I invented (in 1941) is of real and continuing utility to physicists. What further honor is possible?

Thank you again.

Sincerely,

Richard P. Feynman

## LEWIS H. STRAUSS TO RICHARD P. FEYNMAN, AUGUST 2, 1977

Dear Dr. Feynman:

This letter is written to you as a former recipient of the Einstein Award. This award was presented in the following years and to the persons listed below:

| | |
|---|---|
| Professor Julian Schwinger | 1951 |
| Professor Kurt Godel | 1951 |
| Dr. Richard P. Feynman | 1954 |
| Dr. Edward Teller | 1958 |
| Dr. Willard F. Libby | 1959 |
| Dr. Leo Szilard | 1960 |
| Dr. Luis Alvarez | 1961 |
| Dr. Shields Warren | 1962 |
| Dr. John A. Wheeler | 1965 |
| Dr. Marshall N. Rosenbluth | 1967 |
| Dr. Yuval Ne'eman | 1970 |
| Dr. Eugene P. Wigner | 1972 |

The officers of the Lewis & Rose Strauss Memorial Fund have elected not to present this award since the passing of the award's originator, Lewis L. Strauss.

Now, however, sufficient time has passed so that perhaps some of the peripheral problems associated with the award have had a chance to subside. We have taken note of the work of Dr Stephen W. Hawking of Cambridge University who appears to have made major theoretical

strides in the areas where Einstein labored so long. Do you think that Hawking's work is of sufficient importance to justify the Einstein award, quite apart from the humanitarian aspects of the matter? Is there anyone else whose name comes to your mind as more worthy? Your advice is appreciated.

Sincerely,

Lewis H. Strauss

## RICHARD P. FEYNMAN TO LEWIS H. STRAUSS, AUGUST 9, 1977

Mr. Lewis H. Strauss
Washington, D.C.

Dear Mr. Strauss:

In reply to your letter of August 2, I would most certainly agree that Dr. Stephen W. Hawking's work is deserving of the Einstein Award.

Sincerely,

Richard P. Feynman

## RICHARD P. FEYNMAN TO MICHAEL E. FISHER, NOVEMBER 15, 1977

*Dr. Michael Fisher was working with Christopher Longuet-Higgins on a biographical memoir of Lars Onsager and asked Feynman for personal recollections of Dr. Onsager.*

Dr. Michael E. Fisher
Department of Chemistry
Cornell University
Ithaca, New York

Dear Dr. Fisher:

You wanted some information on my interactions with Lars Onsager. I will give them by memory not being sure of dates, etc. Nor will I refer to indirect interaction via reading his papers, etc., but only my direct personal contacts with him.

My first (probably?) contact was in about 1951 at a theoretical physics meeting in Japan. I had worked out a theory of liquid helium, which was a

new field for me so at that meeting I met many of the people in that field for the first time. At a dinner the day before I was to speak, I was seated next to Onsager. He said, "So you think you have a theory of liquid HeII?" I said, "yes, I do," to which he simply answered "hmpf" and said nothing more. I took it to mean he didn't believe my theory could be anything but nonsense.

The next day at the meeting I presented my paper, in which I claimed to explain everything except, I made clear apologetically, there was the serious flaw that I did not understand adequately the detailed nature of behavior of the thermodynamic functions right at the transition. In the period for questions Onsager spoke first. "Mr. Feynman is new in our field, and there is evidently something he doesn't know about it, and we ought to educate him." I was petrified; this was even worse than the grunt last night—what had I left out—what stupid mistake had he found? He continued, "so I think we should tell him that the exact behavior of the thermodynamic functions near a transition is not yet adequately understood for any real transition in any substance whatever. Therefore, the fact that he cannot do it for HeII is no reflection at all on the value of his contribution to understand the rest of the phenomena."

Thus it was that I learned that he was a kindly and generous man whose habit of saying as little as possible had given me the wrong impression. After that I met him at several meetings and we had many conversations about physics problems (transitions, turbulence, helium, superconductivity, etc.). He would speak little (especially relative to me) but the little contained a great deal of thought—and his innate friendliness always showed through.

On one occasion when we were standing together, a young man came up to explain his ideas on superconductivity to us both. I didn't understand what the fellow was saying—so I thought it must be nonsense (a bad habit I have). I was surprised to hear Onsager say, "Yes, that seems to be the solution to the problem." Did he mean the puzzle of superconductivity was solved—and I didn't even know what the young man said? I guess so. I have never been sure—I think the young man could have been Cooper. Could you check?

More specifically, work which Onsager did influence me in at least the following ways. His work on interacting dipoles in liquid dielectrics answered a very puzzling physical question and I studied it carefully to understand rotor interactions in helium. I have had great interest in his theorem about the non-existence of turbulence in two dimensions—and the

Hamiltonian he derived for this. Finally, and most directly, I independently discovered quantized vortex lines without knowing of his previous discovery of them. For, after all, in typical Onsager fashion, their description was not published directly by him, but appears in the records of a meeting in a remark by him during a discussion of somebody else's paper on a different subject. But, of course, finding that he had come to the same conclusions was, for me, strong encouragement that I might be on the right track.

Sincerely,

Richard P. Feynman

## RICHARD P. FEYNMAN TO TEPPER L. GILL, OCTOBER 10, 1977

Dr. Tepper L. Gill
Department of Mathematics
Howard University
Washington, D.C.

Dear Dr. Gill:

Thank you for sending me your research paper on infinite tensor products. I am too ignorant of modern mathematics to understand it, but I am happy to learn that my work is beginning to give pleasure even to professional mathematicians. It was lots of fun to discover ordered operators (time is too special an ordering parameter). I realized from the start their possible application to equations with random operators and spent some time formulating and trying to solve problems on the rate of mixing paint by random stirring, or on the magnetic field generated by random convection currents in the earth's core—as well as problems in the theory of turbulence. I published nothing on these problems because I did not carry them far enough to satisfy me—but I knew the ordered operator stuff would be of great importance someday.

I am pleased to hear that mathematicians can join in the fun, for fun it must be for you too—as it seems to have all the necessary qualifications, according to your quotes—it is "in intimate contact with sin" and "causes mathematical heartburn."

Sincerely,

Richard P. Feynman

## RICHARD P. FEYNMAN TO MAX M. LEVIN, NOVEMBER 21, 1977

*Dr. Levin wrote Feynman from the University of California at Santa Cruz at the suggestion of Chancellor Sinsheimer and Professor Matt Sands. He was developing a lecture series on the subject of the relationships between Sciences and the Arts. Dr. Levin thought Feynman was interested in this topic and wanted to know whether he would be interested in participating in the symposium.*

Dr. Max M. Levin
Deputy Provost
University of California
Santa Cruz, California

Dear Dr. Levin,

After a small amount of careful thought I could think of no recent significant influence of the Arts on at least Physical Science. It is likely that the influence is there—so my inability to come up with an example shows that I am not a good candidate to lecture for your symposium.

It is possible that Matt Sands and Sinsheimer thought of me because I once learned to draw, and they think of that as Art.

I am now playing bongo drums for a small San Francisco ballet company—so if this gets around my scientific friends will be recommending me as a lecturer on Music and Dance in Science (about which I know nothing).

Sincerely,
Richard P. Feynman

## RICHARD P. FEYNMAN TO FRANCIS CRICK, MARCH 7, 1978

Francis Crick
The Salk Institute
San Diego, California

Dear Francis,

I regret having to do this, but I'm returning this paper to you unread. My schedule is such lately that I must refuse to get bogged down reading

someone else's theory; it may turn out to be wonderful and there I'd be with something else to think about.

Sincerely,

Richard P. Feynman


## FRANCIS CRICK TO RICHARD P. FEYNMAN, MARCH 10, 1978

Dear Dick,

I would have done the same! The usual expression used in Molecular Biological circles is due to Frank Stahl: "Don't tell me—I might think about it!"

Yours ever,

Francis

Kieckhefer Distinguished Research Professor

The Salk Institute


## RAFAEL DY-LIACCO TO RICHARD P. FEYNMAN, MAY 24, 1978

Dear Dr. Feynman,

Please forgive my presumptuousness in sending my ideas to you. I have an idea on the origin of cosmic rays boiling in the back of my head but I do not know who to go to. I'm mailing the idea to *Scientific American* for the "Letters" section in the hope that it will be published and that scientists who know more of the physics and mathematics involved in such a thing will be able to criticize it and either disqualify it or say it is possible. The only physics I've taken is what I'm taking now—high school physics (I'm 16).

My father is driving me crazy by suggesting that *Scientific American* will plagiarize the idea. I saw you in a film about the universe and you impressed me as being the most sincere man I ever heard (though you were only talking physics). So I ask for your time (if you will) to criticize this idea (if you do, please do so in terms easily understandable). I do not want to go to my physics teacher for many reasons. An important criticism I'm asking for is this: Is this a new idea? This is the idea:

Assumption: Antimatter exists at least in the outer layers of the sun.

The random collision of particles with their antiparticles provide a small but steady source of cosmic rays.

In an extremely strong magnetic field, like that causing sunspots, charged

*As Frankie Scarpini in campus production
of* Fiorello, *1978.*

particles will move in a less random fashion—they will be accelerated toward one of the poles. Oppositely charged particles are likewise accelerated toward the outer pole. Thus a charged particle will have a greater chance of colliding with its antiparticle. The result is an increase in the rate of matter-antimatter collisions and so a higher cosmic ray output is observed during intense sunspot activity.

Sincerely,

Rafael Dy-Liacco

*Lucille, Gweneth, Richard, Michelle, and Carl in front
of the Feynman van, Mexico 1978.*

## RICHARD P. FEYNMAN TO RAFAEL DY-LIACCO, JUNE 6, 1978

Dr. Rafael Dy-Liacco
Kobe, Japan

Dear Dr. Dy-Liacco,

There are several reasons why your suggestion that cosmic rays come from antiparticles from the sun is probably wrong.

First, in the presence of all the matter in the sun, such antiparticles would all be annihilated in a very short time. Since they are used up so rapidly your problem would be where do new ones keep coming from?

Next, cosmic rays often have very much more energy than would be available from annihilation (of say, proton and antiproton), so at least you could not explain the source of the higher energy rays are known not to come from the sun but from outside the solar system (because of the way they behave during magnetic storms).

Third, annihilating matter and anti-matter make gamma-rays, not fast protons; but the incoming cosmic rays are fast protons and not gamma-rays.

The usual supposition is that low energy cosmic rays do not come from the sun but from acceleration of hydrogen ions (i.e., protons) by the changing solar magnetic fields. There does not seem to be any serious difficulty with this explanation. The origin of the higher energy rays is not as clear.

Keep up your interest in these things. If at first you don't succeed, try again. And tell your father not to worry about plagiarism—because in science (unlike business or many other pursuits) we are all working together cooperating to try to understand Nature and we have learned to be very careful to recognize and commend anybody who gets a really useful new idea—we are glad to have him aboard, so to speak. We haven't had serious trouble with that sort of thing since the seventeenth century.

Sincerely,

Richard P. Feynman

## LINUS PAULING TO RICHARD P. FEYNMAN, JUNE 28, 1978

Dear Dick:

I have learned from Linda that you have had a malignant tumor removed.

These abdominal malignancies are serious. The 5-year survival fraction is rather small. Chemotherapy has little value—in Britain it is rarely used for these cancers.

I think that the best thing to do is to begin immediately a high intake of vitamin C—20 g. per day or more. I am corresponding with a man who had extensive abdominal cancer, and who took 60 g. per day for 3 months. He is now much better, and is down to 35 g. per day.

Enclosed are a couple of papers, with references to more. Linda can tell you where to get pure ascorbic acid and sodium ascorbate and how to take it.

Vitamin C works largely by potentiating the body's immune mechanisms. The cytotoxic drugs destroy them, and probably decrease the effectiveness of the vitamin C. On the other hand, immune stimulants, such as BCG may well be compatible with vitamin C.

It is very important not to stop the intake of vitamin C, once you have started.

We have another paper in press in PNAS. Also, Morishita and Murata in Japan have got similar results.

Best wishes,

Linus

P.S. Also no sugar, little meat, lots of fresh vegetables, vegetable juice & fruit juice.

## RICHARD P. FEYNMAN TO LINUS PAULING, JULY 7, 1978

Dr. Linus Pauling
Linus Pauling Institute of Science and Medicine
Menlo Park, California

Dear Linus,

It was very good to hear from you. Thank you for your special interest in my problems.

It turns out my cancer is a very unusual kind of abdominal cancer called myxoid liposarcoma—a soft tissue cancer and although it weighed over 2800 grams it still seemed to be nicely encapsulated. It was apparently neatly removed in its totality and the pathological laboratory can't even find any apparent invasion of the blood vessels by the cancer cells. So my oncologist (Dr. Thomas C. Hall, a man introduced to me by Benzer) suggests even no chemotherapy at all but of course a very careful periodic thorough search with x-ray for metastases. At any rate, I have given him your letter to be sure he is thoroughly familiar with the results given in your references.

Linda has already offered the information, etc., that you mentioned in a very kind and lovely note. One of your great accomplishments, Linus, has been to help to produce such a lovely daughter.

Thanks again for your interest.

Sincerely,

Richard P. Feynman

.  .  .  .

## RICHARD P. FEYNMAN TO R. WAYNE OLER, MARCH 2, 1979

*A letter arrived with a royalty statement from Mr. Oler, Vice President and Publisher at the Benjamin/Cummins Publishing Company. Mr. Oler was concerned about a problem that he felt might affect royalties and incur losses for the entire college textbook industry. It was common among textbook publishers to send complimentary copies of textbooks to professors teaching courses for which the books might be used. Apparently, professors were now known to be selling these books for personal profit.*

Mr. R. Wayne Oler
Vice President & Publisher
The Benjamin/Cummins Publishing Company
Menlo Park, California

Dear Mr. Oler,

Let us not be so greedy. (Unfortunately, I write my books because I want to spread knowledge—not to make money, so my opinion differs.)

It seems to me that if you intend the books you send out free to promote the books and create good will, they have presumably served their purpose. If they can be sold for others to read, good. If you don't like it, just don't send them out.

If you send them out because you get value from doing so, so let it be—you have already been "paid off" by whatever value you imagine you get by sending them. If the receiver can get further value, we have no claim to stop him.

If the net result is a net loss (the loss of sale offsetting the value of advertising and good will) simply stop sending them and stop the loss.

Previously I have always returned, unopened, unsolicited books from publishers (I dislike advertising). But now you have given me a better idea.

Sincerely,

Richard P. Feynman

.  .  .  .

## RICHARD P. FEYNMAN TO G. C. DIJKHUIS, JUNE 13, 1979

Dr. G. C. Dijkhuis
Zeldenrust College
Zeldenrustlaan 2
Terneuzen
The Netherlands

Dear Dr. Dijkhuis

Thank you for your note on the signs in Section 21–8 of Vol. III. As I said in the earlier part of the lecture, this is like a seminar. And now you see this is realistic even to the signs in the equations—which, in seminars, are notoriously wrong. Have no faith in my signs. Whatever you carefully derive is right. (I'm sorry but I don't have the patience to check it all over now—but you can bet 50% of my signs are wrong.)

Sincerely,
Richard P. Feynman

## RICHARD P. FEYNMAN TO ROBERT ILSLEY, JUNE 13, 1979

Mr. Robert Ilsley
Optical Coating Laboratory, Inc.
Santa Rosa, California

Dear Mr. Ilsley:

This letter is to confirm our telephone conversation of Wednesday, June 13, during which I told you I would be unable to serve on the Board of Directors of your photoelectronics company. The reason is my lack of business experience. I did not feel that the depth of business experience of the other outside members was sufficient to compensate for my lack. Therefore, I could not feel confident that I would be adequately discharging the responsibilities of a member of a Board of Directors. Because of this discomfort I cannot serve.

I appreciate the very gracious way in which you accepted my vacillating position, after what must have been considerable trouble and expense involved in considering me.

As a result of all this, photovoltaics and their future have become very interesting to me. I expect to watch with great interest the development of your company, and hope we can stay in contact from time to time.

*National Science Medalist Feynman with President Jimmy Carter, 1979.*

It was good to get to know you, and other members of the staff at the plant.
Good luck,
Richard P. Feynman

## RICHARD P. FEYNMAN TO GWENETH, MICHELLE, AND CARL FEYNMAN, JUNE 29, 1980

Sat. June 29, 1980
3 P.M.
Royal Olympic Hotel
Poolside

Dear Gweneth, and Michelle (and Carl?)*
This is my third day in Athens.

* Carl was expected home shortly, after spending part of the summer working in Massachusetts.

I'm writing by the side of the hotel pool with the paper in my lap because the tables are too high and the chairs are too low.

The trip was all on time but uncomfortable anyway because the plane from N.Y. to Athens was absolutely full—every seat. I was met by a Prof. Iliopoulos, a student, and his nephew, who is just Carl's age. I was surprised to find the weather here just like in Pasadena (but about 5° cooler)—the vegetation is very similar—the hills look bare and desert like—same plants, same cacti, same low humidity and same cool nights. But there the similarity ends. Athens is a sprawling, ugly, noisy, exhaust-filled mess of streets filled with nervous traffic jumping like rabbits when the lights go green and stopping with squealing brakes when they go red and blowing horns when it goes yellow. Very similar to Mexico City—except the people don't look as poor as in Mexico and there are only occasional beggars in the streets. You, Gweneth, would love it because there are so many shops (all small) and Carl would love walking around in the arcades and rabbit warren twists and surprises, especially in the old part of town.

Yesterday morning I went to the archeological museum. Michelle would like all the great Greek statues of horses—there is one big one especially of a small boy on a large galloping horse, in bronze, that is a sensation. I saw so much stuff I got all mixed up. It is not labeled well and my feet began to hurt. Also it was slightly boring because we have seen so much of that stuff before. Except for one thing—among all those art objects there was one thing so entirely different and strange that it is nearly impossible. It was got from the sea in 1900 and is some kind of machine with gear trains—very much like the inside of a modern wind up alarm clock. The teeth are very regular and many wheels are fitted closely together. There are graduated circles and Greek inscriptions. I wonder if it is some kind of fake. There was an article on it in the *Scientific American* in 1959.

Yesterday afternoon I went to the Acropolis—which is right in the middle of the city—a high rock plateau on which was build the Parthenon and other shrines and temples. The Parthenon looks pretty good—the Temple at Segesta that Gwen and I saw in Sicily is just as impressive because you are allowed to walk around in it—but you can't go up and around among the Parthenon columns. Prof. Iliopoulos' sister came with us and with a notebook she had—she is a professional archeologist—and guided our tour with all kinds of details, dates, quotations from Plutarch, etc.

It appears the Greeks take their past very seriously. They study ancient Greek archeology in their elementary schools for six years, having to take

*10 hours* of that subject every week. It is a kind of ancestor worship for they emphasize always how wonderful the ancient Greeks were—and wonderful indeed they were. When to encourage them by saying yes and look how modern man has advanced beyond the ancient Greeks (thinking of experimental science, the development of mathematics, the art of the renaissance, the great depth and understanding of the relative shallowness of Greek philosophy, etc., etc.)—they say, "What do you mean—what was wrong with the ancient Greeks?" They continually put their age down and the old age up, until to point out the wonders of the present seems to them to be an unjustified lack of appreciation for the past. They were very upset when I said that the thing of greatest importance to mathematics in Europe was the discovery by Tartaglia that you can solve a cubic equation—which, altho it is very little used, must have been psychologically wonderful because it showed a modern man could do something no ancient Greek could do, and therefore helped in the renaissance which was the freeing of man from the intimidation of the ancients—what they are learning in school is to be intimidated into thinking they have fallen so far below their super ancestors.

I asked the archeologist lady about the machine in the museum—whether other similar machines, or simpler machines leading up to it, or down from it were ever found—but she knew nothing so I had to meet her and her son of Carl's age—who looks at me like I was a heroic ancient Greek, for he is studying physics—at the museum to show it to her. She required some explanation from me why I thought such a machine interesting and surprising because didn't Eratosthenes measure the distance to the sun, and didn't that require elaborate scientific instruments, etc. Oh, how ignorant are classically educated people. No wonder they don't appreciate their own time. They are not of it and do not understand it. But after a bit she believed maybe it was striking, and she took me to the back rooms of the museum—because surely there were other examples and she would get a complete bibliography. The complete bibliography was a list of three articles (one in the *Scientific American*) all by one man, and he is an *American* from Yale!

I guess they think all the Americans must be dull, only interested in machinery when there are all those beautiful statues and portrayals of lovely myths and stories of gods and goddesses to look at. (In fact the lady from the museum staff remarked when told the Prof. from America wanted to know more about item 15087, "Of all the things in the museum why does he pick out that particular item, what is so special about it?")

Everyone here complains of the heat—and is concerned about whether you can stand it, when in fact, it is just like Pasadena but about 5° cooler on the average. So all stores close from perhaps 1:30 P.M. to 5:30 P.M. ("because of the heat"), which turns out to be really a good idea (everyone takes a nap) because then they go late into the night, supper between 9:30–10 P.M. etc. when it is cool. People are *seriously* complaining of a new law. To save energy all restaurants and taverns must close at 2 A.M. This, they say, will spoil life in Athens.

Now is the witching hour between 1:30 and 5:30 and I am using it to write to you. Honest tho, I miss you and I would really be happier at home. I guess I really have lost my bug for traveling. I have a day and a half yet here and they have given me all kinds of literature about a beautiful beach (of pebbles) here, of an important ancient site (altho in rather complete ruins) there, etc. but I will go to none of them, for each—it turns out—is a long (2 to 4 hours, only each way) bus ride on a tour bus. No. I'll just stay here and prepare my talks for Crete (they have me giving an extra *three* lectures to some twenty Greek university students who are all coming to Crete just to hear me. I'll do something like my New Zealand lectures—but I haven't got my notes! and have to work them out again.)

I miss you all, especially when I go to bed at night—no dogs to scratch and say good night to!

Love
Richard. Poppa.


P.S. IF YOU CAN'T READ THE ABOVE HANDWRITING, HAVE NO FEAR—IT IS UNIMPORTANT RAMBLINGS. I AM WELL AND IN ATHENS.


## MICHAEL H. HART TO RICHARD P. FEYNMAN, MARCH 25, 1980

Dear Professor Feynman,

One morning last fall, while teaching electromagnetism to some college students, I showed them the formula

$$\vec{F}_{mag} = q\vec{v} \times \vec{B}.$$

I pointed out (as most texts do) that since, by this formula, $\vec{F}_{mag}$ is always perpendicular to $\vec{ds}$, it follows that magnetic forces can never do any work.

A few days later, in the laboratory, I lifted a small iron bar with a permanent magnet. (The bar had been lying on the table, held in place by my finger, while I positioned the permanent magnet above it.)

One of my students (who obviously was not bright enough to properly understand the text) asked me how, if magnets could do no work, the permanent magnet was able to lift the iron bar. Such a foolish question! The correct explanation of this apparent contradiction was, of course, immediately obvious to me. However, at the time I could not think of a way of phrasing the explanation which would make it clear to even a dull student like that one. (It is depressing to see how the high schools have lowered their graduation standards in recent years.) I would appreciate it if you would advise me how I should have phrased the explanation.

Cordially yours,

Michael H. Hart

Assistant Professor

## RICHARD P. FEYNMAN TO MICHAEL H. HART, DECEMBER 4, 1980

Dear Professor Hart,

Clearing up my desk recently, I found your interesting problem about magnetic fields doing work. I must have put it aside to answer later, so please excuse me. (Perhaps I am reluctant to admit that I have been working on it for the last six months and have finally found the answer.)

The idea of "who does the work" is not a very clear or useful one in physics. It does not help our intuition. Suppose I have two blocks A, B, separated by a partly compressed spring ▯·⋯·▯ . A is held fixed and B is
                                    A   B
curved toward A compressing the spring, clearly the spring energy comes from work done by B. But if instead it was A that was moved toward B, fixed, it would be A who did the work. Which is which depends on relative motion of the observer! If we simply took the blocks A and B in our hands and pushed them together we could only say the compressed spring's energy is work done by blocks A and B together and who cares which did what part because it depends on just how my hands moved, a matter that might be irrelevant to any further interesting question. So to give examples:

A is a magnet, B a wheel carrying a charge Q. If A is moved toward B the wheel rotates faster. The extra kinetic energy comes from where? From the work done in pushing them together is what I would like to

say. But we have this theorem which says a particle's kinetic energy cannot be changed by action of a magnetic field. Well in this case, of course, if you insist, it was done by the electric field indeed by moving the magnet. But your dull student wants a reasonable and sensible intuitive answer and will be held back in a flash asking what happens if it (A) is held fixed and B is moved toward it. There is no electric field! It is work done by the force pushing the wheel up acting in a kind of complicated way through the spokes (how exactly?). The dull student loses patience with such fine considerations that depend on which moves toward which because he foolishly thinks nothing essential can really depend on that; and therefore all your fine remarks must be pedantic drivel about something inessential.

Finally, however, and most interestingly, the example you use of a piece of iron lifted by a magnet has a special consideration. The magnetic moment commences from electric spin, a quantum mechanical effect. The forces are not given by E + vxB but, if by anything approximately correct classically, by  where  is the atomic magnetic dipole moment. This time work can be done by the magnetic field! And the paradox here is resolved by the observation that atoms could not behave like little magnets if the electrons in them really obeyed classical physics (see Feynman Lectures on Physics, Vol. II, p. 34–6).

Simple questions with complicated answers are always asked by dull students. Only intelligent students have been trained to ask complicated questions with simple answers—as any teacher knows (and only teachers think there are any simple questions with simple answers).

Sincerely,

Richard P. Feynman

## RICHARD P. FEYNMAN TO GERALD FREUND, APRIL 6, 1981

*Stephen Wolfram was a prodigy in theoretical physics. He earned his Ph.D. at Caltech in 1979 at the age of twenty and continued working there for several years before moving to the Princeton Institute for Advanced Study.*

Dr. Gerald Freund
Director, Prize Fellows Program
John D. and Catherine T. MacArthur Foundation

Dear Dr. Freund,

Looking for something else I found this unfinished recommendation for Stephen Wolfram in a drawer.

I did not consciously remember it when the Zweig request came. I see now my desire to revolt against writing these recommendations has an unconscious basis in previous experience. Anyway, for what it is worth I am having my secretary type the Wolfram recommendation and send it to you in the unfinished state I left it.

You asked for information about Stephen Wolfram in connection with the MacArthur Foundation Prize Fellow Award.

Dr. Wolfram has the truly exceptional qualities you are seeking for your award. He has worked energetically and creatively in many fields. Examples of his creative originality are many. For example, he invented an ingenuous way to make comparison of high-energy experiments to incomplete theoretical predictions of quantum chromodynamics. As another example he has designed and programmed a new system for computer algebraic and symbolic manipulation starting from scratch and in an entirely original and independent manner. His analysis of baryon production in the early universe pointed out significant errors in the work that came before him.

He has made himself fully aware of all the problems in fundamental theoretical physics, from gravitation, cosmology and the new unified field theories, to the detailed problems of hadron physics and the weak interactions. The method he uses in studying each question is not so much to read about it, but to work it all out himself. He works very hard. Such methods and such industry is, of course, the true source of creativity and originality.

There is not a seminar here that proceeds without an important comment, question, or criticism by Wolfram.

I don't know of any others in this field that have the wide range of understanding of Dr. Wolfram. He seems to have worked on everything and has some original or careful judgment on any topic.

Sincerely,

Richard P. Feynman

*Dr. Wolfram received the MacArthur "genius award" in 1981.*

.  .  .  .

*At home with Sandra and Venus, 1981.*

## RICHARD P. FEYNMAN TO RODNEY C. LEWIS, AUGUST 10, 1981

*After saying how much he enjoyed Feynman's colloquium at MIT, Mr. Lewis made the comment that he felt many scientists and educators had lost their thrill of exploration. Mr. Lewis said he was tired of standard textbook examples and was happy with the different points of view Feynman presented.*

Mr. Rodney C. Lewis
Lincoln Laboratory
Massachusetts Institute of Technology
Lexington, Massachusetts

Dear Mr. Lewis:
Thank you for your nice letter. It makes me feel good to hear that you liked my colloquium so much. Don't despair of standard dull textbooks. Just close the book once in a while and think what they just said in your own terms as a revelation of the spirit and wonder of nature. The books give you facts but your imagination can supply life.
My father taught me how to do that when I was a little boy on his knee

and he read the *Encyclopaedia Britannica* to me! He would stop every once
in a while and say—now what does that *really* mean. For example, "the
head of tyrannosaurus rex was four feet wide, etc."—it means if he stood
on the lawn outside, his head would look in at your bedroom on the second floor, and if he poked it in the window it would break the casement on
both sides. Then when I was a little older when we would read that again
he would remind me of how strong the neck muscles had to be—of ratios
of weight and muscle area—and why land animals can't become the size of
whales—and why grasshoppers can jump just about as high as a horse can
jump. All this, by thinking about the size of a dinosaur's head!

Yours sincerely,

Richard P. Feynman

## RICHARD P. FEYNMAN TO DON NEMEC, OCTOBER 19, 1981

*Mr. Nemec and Feynman shared an eight-year correspondence. Mr. Nemec and
Feynman had a mutual friend, the physician John Lilly, who was a pioneer of the
isolation tank method of exploring consciousness (and probably best known for
his interspecies communication research projects between man and dolphin). Their
correspondence is greatly varied; there is a letter concerning Lilly's sense depriva-
tion tank, a page or so of physics written by Mr. Nemec with comments added by
Feynman, and discussions on dreams and OBEs—out-of-body experiences.*

*Most of the letters are Mr. Nemec's, with a few notes from Feynman. The
following is the most substantive letter of the bunch.*

Mr. Don Nemec
Boron, California

Dear Don:

I was very glad to hear from you in your long letter of August 5. You
seem quite a bit happier (and much fatter!) than you were in that shop in
Pasadena. That was a most unideal place for experiments in psychic phenomena.

As you know I still think that these phenomena you are practicing and
working on are simply the wanderings of a relaxed mind. Wanderings which
are entirely internal but partly directed and greatly affected by what your
beliefs and expectations are (that is to say, what you have been thinking of
when your mind was not relaxed but operating actively in the more usual

directed manner of every day). Thus I do not think one is "making contact with a universal intelligence," or "exploring some new aspect of the real world," etc., in the sense that there is a common *external* reality that one is becoming aware of by such experiences. They are more of the kind of private experience of ordinary dreams. Dreams, I think, do not predict the future, nor do they correctly tell one what really may be happening elsewhere, etc. They only might tell one what one is concerned or worried about during the day.

Well maybe, but how to prove or disprove it?

It is good to hear you have found someone (Kim) to work with. I don't know if you are still working with her. She has OBE's, but what do you mean when you say "in no way can she be convinced that the experiences are anything but what they appear to be?" What do they appear to be?

As you can guess—from my point of view it is really sad to hear that such a promising and intelligent girl has given up a college career to become obsessed with these phenomena which occur occasionally inside her head. Couldn't she pursue both, keeping things in better proportion? Because, again from my particular point of view, she may be wasting her time chasing phantoms and dreams while the delightful, exciting real world can no longer serve her as a great source for thought, pleasure, and beauty. I know we disagree on this a bit—and of course it is none of my business.

Thank you for your letter. Write me again, and lose a little weight! Good luck.

Your friend,
Richard P. Feynman

## RICHARD P. FEYNMAN TO P. P. LELE, NOVEMBER 16, 1981

Dr. P. P. Lele
Room 260–23
Massachusetts Institute of Technology
Cambridge, Massachusetts

Dear Dr. Lele:
Thank you very much for your report on your work with ultrasound. The care and analyses and ingenuity used in controlling the charting is in vast contrast to the sloppy way which is used by everyone else. It's very impressive and I hope that ultimately it is as successful as the early tests suggest.

I'm grateful for your offer to try your technique on my sarcoma; I gave it a great deal of consideration but I could not satisfy the condition that every conventional therapy had already been tried. The doctors at UCLA think that surgery is still a reasonable option and I have elected to have it next week (Dr. Donald Morton).

Preoperatively we have tried the usual chemotherapy, some radiation and hyperthermia (with Dr. Sturm). The hyperthermia was 1000 watts of 13.5 megahertz magnetic field to the abdomen. It was given in 35-minute sessions on 10 successive days (nothing has had a very dramatic effect). The purpose was to try to reduce the size to make a difficult surgery possibly easier. No thermometers were put into the tumor. Surely a crude technique compared to yours.

Perhaps when my sarcoma recurs you will consider me again as a possible candidate to use your ultrasound.

Thank you very much for your time and patience in discussing these things with me on the telephone.

Sincerely,

Richard P. Feynman

# 1982–1984

*The Pleasure of Finding Things Out,* a one-hour documentary program featuring Feynman talking to the camera in a tight head-and-shoulders crop, aired on both sides of the Atlantic in the early 1980s. Much to the surprise of some reviewers, it went over very well. More letters from laymen, admirers, and the merely curious soon followed.

A fluctuation between patience and a more brusque stance characterizes some of these later exchanges. Time seemed more scarce. In a letter from early 1982, Feynman claimed he found it very difficult to have discussions by letter, "because I cannot indulge in my habit of immediately asking for clarification of things which at first confuse me. I don't have the patience to figure out what is probably meant." Nonetheless, he would read through pages and pages of often unorganized, rambling thoughts, letters on subjects both technical and not, and in his reply hone in on what the letter was substantively about.

Meanwhile, his health problems continued. His mother passed away in early November 1981, and soon after her funeral, Feynman again underwent surgery to combat his abdominal cancer. During the fourteen-and-a-half-hour operation, his aorta split, and he required over seventy pints of blood. Dozens of Caltech students and faculty, as well as many staff from the nearby Jet Propulsion Laboratory, rushed to UCLA Hospital to donate blood. I remember speaking to a number of them on the telephone at the time, trying to round up donors, and their love and admiration of my father was palpable. It made a difficult time a little easier for my family.

In March 1984, he tripped over a curb in a parking lot on the way to pick up a long-anticipated personal computer. He wiped the blood off his head and continued on to the store. Later, when my mother and I returned home from a day out, we found him happily playing with his new computer, still wearing his blood-stained shirt. A few weeks later it was discovered that he had a subdural hematoma (a collection of blood on the surface of the brain) as a result of the fall, and so he returned to the hospital to have two holes drilled in his skull to relieve the pressure. He then proudly told his friends, "If you think I'm crazy, I now have an excuse: I have holes in my head—you can feel them right here!"

Computers figured prominently in his professional life as well at this time. Thinking and lecturing about computers, their ultimate limits, and the possibility of quantum computers was primarily stimulated by work done by none other than Carl Feynman.

.  .  .  .

## DON WRIGHT TO RICHARD P. FEYNMAN, NOVEMBER 23, 1981

*The Christopher Sykes documentary,* The Pleasure of Finding Things Out, *first aired in November 1981 in England, on a series called "Horizon."*

Dear Professor,

I don't have a lot of time for most Americans (U.S. ones anyway) and I've never written to a T.V. personality before and probably will never do so again. But I was so impressed by your television talk ("lecture", "interview"—whatever) which went out tonight on our BBC2 that I had to write and tell you so.

If you're ever in England and have time to come to Swanage in Dorset—beautiful outside and in—and visit a very ordinary little Englander in a little house you'd be more than welcome.

I'll post (mail) this letter knowing that if you feel that way your waste bin is probably bigger than mine.

Thanks for the talk.

Sincerely,

Don Wright

.  .  .  .

## RICHARD P. FEYNMAN TO DON WRIGHT, JANUARY 13, 1982

Mr. Don Wright
Swanage, Dorset
England

Dear Mr. Wright:
Thank you very much for your kind remarks about my TV program.
Maybe it would help you with your problem about my being an American,
to know that my wife is an Englishwoman from Yorkshire. She has probably
improved me greatly.
Sincerely,
Richard P. Feynman

## RICHARD P. FEYNMAN TO CONNIE GRIFFITHS, JANUARY 25, 1982

*Connie Griffiths also saw the "Horizon" program and wrote a letter proposing
that transcendental meditation might help "fully coordinate" his brain, thereby
saving time and effort. "I can see you must have a very fine, clear mind but
even so I wonder if those ultimate problems you touch on might not be more
easily solved by letting your attention go effortlessly through T.M. to the very
area you are researching—the area next the vacuum state," she suggested. "The
laws of Nature there would be closer to you and reveal themselves—not one by
one but suddenly."*

Connie Griffiths
Essex, England

Dear Miss Griffiths:
Thank you for your note about T.M.—Transcendental Meditation.
I have tried something similar—"sense-deprivation" tanks (made by a
man called John Lilly—no sound, no light, little feeling for you are floating
in salt water). I did it many hours, and had hallucinations, etc.
It is not the same as T.M. of course, but I had the impression that
although such experiences may help your psychological feeling about
yourself, they are no substitute for the hard directed thought that one
apparently needs to get ready to solve a hard problem. After loading up

with all the "facts" it is possible that some change in mental set, like T.M., or simply relaxing, might help one to get things to click all together.

Anyway, thanks for your suggestion.

Sincerely,

Richard P. Feynman

## J. GERARD WOLFF TO RICHARD P. FEYNMAN, NOVEMBER 23, 1981

Dr. Richard Feynman,

c/o Horizon,

BBC,

London

Dear Dr. Feynman,

I found what you had to say on today's BBC Horizon programme very interesting. In particular I was struck by your remarks about the problem of working out the implications of a theory so that it may be tested against observations.

This is precisely the kind of problem which arises in my area of research—how children acquire their first language. It is fairly obvious that a child's brain does a large amount of complicated "computing" in working out grammatical patterns from the language he or she hears. Many theories of language acquisition are fairly simple to state—at least in outline—but the implications can be very complex and the theories can, as a consequence, be difficult to confirm or refute.

My favoured method of tackling the problem is not via mathematics (partly because I don't know that much) but by expressing the theory in the form of a computer program and then allowing the computer to tease out the implications. This kind of technique is common enough in many areas of science and it is rather unlikely that you have not thought of it. But since you did not mention this as a possibility in your interview I thought it was worth making the point to you.

I agree that much social science, including psychology, is a sterile aping of the methods of physical sciences. But there are substantial problems—I think language acquisition is one of them—where precise and falsifiable theories can be devised and where potent insights can be obtained.

Do you think computer simulation would be any good for your problem?

Yours sincerely,

Gerry Wolff

(Dr.) J. Gerard Wolff

## RICHARD P. FEYNMAN TO J. GERARD WOLFF, FEBRUARY 4, 1982

Dr. J. Gerard Wolff
Department of Psychology
The University
Dundee, Scotland

Dear Dr. Wolff:

Thank you for your note about my program.

You guessed right—I had thought about computer simulation of a child acquiring language. I came to it the other way about. In studying the abilities of computers a friend gave me a problem. Computers can be programmed to play chess well, or they can be programmed to play backgammon, or bridge. Can you program a computer that can play any game? You tell it the rules of draughts say—and play some games with it and it gradually gets better and better as it gains experience (as a person might.) As it "learns" more it doesn't run slower and slower while it searches its increasingly large memory storage. That would be an intelligent computer.

In thinking about that I tried to find a simpler system—first I thought of the problem of making a computer that could learn more and more mathematics as you went along—and then I thought of the problem of learning a language or the "simplest" problem of this type. Then I wondered how children do it.

I couldn't figure out a way to store the knowledge so that as you learned more, new grammar or vocabulary, it would not make the computer slower and slower as it searched through its ever greater stores of memory to find the correct meaning.

Do you have some ideas on this subject? Could you send me some reprints if you have any published work?

In my physics problem we have, of course, attempted to work out some details by computer and some things have been learned that way, but so far

the computers do not seem to have a large enough capacity to give clear results.

Sincerely,

Richard P. Feynman

## RICHARD P. FEYNMAN TO H. DUDLEY WRIGHT, FEBRUARY 17, 1982

Mr. H. Dudley Wright
Geneva
Switzerland

Dear Dudley:

Well, the operation was a success—I mean I am still alive! In fact I have been recovering very well and would now be easily able to come to Geneva any time you want. I look forward to resuming work after that interruption.

Nearly every day I come to Caltech to work, but Helen, my secretary, won't let me teach—she says I should take the year off—but she can't stop me from doing research, thinking harder than ever.

It will be great to see you when you come in March—do you want to stay with us? But when will you let me go to Geneva to find out how everything is going, and to be with the gang again?

I got a great long letter with lots of non-technical news from the Hales.

Yours sincerely,

Richard P. Feynman

## RICHARD P. FEYNMAN TO V. TELEGDI, MARCH 30, 1982

Professor V. Telegdi
Zurich, Switzerland

Hi Val:

Thanks for your note. There is nothing to worry about as I'm completely recovered. I had surgery for cancer in November. The Doctor took out everything he could see so I'm missing various organs and pieces of organs. The surgeon said he came very close to losing his first patient during surgery, but everything turned out OK.

Have you any plans to come back to Caltech in the near future? Perhaps I will see you in Geneva—I expect to go there around the first or second week in June.

Sincerely,

Richard P. Feynman

P.S. I am teaching this quarter so I am back to 100% normal activity.

## RICHARD P. FEYNMAN TO FREDERIC DE HOFFMAN, MARCH 29, 1982

Mr. Frederic de Hoffman
The Salk Institute
San Diego, California

Dear Fiddle-dee-dee:

Thank you for your note of January 29 (it takes me a while to get around to my mail basket). I am feeling great and am going to be doing my normal teaching duties starting this week. I also regret that we haven't seen each other for a long time.

Sincerely,

Richard P. Feynman

## RICHARD P. FEYNMAN TO STUART ZIMMER, FEBRUARY 18, 1982

*From a high school for students dedicated to future careers in science, science teacher Mr. Zimmer wrote to request an autograph. The school was creating an exhibit of American science, and Feynman's autograph was to be part of a section of Nobel Prize winners.*

Mr. Stuart Zimmer
Elwood, New York

Dear Mr. Zimmer:

I am not in sympathy with an exhibit which is devoted to "American Science." Science is an international human effort and there would be no "American science" were it not for scientific development: in the rest of the world.

Secondly, it bothers me that everyone always chooses "Nobel prize winners" as important examples of scientists. Why do we pay such attention to the choice of the members of the Swedish Academy? That may be OK for the unknowing public, but surely a science teacher can make his own independent choices of which scientists excite his imagination and which men he would like to call to the attention of his students.

Don't mind my discourtesy—carry on with your good work. I just get tired of being a "Nobel Prize Winner" from time to time, and have to explode at somebody, so here is your autograph.

Sincerely,

Richard P. Feynman

## ALAN WOODWARD TO RICHARD P. FEYNMAN, DECEMBER 29, 1981

Dear Professor Feynman,

I write to you realising that Professors of Physics can be extremely busy people, but in the hope that some reply may be forthcoming at a future time.

I am only an undergraduate physics student at the University of Southampton, but already serious doubts are creeping into my mind. I see all these students, lecturers and professors engrossed in their subject, and not realising what is happening outside their laboratory door.

They do not seem to realise where the world is heading and would literally do anything in the name, and for the sake of their science. Surely this is wrong?

Yet, when one tries to broach the topic with them, everyone of them ignores their responsibility—I feel like a peculiarity.

Now, I read your books for my course, I am taught about your theories, and my Quantum Physics lecturer, Dr. Tony Hey, is constantly telling us what a great force in Physics you are, so you are obviously an authority in Physics. But, have you never had any such fears?

I used to be an officer in the Royal Navy, myself, until I decided to resign my commission and find a more humanitarian career. Unfortunately, the course that Physics is taking me on still seems to have something missing!

I am afraid I cannot explain such fears very eloquently but I hope you understand my meaning. Perhaps one day you might care to reciprocate your ideas as it confuses me terribly?

Yours sincerely,

Alan Woodward

## RICHARD P. FEYNMAN TO ALAN WOODWARD, MARCH 31, 1982

Alan Woodward
Southampton, Britain

Dear Mr. Woodward:

Surely increased knowledge is not incompatible with a humanitarian career—no matter what it is you learn. And surely if your professor and fellow students seem to know some things, but seem to be oblivious to other things ("outside their laboratory door," as you say) that does not exclude *you* from learning what they know whilst remaining deeply aware of what they are blind to.

Of course, the course that physics is taking you has something missing. You cannot develop a personality with physics alone, the rest of your life must be worked in.

Sincerely,
Richard P. Feynman

## RICHARD P. FEYNMAN TO YETTA FARBER, MARCH 30, 1982

*Ms. Yetta Farber wrote to remind Feynman that she had once dated him at Cornell. She also had a story to tell that made her laugh every time she saw Feynman's name in the papers. Immediately after their enjoyable date, word had gone out that there was a rapist, described as wearing "a brown or brown leather-like jacket," loose on the Cornell campus. "A-hah! I said—I went out with this nice fellow and he wore a brown leather-like jacket! Maybe it was he. When you called me for another date, I said, 'No, I'm busy.'"*

*Ms. Farber had thought that Feynman was too young to be an assistant professor (in fact, Feynman was a full professor), so she was rather suspicious.*

Dear Yetta,

Naturally I could never understand why the girls I went out with in Ithaca wouldn't go out with me again. At last I find out—it was my brown leather jacket!

So often, was I thus frustrated by pretty girls (like you) that I came out to California. Since the weather was so much better I threw away my leather jacket and at last found someone who would go out with me more than once—so I married her.

I always thought that the girls in California were more tolerant—but now I know the inner workings of the phenomenon.

Physics is much easier to understand.

Your former date,

Richard P. Feynman

## RICHARD P. FEYNMAN TO RICHARD MARRUS, APRIL 30, 1982

Dr. Richard Marrus
Department of Physics
University of California
Berkeley, California

Dear Dr. Marrus:

A student of mine, Peter Thomas,* asked me to write to you in connection with his application to Berkeley to work in particle physics.

He is a good student (I assume you have his records) who showed considerable independence and initiative. We had many conversations as he learned the details of quantum chromodynamics (on which I was working). Things were going well and he was asking insightful questions. I am sorry to have lost him to you, but I should like to explain the circumstances.

Firstly, he had always been discouraged a bit by the social conditions at Caltech, particularly the lack of women. In addition, in that year we did not have many good students, and he was unable to find students to talk to profitably. Therefore, he often thought of going to Berkeley where the society is (believe it not) more normal. He never got around to it because he wanted to work with me.

In the Fall, however, I became seriously ill. I required extensive surgery, and there was a real chance I might not survive, or might never be able to teach again. (As it turned out I have temporarily recovered completely.)

This uncertainty plus certain personal disappointments (unsuccessful love, etc.) confused and discouraged him badly. He spent time working and taking classes at UCLA (I believe he also taught part time here at Caltech while he lived near UCLA).

He also seems recovered from the shock and now appears definite in what he wants to do—study at Berkeley.

---

* Not his real name.

To summarize. He is a good student, and I am sorry to have lost him. I know that his application is late but I hope you can do something anyway. I think his talent should not be wasted. So I would be delighted to hear you have accepted him.

Sincerely,

Richard P. Feynman

## RICHARD P. FEYNMAN TO BLAS CABRERA, SEPTEMBER 7, 1982

*In 1931 the great physicist Paul Dirac argued convincingly that there may exist in nature magnetic monopoles, fundamental particles out of which emerge radially pointing magnetic fields, analogous to the electric fields that stick out of an electron or proton, and he predicted the strength of such a monopole's magnetic field. These predictions have motivated many experiments, over the seven decades since, to search for magnetic monopoles.*

*In 1981–1982, Dr. Blas Cabrera (a newly appointed assistant professor at Stanford University) carried out a high-sensitivity search for magnetic monopoles using an apparatus he had built that relied on a SQUID (superconducting quantum interference device). On February 14, 1982, Cabrera's experiment exhibited a sudden excitation that was "consistent with the passage" of a magnetic monopole through the apparatus. Many physicists would have touted this excitation as the discovery of a magnetic monopole, but Cabrera did not. Instead, he published a paper in which he presented the details of his event forthrightly and clearly, explained that the sudden excitation was consistent with what one would expect of a monopole, and then used it to place an upper limit on the number of monopoles in the universe. Cabrera's paper created great excitement in the physics community, and many physicists misinterpreted him as claiming a monopole discovery.*

*Cabrera presented the details of his "candidate" magnetic-monopole event in a physics colloquium at Caltech in May 1982. Feynman almost always attended Caltech's physics colloquia and made them lively by his intense questioning of the speaker. At Cabrera's colloquium, Feynman asked piercing questions about the experimental details. Cabrera, evidently dissatisfied with his on-the-spot answers, wrote to Feynman on June 24 with a follow-up clarification: "I want to answer the question you asked at the colloquium several weeks ago a bit more clearly. I first learned the theory of SQUID operation from Vol. III of your Lectures on Physics as an undergraduate at the University of Virginia. . . . Your three volumes had a profound influence on my early appreci-*

*ation for the underlying unity of physics. I thank you." He then went on with three pages of technical explanation, including six diagrams.*

Dr. Blas Cabrera
Department of Physics
Stanford University
Stanford, California

Dear Dr. Cabrera:

Thank you very much for your detailed explanation of the working of your squid. Some days after your talk I realized how things worked and that my comments at the talk were dumb, that the true weakness was small sub-quantum changes in the flux through the squid coil.

(I found this above paragraph of an unfinished letter started months ago—and now I'll finish.) May I say that your own evaluation and attitude toward your experiment is much more scientifically satisfactory and profound than what theorists say about you. "Cabrera says he found a monopole." I defend you at each opportunity to explain that you don't feel you can go so far—that it looks exactly like what a monopole should look like—but there is only one and you are uncomfortable. (Exactly as you say in your next to last paragraphs.)

Some days ago someone came in and said it has been recently shown that the theoretically expected monopole would interact strongly with protons and if there were as many as implied by your experiment we would have already seen them in apparent proton decays. "Won't Cabrera be embarrassed?" he gloated. "Why?" I asked. "Because he claims to have seen the monopole."

I carefully explained your position once again and then asked him whether *he* wouldn't be embarrassed if you saw another one. Because there still might be such monopoles for reasons other than today's brave new theories of unification. Some theorists do not understand the relation of theory and experiment—and where lies the true source and test of all their knowledge?

So keep up the elegant work and send me a telegram if Nature does it again!

Sincerely,
Richard P. Feynman

*Nature did not do it again. Cabrera's event remains today, in 2004, the most convincing example of a possible magnetic monopole. Cabrera continues his*

*experimental research on fundamental physics as a highly respected full profes-*
*sor at Stanford University.*

## RICHARD P. FEYNMAN TO GWENETH AND MICHELLE FEYNMAN, SEPTEMBER 12, 1982

*The following letter was written during a trip to Switzerland. Names have been changed.*

Dear Gweneth and Michelle,

"The Curse of Riches"
Yesterday Donald took me to see a man he said I would find very interesting and I will tell you all about Jorge Rueda Orozco, one of the sons of
a great Bolivian tin magnate, who inherited great wealth from his father.
We drove over there with me driving Donald's new Rolls Royce—
because for safety sake on previous occasions where he was drunk I asked
him to let me drive it, insisting it was not because he was incapable but
because I enjoyed driving it so much, so to keep up appearances I asked to
drive it that morning also even tho he was sober. Mr. Orozco had bought
an estate—a farm really—nearby and was making over the old farm buildings into beautiful new buildings. So far he had been working on it seven
years.

Expecting a beautiful entrance gate I was surprised when Donald
announced suddenly that we had arrived—opposite a rickety jerry built
gate of planks and chicken wire you might expect at the entrance to a construction site—but there was Mr. Orozco, a small thin but mildly handsome
man of about 55 who appeared very gracious and intelligent and laughed
when Donald told him of my comment—"Not a very elegant entrance."
We drove away to a nearby restaurant—where with his charm and good
manners he began to impress me with his wit and intelligence. (While I
was parking the car at the restaurant he asked Donald what I did—when
told I was a professor he remarked in surprise, "How does a Professor come
to be driving a Rolls?" not knowing it was Donald's car.)

But soon things began to deteriorate slightly with his statement that it is
not good to eat light lunches, for it is not natural, the argument being bol-
stered by some vague reference to the habits of animals. It soon appeared
that Mr. Orozco was a principled and moral man, with various definite

pronouncements about the need for forbearance and sacrifice before gratification of desires, about the dangerous slippage of morals in America,
about the fact that the main cause of all this is the press, and such things.
Donald told him about my recent operation—which he found surprising
because, he said, I looked so well. Then looking at me he very seriously told
me that if I kept up my good attitude he was "absolutely certain" that the
cancer would not return. He asked me why was I smiling—whether I was
laughing at him. I said, "Yes, I feel like I am in the tent of a fortune-teller,"
and explained I have found the world more uncertain. He countered with
an example of his mother who was in perfect health at the age of 80 but
who died then because she was hit by three blows. One I can't remember.
One was that he and his brother had a falling out and would no longer
speak to each other. His brother filched all his fortune from him (and
explained to me that "filch" is a slang expression meaning to steal). (His
brother was an inefficient "filcher" because as I will tell you below, Jorge
had a considerable fortune still.) Another blow was the kidnapping of his
(Jorge's) daughter Mercedes some years ago (she is now 15).

The latter is a tragic story. She was held several months, for a 2,500,000-
franc ransom. The money was to be delivered in notes none larger than 100
franc and none with consecutive numbers. Checking and counting the 25
thousand notes for delivery to the kidnappers was so difficult his hands
were bleeding after a while (Donald had told me all of this ahead of time).
When he got his daughter back she would tell him nothing of her ordeal,
except that they had warned her that if she told anything they would kill
her father. He took a long time to get over the experience.

Mr. Orozco told me many things of his early life—of his governess who
would berate him because he was small, partly Indian—and particularly how
uncomfortable he has always felt about the fact that he inherited wealth but
has done nothing to earn it. There were confessions of courses with a psy-
choanalyst to help him. There were stories of his Catholic bringing up—of
his turning toward Protestantism as an improvement—but now of a disbelief
in all religion—for how could a good God put men to test knowing before-
hand who would fail, and yet it still bothered him—and he didn't understand it—that during the kidnapping at one point he found himself on his
knees and spontaneously praying. What could it mean, he asked me, was it
proof it was really true in spite of his intellectual misgivings? I told him that
I was as strong an atheist as he was likely to find—but that I could conceive
of myself praying in such a situation. That it was the cry of anguish of an

utterly helpless man in the face of a threatening tragedy. He seemed genuinely pleased with my answer and muttered, "Yes, that's all it was."

Well after lunch we returned to his house—the rickety gates were
opened and we drove in past stoves, lumber, and operating cement mixers
to a driveway in front of a large square building—which had previously
been a great barn but now looked well enough but somewhat uninterest-
ingly rectangular. The work going on was on other nearby buildings yet to
be "restored." Two massive snarling police dogs came out barking—but
they only reminded me of Pasha—yet Donald was genuinely afraid. Mr.
Orozco assured us there was nothing to worry about, and tried to hold
their chain collars, but they were not particularly obedient. We walked out
into a large area of the "estate" behind the house, an enormous well kept
grass lawn on part of which was an orchard of fruit trees. Mr. Orozco led us
out there, the dogs circulating around and barking loudly while Mr.
Orozco only partly successfully held their collars with one hand while he
picked plums and "rare mirabelles" for us with the other assuring us always
that the dogs were no concern—a thing I fully believed for they looked
playful and puppyish. I put out my hand for one to smell—and was warned
frantically not to try to touch them. When one circled in front of Donald
he stopped short in fear, but Mr. Orozco said, "Don't stop suddenly" and
"Don't show fear because the dogs smell and sense that and become
enraged." The other escaped him and came up behind me and snuggled my
leg with his snout—he only wanted to play—but Mr. Orozco continued
his inconsistent patter about how there was nothing to worry about and
don't pay any attention to them or put your hand out to them or they may
become vicious. I think he knew the former was true—they were gentle
altho loudly barking dogs looking for a little love—and the latter he wished
were true—that they were dangerous guard dogs.

We entered the house—marble floors—marble pillars—marble staircases
and into an enormous "living room" ceiling 35 feet high! With great single
pane glass opening (a better word than mere "windows" unless you thought
of store windows—and even those are too small) at each end. The walls—
flat fine cement or plaster—had six great medieval tapestries, and two large
very strange gold plated chandeliers hung down. So far I had seen no furniture in the great empty barn of a marble house—and yet there was furniture everywhere for his mother's collection was in large crates about in the
"living room," but could not be seen. There was much talk about how
noisy the room turned out to be (before the crates. He has been working

on this project 7 years—living in the house now for a year and a half) and what he might do about it. It probably bounced from flat ceiling to flat floor—which floor by the way was all in old worn terracotta tiles.

We then entered a normal sized but abnormally long but narrow room, in the center of which was a long antique 15 ft. narrow table ("feel the edge marks on the underside") with six ancient chairs—a similar antique sideboard and an odd and ancient candlestick to hold 3 candles on the table. The walls of this room were grey and unpainted still carrying the pencil marks of the workmen who had lined up the pictures which hung on the walls. They were Roman paintings from the walls of Roman houses from Herculaneum or Pompeii. They were very strange "of mythological figures" he said. The nicest was simply a dove eating a fig. They all had weak flat colors surrounded by a border of yellow orange—the outside of which was irregular in shape as they had been removed from their original houses. On one wall was a larger fuller painting of three naked women at a fountain. There was very considerate discussion between Donald and Jorge as to whether the pictures were too high and should be moved down. I was surprised to realize that they were discussing only the motion of an inch and a half.

"Let us go into the kitchen," said Orozco as he moved to one wall, which I now realized, contained a door, flat and unpainted like the walls (everything, so far, had this unfinished look) and with no door handle. Pushing thru, with amazing contrast we came into a modern, fully equipped kitchen with Formica counters, copper sinks, beautiful plastic faucets, etc. Everything "normal" except if you discount three large (4 ft. diameter) wooden barrels covered roughly with a piece of oilcloth (on which there was a partially eaten apple) and standing directly in the center of one of the kitchen rooms where you might expect, if anything, a table. There was a half loaf of bread—his wife made it—from which he cut and offered me a piece. I had it with jam—or rather with the foam of jam that forms while it was being made—his wife makes jam too—from the fruit trees. It was delicious.

Then to the library. To get there you must climb a set of very high and particularly narrow marble steps that reminded me of the Mayan pyramids. It was an enormous room with brown carpeting on a floor of different levels high in the barn—bookshelves all around with many books. On antiquities "Animals in Roman Art," "Greek Vases," about things of all times—of all ages in several languages. The house had been finished thus now for a

year and a half. No, he had not gotten a chance to sit down in his library (there were no chairs—but you could sit on steps) to read any of the books. I suggested to him that there he might find the answer to a question he asked me about what the writing might be about on a door he had showed us that had a circle around which were written 1562, 1563, etc. and other things we couldn't make out.

During all this time there was a perpetual tapping of a chisel on plaster coming from somewhere in the house. Well, there was a (marble, or course) stairway leading to a great (now empty) vault with two steel doors 20 inches thick one at each end, each containing the locks and equipment of a bank vault door (but these were wider than any vault door I had previously seen). There was a workman chiseling into the concrete wall a long channel for the drain from the air conditioner that was to be installed in the vault. On passing thru the vault you climbed up another set of marble steps to another part of the house.

There were rooms full of crated and partly crated material. "What do you think these are?" he asks pointing to some oddly carved wooden panels. I guess wrong. "No, Polynesian" but he is very upset that he cannot remember the name of the island, except that it begins with T, but although we guess Tahiti, Truk, Torya and the Trobriand islands he knows that it is not one of those.

I will not lead you up and down thru any of the other rooms—but just as we are about to leave and he has given me some plums (which I am eating now) he suddenly remembers that he has forgotten to show me the Mayan statue (he knew from our lunch conversations that I had Mayan interest). Another room, under a semitransparent plastic sheet covering there it is. With the cover off I see it is a nearly (perhaps 2/3) full size sculpted figure of a man in a somewhat curved odd position with one hand gracefully curved out above his head—and the neck twisted to one side as if it is too weak to hold the head straight up. The thing is of white plaster-like stone (probably soft sandstone) evidently late for altho there are earplugs there is none of the elaborate carving of headdresses, etc. of the usual stelae. But the most remarkable unusual feature is the large patches here and there still remaining of the original painting—all dark red— which presumably covered it. How could it have been preserved? There was only one possibility—it was from Bonampak (or possibly some more recently discovered cave) where actual wall paintings have been found. And the figure—of course—was exactly the pose of the prisoners depicted on

the Bonampak painting. I asked him where it was from, suggesting Bonampak, but he didn't know—had it written down somewhere, he said. But maybe he knows but didn't want to tell me because it's all illegal to take that stuff out of Mexico (Donald says it is equally forbidden to take Roman paintings out of Italy—any he often hears Orozco on the phone talking to the lowest of the low making deals. He adds that Jorge's wife and daughter Mercedes are friends of Maya and Donald's daughter—and that Jorge insists that his wife make bread and the jam—I guess such activity is good for her, keeps her closer to Nature).

As I left and thought about it later I realized how my first favorable impressions of Jorge Orozco had turned into a vision of surrealistic horror—as I imagined the three—he, his wife and daughter (the other children are off at college) eating alone in that long room with unpainted walls, with the Roman paintings looking down on the dark scene lit only by candles (electricity isn't installed in that room yet) in an ancient candlestick. For such they do. This is the pleasure of the autocratic, self-righteous, moralistic man with his principles—which include the requirement that his wife bakes the bread and serves the dinner. Meanwhile his guard dogs are outside and his vaults are being readied to contain the robbed remnants of other people's culture. The tin mines of Bolivia slaved for this—that money he did not earn is spent in his "collection" in a crescendo of single mindedness to try to expiate the guilt he feels from not having earned it.

He has no friends. His world is so distorted as to border on the insane. Such, in this case, is the curse of wealth.

Love,
Richard

Which reminds me—I hope you got our bank balance problem straightened out?

## RICHARD C. HENRY TO RICHARD P. FEYNMAN, DECEMBER 2, 1982

Dear Professor Feynman:
I am an astronomer located in a Physics department, and experiencing ever increasing joy at teaching physics (as opposed to astronomy). Some of that joy expressed itself in an article for "The Physics Teacher" (copy enclosed); if you have time to read it and have any corrective comments, I would be very glad.

The reason I write to you is that I vacationed in Florida over Thanksgiving, and for light reading had Eddington's *The Mathematical Theory of Relativity.* He puts words to equations extraordinarily well; almost as well as the Feynman Lectures. But even he is not able to verbally discuss, and does not try to mathematically discuss, a fascinating question: what would it be "like" to inhabit a world with 3 space dimensions and two time dimensions (see enclosure from Eddington's book)?

This was a question I asked myself in grad school, and that I once asked Charlie Misner at a party; I've never gotten an answer. At first I asked myself whether there might not be two time dimensions, and that's where the quantum-mechanical funny business comes from! More lately I think the answer might be much duller: identical clocks running at different rates, and changing rates if they collide.

If the answer is interesting, would it make a good article (for you to write!) for the *Physics Teacher?* Just a suggestion!

Thanks for your attention.

Sincerely,

Richard C. Henry

Professor of Physics

P.S. I am not making the Editors of the Physics Teacher aware of this letter, so no disappointment is possible.

## RICHARD P. FEYNMAN TO RICHARD C. HENRY, JANUARY 7, 1982

*This response to Dr. Henry is perhaps the only letter that hints at the unique relationship between my father and my brother. While always close, during Carl's teenage years they became collaborators. They went for long walks together, discussing highly technical ideas (I know, for I sometimes tagged along and regretted it).*

*Richard's renewed interest in computers was sparked by Carl's passion, and the two kept a notebook together of computer problems and solutions they had tried. When Carl was at MIT, he sent home handouts from his computing courses for Richard to read. In the Caltech Archives, there is a long, completely technical letter about algorithms from Richard to Carl during this time. At the end of the typed letter, Richard wrote in his own hand, "TO BE CONTINUED. Love, POPPA FEYNMAN." It was one of the deep joys in our father's life to have a son like Carl, who spoke his language. To say he was proud of Carl would be an understatement.*

Professor Richard C. Henry
Department of Physics
The Johns Hopkins University
Baltimore, Maryland

Dear Professor Henry:
My son Carl Feynman and I discussed the case of two time and two space dimensions for a few days at our beach house in Mexico. It certainly is fun to think about. He developed a way of geometrically visualizing things—involving something like a small two-dimensional picture at each point of a plane that we found useful for kinematic considerations. I forget exactly how it worked, and I don't know if he pursued it any further. If you want to ask him his address is below.
    Sincerely,
    Richard P. Feynman

*Professor Henry did not pursue this question with Carl.*

## RICHARD P. FEYNMAN TO BEATA C. KAMP, FEBRUARY 28, 1983

*Beata Kamp wrote to say she had watched* The Pleasure of Finding Things Out *all three times that it aired on her local PBS station. Other than expressing her appreciation for the show, she explored the idea that an "explorer of the spirit" like herself (and her cousin) and someone like Feynman, who pursued hard knowledge, were doing much the same work. "You unravel mysteries contained in matter, thinking you do it all with your own brainwork, while I listen for the soundless voices that teach me the mysteries behind matter."*

Mrs. Beata C. Kamp
Chico, California

Dear Mrs. Kamp:
Thank you for your fan letter.
    There certainly are more mysteries than knowledge and, perhaps, more ways of finding things out than science. I like science because when you think of something you can check it by experiment; "yes" or "no", Nature says, and you go on from there progressively. Other wisdom has no equally certain way of separating truth from falsehood. So I have taken the easy

course with easy methods, while you and your cousin are pursuing far more difficult matters with less to guide you.

Good luck in your endeavors.

Sincerely,

Richard P. Feynman

## RICHARD P. FEYNMAN TO CHRISTOPHER SYKES, MARCH 11, 1983

Christopher Sykes
BBC Television
London, England

Dear Christopher,

Well, they showed it on NOVA, and we wowed them!

It was a big success. I got all kinds of great comments from my friends, and very many fan letters. . . all good. I get the compliments, and all you get it some remark like "and there was no buffoon commentator asking you dumb questions," or the like. Little did most people (even some in the business) realize how it was actually done. They all believed the illusion that all I had to do was open my mouth and talk for an hour. Like all true art, the artist disappears and it looks natural and wonderful.

You and I know better. Three days of interviews, and four hours in the can to make one hour of program. But your original idea, so carefully considered and worried about, to make a talking head seems to have made it an unusual program. Congratulations on the American success of "your program."

The people at Nova claim it made a record. There were more requests for transcripts than for any previous Nova program. But I saw one of these transcripts. Boy, are they a disappointment. They are hard to read, the sentences are incomplete, the grammar is murderous, etc.

I don't think much of the new little programs. But I have learned to trust your judgement far above my own. So if you say they are O.K. they are O.K. by me. All the sweat is in the cutting room. Good luck!

Sincerely,

Richard P. Feynman

.  .  .  .

## RICHARD P. FEYNMAN TO DOROTHY W. WEEKS, FEBRUARY 25, 1983

*"Your NOVA program was superb," Dorothy Weeks wrote. In response to a seg-ment in which Feynman spoke about the different ways Carl and I responded to his stories, she added that she too had noticed differences in the ways boys and girls learned about the physical world. She also wanted to know if he had been acquainted with Norbert Weiner, founder of cybernetics and eccentric genius in his own right.*

Dr. Dorothy W. Weeks
Wellesley, Massachusetts

Dear Dr. Weeks:

Thank you for your note.

The way my two children responded to my stories was very different but I don't know that it was because one was a boy and the other a girl. I just think people are very different and that if I had two sons they would respond differently too—maybe.

Norbert Wiener was at MIT when I was a student. I saw him often, but didn't know him very well.

Sincerely,

Richard P. Feynman

## ROBERT L. CARNEIRO TO RICHARD P. FEYNMAN, FEBRUARY 1, 1983

Dear Dr. Feynman,

A few days ago I saw the Nova television program devoted to your life and work in physics and found it most absorbing and enlightening. Having a one-year-old son whose education I must now begin to consider, it was of great interest to me to hear how your father directed you toward scien-tific thinking and a career in science. His methods worked so well in your case that I am thinking of borrowing and applying them myself!

But now to the substance of this letter. At one point in your conversation you belittled the idea of a social science. This was nothing new to me. Over the years I have heard this feeling expressed by a number of physical scien-tists. However, I would like to see if I can make you reconsider this attitude.

The rejection of social science by physicists (or anyone else) must have at least one of two bases: either a social science is inherently impossible because of the nature of man and society, or else it does not exist because

those who have practiced it have done so incompetently. Let us briefly examine both these assumptions.

It would be very strange indeed if evolution, having produced a series of levels of organization of increasing complexity, should, once it reached the level of culture, suddenly suspend the operation of cause and effect and preclude pattern, order, and regularity—the stuff of which science is made at all other levels. I don't think evolution has done this, and I daresay you don't think so either.

Now, let us consider the other possible basis for denying social science. It is perfectly true that much of what is called by that term is not science at all. But can you, or any other physicist, necessarily acquainted with but a tiny fraction of what is being done in the field of social science, really assert that none of it can pass muster as genuine science? I don't think you can.

Having said all this, I am emboldened to enclose reprints of a few articles of mine in which I have tried to look at certain aspects of social systems from the point of view of science. In my opinion, some of the regularities uncovered during the course of this work and expressed as scientific generalizations, have a valid claim to being considered science.

I know you are exceedingly busy with more important things, but perhaps you have enough curiosity about what is happening in newer disciplines with scientific pretensions to look briefly at what I am sending you. Then you may conceivably find reason to reexamine the question of whether social science is possible or not.

Needless to say, whatever your conclusion, I would be most interested in your reply.

Sincerely,

Robert L. Carneiro

Curator of Anthropology

American Museum of Natural History

## RICHARD P. FEYNMAN TO ROBERT L. CARNEIRO, FEBRUARY 28, 1983

Dr. Robert L. Carneiro

American Museum of Natural History

New York, New York

Dear Mr. Carneiro:

You are, of course, completely correct.

My reference to "social sciences" just before I talked about pseudo-science was inappropriate. I was thinking as I spoke and had in mind the "much of what is called by that term which is not a science at all," and forgot things like anthropology, history, archaeology, etc., which I admire and would not wish to criticize in such a blanket manner. By such carelessness, I have done your field and others a considerable disservice. I apologize (but it does no good now!)

With some remorse, yours sincerely,
Richard P. Feynman

P.S. This evening I am going to relax & read the articles you sent me.

## JUDAH CAHN TO RICHARD P. FEYNMAN, JANUARY 26, 1983

*In 1946, Dr. Cahn presided over the burial of Melville Feynman, Feynman's father.*

Dear Dick:
It is many, many years since we have written to each other. Last night however I was fortunate enough to hear and see you on television and you were superb. I was enthralled not merely with your recollections of your father, but with the manner in which you presented your thesis and philosophy of education and life.

I am happy to tell you that my own sons are following in your footsteps. My elder son, Steve, was Chairman of the Philosophy Department at the University of Vermont and he is now a Director of the National Endowment Foundation for the Humanities in Washington. My younger son is teaching English at Skidmore.

I am enclosing a copy of a book I wrote. It isn't very much but it indicates my own thinking on some of the matters I deal with. I hope you find the time to read portions of it and perhaps let me know what you think of it.

The last time I wrote to you I told you that I had traveled extensively in the Soviet Union and had held seminars with faculty and students at various universities. I did not mention that I had the opportunity of meeting with Lev Landau for a short period of time. It was before his terrible accident and I just felt privileged to be in his presence. I don't know whether you ever met him and I really find it impossible to classify him in any way.

I'm sure he had a sense of humor or else he couldn't possibly have lived through the trials he experienced. The enclosed book tells his story.

I have not retired from the active Rabbinate and am enjoying the pleasures of reading and writing. I shall never forget the evening you spoke to the congregation at Temple Israel. I have often referred to it. I wonder whether you remember it. You were speaking about the atom bomb. Your notes were well prepared. At some point in the sermon you wandered from your notes and you began to talk not to the congregation but simply aloud, to no one in particular. I still remember your words not literally but approximately. You said, "I was asked to assist in the creation of the world's most destructive machine but I was never asked how to use it. Now I realize what I have done and what that machine could do, and I am afraid." With those words, "I am afraid" you sat down. Dick, I will never forget the look on your face when you came back to the seat and sat down next to me. You are not a religious man and in the acceptable term, neither am I. But you were quite prophetic. Except that now not only you are afraid but all of us are afraid.

I have reread your letter in which you said that you could understand religion perhaps if there were no God in religion. It is for that reason that I send you a book by my elder son containing answers to your questions. Incidentally he possibly will be out in California some time in the future and if he has the time and you have the time, I would be very pleased indeed if he were able to call you and spend a little time with you.

My younger son, Victor, who teaches English at Skidmore, will not be happy with the choice of book I sent to you. It was an early work of his but I enjoyed it so much that I felt that someone with your capacity of laugh would enjoy it equally as much. His other books are on Tom Stoppard, the field of Absurd Drama, etc.

I hope you and your family are well and that you will continue to enjoy the success which obviously you have attained.

With my best wishes.

Sincerely yours,

Dr. Judah Cahn

. . . .

## RICHARD P. FEYNMAN TO JUDAH CAHN, MARCH 15, 1983

Dr. Judah Cahn
New York, New York

Dear Judah:

How very wonderful to hear from you! You are rightly proud of your children—and so am I. I also have two children, a boy (21) and a girl (14). The boy studies computers at MIT and will soon graduate. My daughter is "into" horses and cello.

You probably remember my mother. She died a year ago. She liked you very much and we often reminisced about you and Temple Israel.

I am most curious and interested in your remembrance of my Temple Israel talk. I don't remember the exact remark—had you asked me I would have said that what I said was something like this: it was "brotherhood week" and I had explained how serious the bomb was and therefore on how important it was to try to get along with our enemies. Then I said "but there is something wrong about all this—for brotherhood should be based on love and not fear," and sat down.

I am very curious on how reliable old memories are and wonder how much we make up in our own mind when we review events. Maybe we remember saying what we would have liked to say. Your memory is probably more accurate—and there is no doubt I could have said things just as you say—because it does express my sentiments then.

Thank you for writing.

Sincerely,

Richard P. Feynman


## HEIDI HOUSTON TO RICHARD P. FEYNMAN, MAY 2, 1983

Dear Dr. Feynman,

I believe that your comments at last week's Physics Colloquium were arrogant, rude, and disruptive. In addition your attitude appeared to encourage the students (or post-docs?) sitting near you. Their persistent giggling and snickering was annoying and rude. Please consider this.

I remain, however, a great admirer of yours.

Sincerely,

Heidi Houston

## RICHARD P. FEYNMAN TO HEIDI HOUSTON, MAY 13, 1983
## INTEROFFICE MEMORANDUM

Thank you for your observations on my behavior at the Colloquium. You are probably right.

## RICHARD P. FEYNMAN TO BOB VALLEY,
## OCTOBER 14, 1983

*Mr. Valley, my Algebra II teacher, had marked me down for using "incorrect" methods in his math class (the resulting encounter is discussed at greater length in the introduction).*

Mr. Valley
John Muir High School
Pasadena, California

Dear Mr. Valley:

I should like to apologize for the personal remarks I made about you last Tuesday. They were entirely uncalled for, and as I find from speaking to others who know you better, entirely unjustified. I was quite wrong. I hope you will accept this apology for my unwarranted remarks.

If you have any mathematical interest in the point I was trying to make, it is explained on the accompanying sheet.

Sincerely,

Richard P. Feynman

## RICHARD P. FEYNMAN TO FRANCES BEST,
## NOVEMBER 2, 1983

*Nineteen-year-old Miss Best had recently started reading* The Feynman Lectures on Physics *and discovered that she loved the subject. Her parents arranged for an hour per week with a physics tutor, which she thoroughly enjoyed. Unfortunately, she had to give that hour up to cram for finals, finals that she did not do very well in, and as a result she was not able to go to university as she had hoped. She wrote to Feynman, seeking some advice.*

Miss F. Best
London, England

Dear Miss Best:

I received your letter with its unhappy cry about the difficulty of entering the University to study physics. It is all to the good that you find fascination in physics and enjoy studying it so much. Nature is wonderful indeed.

The best place to study it is a University of course. It must have been devastating to discover you didn't get in.

I am not familiar with the system in England. Here in this country we have very many schools large and small, state financed or private, specializing in this and that. Therefore even if someone doesn't get accepted by the school he chooses, he can find another. I'm afraid I cannot help you. I can only give the obvious advice that all is not yet lost, you are still young and strong, and persistence will win out in the end. I'm sure you have thought and heard all that before—and youth is impatient and anxious with time.

I am very sorry that I cannot help you with more than sympathy and platitudes. I am very glad to hear you found my book useful, but hope that it has not simply led you into desperation and frustration—but rather introduced you to the delight we can all get from a close scrutiny of Nature's patterns.

Sincerely,

Richard P. Feynman

## RICHARD P. FEYNMAN TO PAUL PRIVATEER, NOVEMBER 9, 1983

*After seeing* The Pleasure of Finding Things Out, *Dr. Privateer wrote to tell Feynman how delighted he was by the program. He wondered if the rift between science and literature was a true one, citing, as a historical example, "William Blake's open hostility to Newton who he saw as. . . mechanically defining the universe in laws that denied the imagination its supreme place in human experience."*

*Dr. Privateer believed there were similarities between the language of science and the language of literature, and he wanted Feynman to speak at a conference on the subject. "I would personally be elated to have you as a guest speaker on this topic and even if you fain a complete lack of knowledge about literature I will know you are being unnecessarily humble."*

Dr. Paul Privateer
English Department
San Jose State University
San Jose, California

Dear Dr. Privateer:

Thank you very much for your long letter commenting on my NOVA program and inviting me to a conference on science and literature. I am, of course, dishonored for I cannot accept, for as you guessed, I fain a complete lack of knowledge of literature. But not, as you guessed, because I am humble. The faining is not feigning. I heard about Blake's view of Newton only because this professor of literature here is such a wonderfully attractive woman that I took her out to lunch. She is a Blake fiend.

Sincerely,

Richard P. Feynman

*The professor of literature at Caltech that Feynman referred to was Jenijoy La Belle. She discussed Blake's view of Newton with Feynman and showed him Blake's color print of Newton under the sea, which he loved.*

## RICHARD P. FEYNMAN TO JACK M. RICE, JR., NOVEMBER 11, 1983

*After reading the* Los Angeles Times *article, "Nobel Prize: Another Side of the Medal," in which Feynman expressed a certain ambivalence about having received the award, Mr. Rice wrote a blistering letter to Feynman. He concluded: "You may wish, in saying 'to Hell' with Nobel and his prize, to come across as a curmudgeon. You succeed instead at demonstrating what a jerk you are."*

Mr. Jack M. Rice, Jr.
Los Angeles, California

Dear Mr. Rice,

Thank you for your letter concerning my remarks in the L.A. Times. You are right, I am a jerk for telling the reporter my personal feelings and reactions to the Nobel Prize. I realized this after I spoke to him and tried to call him back to ask him not to print the interview—but couldn't reach him.

My feelings are perhaps childish or foolish but they are sincere. I was first notified of the prize at 4 A.M. by a New York newspaper. I was not asked by the Nobel committee whether I wanted to receive it. I wanted to quietly demur the honor, but it was already too late to be possible. It would have been an even greater publicity annoyance if I said no in public after newspapers knew I'd won it. It would be a worldwide sensation.

It has been a mild annoyance to me ever since. Never mind whether it should be or not—or whether you think it would not be to you—or whether it is ungrateful of me—it is, to me, an annoyance. On the other hand, as you imply, I am a jerk to say so publicly. An illogical jerk, certainly, to complain to a newspaper publicly about not wanting publicity.

On the other hand you seem to think my criticism of the prize is a criticism of the Swedish people, or of Alfred Nobel himself. I didn't mean to imply that. The one delightful compensation for the trouble of the prize was the wholehearted and open friendly welcome of all the Swedish people when my wife and I went there. I have many Scandinavian friends and students and they are marvelous people. I am sorry if you thought I believed otherwise.

I don't know Mr. Nobel's life or motives for giving the prize, so if I criticized him personally in my remarks I didn't know what I was talking about.

May I wish you many honors in your life, for I know you will accept them far more graciously than I.

From the curmudgeon,

Richard P. Feynman

## RICHARD P. FEYNMAN TO DAVID MERMIN, MARCH 30, 1984

*David Mermin is an eminent physicist and professor at Cornell who, like Feynman, takes delight in finding simple, elegant explanations for surprising physical phenomena. In 1981 he published an article on quantum mechanics that delighted Feynman and moved him to write the following letter.*

Dr. N. David Mermin
Lab. of Atomic and Solid State Physics
Cornell University
Ithaca, New York

Dear Dr. Mermin:

One of the most beautiful papers in physics that I know of is yours in the American Journal of Physics 49 (1981) 10.

All my mature life I have been trying to distill the strangeness of quantum mechanics into simpler and simpler circumstances. I have given many lectures of ever increasing simplicity and purity. I was recently very close to your description (down to six states, instead of three, etc.) when your ideally pristine presentation appeared.

I have since copied it almost exactly (with attribution, of course) in several recent lectures on the subject. Thank you.

I have been making a similar series of attempts to explain the relation of spin and statistics. Can you do as well there? Perhaps if we meet someday we can discuss it together and create a clear explanation of why exchanging two particles implies a tacit rotation of the axes of one by 360 degrees relative to the other.

Sincerely,

Richard P. Feynman

*The "relation of spin and statistics" to which Feynman refers is the fact that fundamental particles whose spin angular momentum (in units of Planck's constant) is an integer want, in a statistical (probabilistic) sense, to be in the same quantum state (do the same thing at the same place at the same time as each other), while half-integer-spin particles can never be in the same state. This relation between spin and statistics underlies many important phenomena— such as the lasing of lasers and the fact that solid objects cannot easily be squashed—and so Feynman wanted to find a simple explanation for them.*

*Feynman and Mermin knew an explanation that relied on one simple property of all fundamental particles: when two particles, with any spin, are interchanged, the result is the same as not interchanging them but instead rotating one of them through 360 degrees. Feynman hoped Mermin could explain that weird rotational property in a simple way.*

## DAVID MERMIN TO RICHARD P. FEYNMAN, APRIL 11, 1984

Dear Dr. Feynman:

Thanks for the letter about my gedanken demonstration. I'm fond of that paper myself, but I've learned that there are two kinds of physicists: those who enjoy it and those who utterly fail to get the point. I thought

you'd be the first kind, but I'm glad to know for sure... I have nothing simple to say about why interchanging two particles involves a tacit rotation of one by 360° relative to the other. I don't even have a satisfactory complicated understanding. If you ever write anything about that please send me a copy...

Thanks again for the very nice letter. Since you, through your writings, have influenced the way I try to write and think about physics more than anybody else, I'm delighted to have had at least one chance to return the favor.

Yours,
David Mermin
Cornell University Laboratory of Atomic and Solid State Physics

## RICHARD P. FEYNMAN TO WILLIAM G. BRADLEY, JULY 13, 1984

*As a reminder to the reader, the "event" in question is the examination following the nasty bump on the head that the excited customer received on his way into the computer store.*

Dr. William G. Bradley
Director, NMR Imaging Laboratory
Huntington Medical Research Institute
Pasadena, California

Dear Dr. Bradley,
Thank you so much for sending the NMR pictures of my brain. The instrument is sensational in the detail and resolution you can see.

But you can't see what I am thinking—for apparently I still have some functional failure as I remember the event as on 25-Jun–19:33:24 whereas your instrument says it occurred on 06-Jun–19:33:18. The 6 second doesn't concern me, for my time errors already averaged 10 sec. standard deviation before the accident (a deleterious effect of age, I think) but the 19 day error is evidence of severe functional disability (resulting from hematoma, probably).

Sincerely,
Richard P. Feynman

. . . .

# RICHARD P. FEYNMAN TO ERIC W. LEULIETTE, SEPTEMBER 24, 1984

*Mr. Eric Leuliette, a sixteen-year-old high school student, wanted advice on how to prepare for college and a career in physics.*

Eric W. Leuliette
Clarksburg, West Virginia

Dear Eric:

Among the many things I know very little about, one is what one should do to prepare oneself to be a theoretical physicist. My best guess is to do with energy and zest whatever interests you the most. If it turns out not to lead you toward theoretical physics but to, say law, or electrical engineering instead, go that way. It is wonderful, if you can find something you love to do in your youth which is big enough to sustain your interest through all your adult life. Because, whatever it is, if you do it well enough (and you will, if you truly love it) people will pay you to do what you want to do anyway.

About college financial support questions, I am asking the proper office to send you any information they have available.

Sincerely,

Richard P. Feynman

P.S. Try taking my Feynman Lectures out of the library. You might enjoy parts of them—if not they will help you decide what you want to do.

# FRANK POTTER TO RICHARD P. FEYNMAN, NOVEMBER 15, 1984

Professor Feynman.

I wish to sincerely thank you for your influence on my life and my career. You may not remember me well, but I was that ignorant Cal Tech undergraduate who often rode to the Hughes Research Labs with you and Bruce Winstein each week in 1965 through 1967. You were then lecturing on the fundamentals of physics and later, I think, on astrophysics. During these journeys in your car or van you told anecdotes and discussed physics concepts—a short "trial run" for your two hours at Hughes.

The excitement and the genuine interest you displayed in posing the problems and then immediately attacking them eventually affected me so

much that I changed objectives. From an EE major as an undergraduate, I earned a Ph.D. in Physics in 1973. That spirit which I saw you display continues to infect me in my career as a Lecturer in Physics here at UC Irvine.

There is one conversation in the car which I distinctly remember because I have tried to practice the ideas we discussed. Bruce asked you: "If you could do anything different in your career, what would it be?" You replied with very little hesitation: "I would try to forget how I had solved a problem. Then, each time the problem arose, I might solve it in a different way—I wouldn't be thinking about how I had solved it before."

I can honestly report to you that I have attempted to achieve that goal—at least as far as physics is concerned. At first, as a graduate student, this manner of doing things was a great burden, since I ended up starting from the very fundamentals quite often. Eventually, after years of practice, I have grown to enjoy thinking freshly each time a problem arises. In fact, it is a real challenge to find a variety of ways to solve problems. It has even led to my own alternative ways to think about traditional physics as well as contemporary ideas. The pleasures I have enjoyed from this one idea of yours has made me realize how indebted to you I really am.

I may not make any significant contributions to physics in my career, but I have no great concern over that. I find Nature fascinating and challenging, and I have some of that spirit which I saw in you. I have a family which I enjoy, and I have the freedom and the time to think about anything I wish.

I hope that you receive other letters of this nature, for you truly deserve them. You have strongly influenced many people, and you have given us remarkable guidance and instilled a spirit which will last forever.

With great appreciation,

Frank Potter

## RICHARD P. FEYNMAN TO FRANK POTTER, NOVEMBER 21, 1984

Dr. Frank Potter
University of California
Irvine, California

Dear Frank,

Gee, Frank, that was a nice letter.

Of course I remember you and Winstein (and Schlichter) going to

Hughes. I don't remember that particular remark, about wanting to forget solutions, but I agree with it. Your letter surprised me, and I am often surprised, to discover that other people don't ordinarily do things the way I do. It is my greatest pleasure to think anew about things and I am delighted to discover that I have infected you with the same pleasure.

Naturally I have read your letter over several times—it makes me feel good. Thank you.

Sincerely,
Richard P. Feynman

# 1985–1987

In January of 1985, a collection of Feynman stories appeared under the title *Surely You're Joking, Mr. Feynman!: Adventures of a Curious Character* (triple entendre intended). Far exceeding both my father's own expectations and that of his publisher, it went on to spend fourteen weeks on the *New York Times* best-seller list. That same year also saw the publication of *QED: The Strange Theory of Light and Matter*, Feynman's painstakingly accurate and detailed explanation, using very little math, of quantum theory for a lay audience (if you want to understand what he got his Nobel Prize for, this book is for you).

In 1986, at the invitation of former Physics X student Bill Graham (then acting administrator of NASA), Feynman reluctantly agreed to serve on the Presidential Commission on the Space Shuttle Challenger Accident, an experience he recounted in depth in *What Do You Care What Other People Think?* At a critical moment in the hearings, Feynman dipped O-ring material from a rocket booster into a glass of ice water, dramatically demonstrating the technical cause of the accident.

Richard Feynman died less than two years later, on February 15, 1988. When he briefly emerged from a coma induced by kidney failure, his last words, spoken to the three women by his side (wife Gweneth, sister Joan, and cousin Frances), were: "I'd hate to die twice. It's so boring."

·  ·  ·  ·  ·

## RICHARD P. FEYNMAN TO SILVAN SCHWEBER, JANUARY 28, 1985

*Dr. Schweber sent a draft of his chapter, "Feynman and the Visualization of Space-Time Processes," which he had written as part of his book* Quantum Field Theory, 1938–1950. *Dr. Schweber assumed that Feynman would find the chapter boring, but he hoped there were things that would be new to him. Dr. Schweber was striving for an accurate account and welcomed any comments or criticisms from Feynman.*

Dr. S. Schweber
The Martin Fisher School of Physics
Brandeis University

Dear Schweber:

Well, I am sorry for the delay. We had a calamity here and lost everything pertaining to you—your manuscript and my notes on it to you. So I'll try again.

First, I didn't find it boring but very exciting and surprising to find things I thought gone forever, like the page of typing on complex numbers that I remember seeing before me in my almost toy typewriter—but I didn't think still existed! Also I didn't know I wrote so many letters so it was fun to see what I was thinking. You historians have a way of recreating the past until it appears almost real.

Here are some comments on possible corrections but I know it is only my memory against 'facts.' There were several typos but I didn't bother with those.

p. 4 last para. I think I entered MIT in Math (course XVIII). After a bit I went to Franklin (then head of Math Department) to ask "what is the use of higher mathematics beside teaching more higher mathematics." He answered, "if you have to ask that then you don't belong in mathematics." So I changed to practical, Electrical Engineering (course VI), but soon oscillated part way back to Physics (course VIII) and stayed there.

p. 9 central para. (Comment) I didn't know that they thought of three years instead of four. How fortunate that they didn't do it!

p. 9 top para. The scholarship to Harvard was the prize for winning a country-wide math contest (Putnam?). The math department asked me even though I was in physics, to enter because they didn't have the four men needed to enter a team—but looking at records had found I had been in math. I was unsure, but they gave me old exams to practice on.

p. 14 2nd para. Among the "others" was John von Neumann.

p. 17 3rd para. Should read "In this newer version, they gave. . ."

p. 24 last para. (Comment) My interest in this came from this: In high school I had a very able friend Herbert Harris who, when we graduated, went to Rensselaer Polytechnic to become an electrical engineer, while I went to MIT. One summer (end of freshman year?) he returned to Far Rockaway, we friends took a walk, and he told me about the then new feedback amplifiers. He tried to design them in different ways avoiding oscillation and said he was very convinced that there was some law of nature that made it impossible to make the impedance fall off too fast without inducing a large phase shift. I proposed it might be a reflection in the frequency response domain of the fact that signals cannot come out before they come in, but neither of us was, apparently, sophisticated enough to work this out mathematically—but you see why four years later I would find Bode's paper so interesting.

p. 23. I find no reference to fig 1 in the text. The figure (if I guess what it refers to) should be more like

or other more complex knot.

p. 25 line 12 (also p. 36 top). I joined Wilson's group for war work before I finished my thesis—and stopped working on it. After a time I did ask for some weeks off to write my ideas on it myself so I wouldn't forget them. But while doing that I saw (probably erroneously) a way to solve a problem that was holding me up, and Wheeler suggested I quickly write it all up and finish getting my degree.

Comment. Generally, I am not bothering to check all the equations, etc., on these pages.

p. 32 top line 6. I didn't know how to calculate a self-energy in the conventional way with the Dirac theory and holes because I had never studied it carefully, and my path integrals hadn't been clearly developed for the Dirac electron going backward in time yet. But I did know how, in an obvious way, to reinstate the terms representing electrons interacting with themselves. I tried to translate the modification I was proposing into a rule (the scheme of F60, p. 65) for integrating over frequencies or photons of various masses and returned next day to ask Bethe to try it on the conventional calculation that he, but not I, knew exactly how to do.

p. 40 top. It would be nice to get Aage Bohr's recollection of all this, for

I only saw it from one side and may have been partly guessing what went on between his father and him.

p. 66 top para. No, it was *other* people I was referring to who had to keep transverse and longitudinal terms separately—I had known since the time of my thesis, 1942, that they went easily together. In the quote what I meant was ". . . had been worked out so patiently (by everybody else) by separating. . . "

p. 66 line 16. Possibly error for "(except for some nice simplification ways)". Next line, possibly "somewhat less sharp."

p. 70 top para. The story of this mistake is interesting. As near as I can remember it, I first got a relativistic result (we were only working to order $v^2/c^2$). A student found an error in an early line and concluded it would not be invariant—when I wrote the first letter p. 66★. But later on several pages later he found another error where I canceled two equal complicated terms that I should have added. The original answer I had gotten was right—it was relativistic. This miracle of two canceling errors was probably the result of a mixture of having a strong feeling for what the answer must be and algebraic carelessness.

p. 71 2nd para. It is possible I remember things the way I would like but I suspect this Eyges story never happened. Could you get Bethe to confirm it? Schwinger and I compared notes and results and we were good friends. We discussed matters at Pocono and later also over the telephone and compared results. (We did not understand each other's methods but trusted each other to be making sense—even when others still didn't trust us. We could compare final quantities and vaguely see in our own way where the other fellow's terms or errors came from. We helped each other in several ways.+) Many people joked that we were competitors—but I don't remember feeling much that way. And in my letter, p. 70, one reads my thoughts (in the last para.) that I am excited because an old problem may have been solved either by me or Schwinger. Doesn't sound too competitive.

+ For example, he showed me a trick for integrals that led to my parameter trick, and I suggested to him that only one complex parameter function ever affected rather than his two separate real functional (i.e.: they would always end up in the combination $D_1 + iO^2$ in the final answer).

p. 76 line 14. The "bombardier metaphor" was suggested to me by some student at Cornell (who had actually been a bombardier during the war) when I was writing up my paper and was asking for opinions of how to explain it and only had poor and awkward metaphors.

p. 76. Why is "phantezising" spelled wrong? Can we trust the typist recording the taped oral interview to know how badly I spell?

p. 87 Actually I also had the impression that Aage expressed, that Schwinger had more complete results because he had a charge renormalization whereas I hadn't yet done vacuum polarization satisfactorily. As you know there are four diagrams.

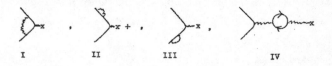

We (Schwinger and I) later found that I had I + II + III (no charge renormalization) and he had I, without II, III and so with a charge renormalization that I (confusedly) thought was due to the vacuum polarization IV. I didn't have IV at the Pocono, and I thought he did. (I do not now know whether at Pocono he really did have IV satisfactory or not.)

p. 100 4th para. From end. I think I expressed myself badly in "that was the moment I got my Nobel prize." I didn't mean that was when I knew I would win the Nobel prize—which never entered my head. What I meant was that was the moment I got a "prize" of thrill and delight in discovering I had something wonderful and useful.

p. 106 Is it necessary to include these religious views? People are often sensitive about such things and they are best presented in a quiet gentle way as I tried to do in my full talk. The short quotations may be more shocking (Oh well, not, I guess, to your readers!)

p. 107 Professor Morgan Ward pointed out to me that the same argument would show that an equation like $x^7 + y^{13} = z^{11}$ (powers prime to each other) would be unlikely to have integer solutions—but that they do, an infinite number of them!

p. 111 1st para. There is no reference for footnote 196.

p. 111 last para. It is not proper for an historian who is writing for posterity to tell us what posterity will think. He should be content to present the evidence to permit "posterity" to come to an opinion, but not to formulate that opinion for it. In all other remarks you give full references in footnotes. Where is your reference for this paragraph?

Almost my entire knowledge of Quantum electrodynamics came from a simple paper by Fermi in the Reviews of Modern Physics (1932).

Sincerely,

Richard P. Feynman

## RICHARD P. FEYNMAN TO BERNARD HANFT, FEBRUARY 4, 1985

*Mr. Bernard Hanft sent a washer with thread attached to demonstrate what he called a new physical force, "The Hanft Force." The Hanft Force operates as follows, he wrote: "It will cause a suitably suspended object, of any substance, and of any configuration, to spin, on its axis." Mr. Hanft felt that since there was no energy used, the Hanft Force could be developed into a source of unlimited free power.*

Mr. Bernard Hanft
Rego Park, New York

Dear Mr. Hanft:

Thank you for sending me your note about the spinning force and the perimeter force, as well as the washer and thread with which to demonstrate it.

The spinning force is a delightful phenomenon and appears quite puzzling at first. However, I have done some experiments and believe I know how it works in spite of the apparent violation of the laws of energy conservation. I shall suppose that the fibres of the thread have a natural tendency to be twisted in the state of lowest energy (presumably because of twists put in manufacture). More simply, I mean when there is no pull along the thread, the thread is twisted. Then when you pull it—for example by the weight of the washer—it tends to untwist at least partially—for it can become longer—giving in to the force by untwisting.

So, when you hang the washer on the thread, the thread untwists rotating the washer. The energy comes from the fact that the thread becomes longer as it untwists—and lowering a weight (of the washer, in this case) can supply the energy of rotation.

To verify my prediction, that the string lengthened, I hung the washer on the thread from a fixed point and carefully marked the position of the top of the washer. When it was in full spin it was indeed lower but by only a little over 2 millimeters! At first I was surprised because I didn't think the energy released in such a short distance could account for such a vigorous spinning, as the washer acquires. But a short calculation showed that my intuition was wrong. The energy released in falling two millimeters is about the same as the kinetic energy of a disc spinning at a rate of three full rotations per second. (I judge the disc spin to be not quite as fast as that—but we lose energy in friction in the thread and in air resistance.)

When the thread is released, the tension of the weight is taken off, and the thread is free to spring back to its original more twisted and shorter configuration. Thus the experiment can be repeated again and again.

The perimeter force effect is less interesting because it is more well known. There used to be a parlor trick using a girl's ring on a thread held by the girl to answer questions yes or no depending on whether the ring goes in a clockwise or counterclockwise ellipse. The motion is caused by inadvertent and unconscious slight motions of the hand. That is why, as you report, it works only if the thread is held in your hand. Try holding it in your hand, but a hand which cannot move because it is held up against a shelf or other rigid object (be careful your fingers don't move also).

Thank you again for calling my attention to these entertaining phenomena.

Sincerely,

Richard P. Feynman

## ROBERT F. COUTTS TO RICHARD P. FEYNMAN, APRIL 1985

Dr. Feynman,

I have been nominated for the Presidential Award for Excellence in Science and Mathematics Teaching. Reception of this award unfortunately requires various recommendations and I decided quite presumptuously to request that you write one for me. I would very much appreciate a brief note, perhaps mentioning the many years that you have been coming to Van Nuys High and the joy we share in a good Physics lesson. I know that writing recommendations is a pain in the neck and I shouldn't ask, but I decided to let the prospects of fame and glory cloud my reason. Thanks for your consideration.

A note to your trusty secretary, the Presidential Award program says I must have it all post marked by April 19th.

My students and I look forward to your visit on the 24th.

Gratefully,

Bob Coutts

.   .   .   .

## RICHARD P. FEYNMAN TO MELINDA JAN, APRIL 16, 1985

Ms. Melinda Jan, Science Chairperson
Presidential Awards Program
California State Department of Education
Sacramento, California

Dear Ms. Jan:

I was pleased to hear that Robert Coutts has been nominated for the Presidential Award. One of my little pleasures in life is to go to the Van Nuys High School once or twice a year to answer questions for the science students of Mr. Coutts' classes. This activity was initiated many years ago by Mr. Coutts and I look forward to doing it every year. The questions are on anything, relativity, black holes, clouds, spinning tops, magnetic force, you name it. The class is alive and very interested and seems to enjoy it as much as I. That they are so alive and ask so many questions, is, I have always supposed, a result of Mr. Coutts' love for science and education (for he never fails to want to show me some new experiment, device or original way of explaining something in the few minutes before the class begins).

If you select him you can be proud of your selection.

Sincerely,

Richard P. Feynman
Richard Chace Tolman Professor of Theoretical Physics

## DEBBIE FEYNMAN TO RICHARD P. FEYNMAN, JANUARY 20, 1985

Dear Dr. Feynman,

I am prompted to write to you at this time for various reasons you will understand as this letter continues.

First, let me explain our relationship. My father is Bert Feynman, son of Frank Feynman, who passed away in 1966. My grandfather, Frank Feynman, is the son of Harry and Sarah Feynman. I believe, correct me if I am wrong, you are the son of my great grandfather's brother. That makes me removed about four times, and my father, Bert, about three times as cousins.

My name is Debbie. I am going to be 17 years old in this April. I go to Forest Hills High School, half way through eleventh grade. My curriculum at school is called the Math-Science Honors Program. My science teachers in the past have asked me if I was related to you. Of course, my answer is

yes, as you are a cousin to my father. I felt very proud of this, as you are held in high esteem as a scientist and as a person amongst the scientific world. I have seen you on T.V. on our channel 13 early last year, when you made an hour long interview program.

This summer my mother, whose name is Audrey, and my father are sending me on a teen tour, which will include going to California. How close I will be to where you are I can't determine yet, but it would be a great treat to meet you if possible. It certainly would make my father proud, also.

Since this is our first correspondence, it is a little difficult to be definite, not knowing about your immediate family and etc.

You should know that my father and mother have great interest in my writing you as I have. You should feel free to write or question anything you might have in mind at the time.

I will be anxiously waiting for your reply, and send my very best to your family.

Sincerely yours,

Debbie Feynman


## RICHARD P. FEYNMAN TO DEBBIE FEYNMAN, MAY 7, 1985

Ms. Debbie Feynman
Forest Hills, New York

Hi Debbie:

It is fun to find someone signing letters with my name, that isn't my wife, child, or sister. Of course, it is because it is your name too. It is such a crazy name that we must be long-lost relatives—nobody could have invented it twice.

But how are we related? It will require more detective work. All I know is that my father Melville, had a brother, Arthur, who died without children, and two sisters who changed their name upon marriage. My grandmother's name was Anna. She lived in Brooklyn. Her husband's name was Jacob Feynman. They had trouble and he ran off to California and remarried (some "Feynmans" exist in California in Long Beach).

The story, as far as I or my sister knows, was that Jacob's name was Pollock but, when they immigrated here, he took the name of his wife (or an approximation of it, maybe it was Feynamonavitchinsky or something). They came from Minsk.

*And,* my sister tells me, he later brought over two brothers, who on their arrival took the same new name, Feynman! That is all we know—we don't know the first name of Jacob's brothers or brother. Could it be Harry? If so, I am the grandson of your great grandfather's brother. My daughter, Michelle, is 16. That would mean that you and she have the *same* great great great grandfather.

Anyway, we must be related somehow—and if we can't prove it, let's assume it—it is more fun that way.

So we are all waiting to see you when you get out here on your teen tour. Please call us when you know when you are coming, our number is [withheld].

Your relative (probably),

Richard P. Feynman

P.S. By the way, our name is all over the place on a new book, *Surely You're Joking, Mr. Feynman,* published by Norton.

*Unfortunately, we were out of town when Debbie Feynman came to California and were unable to meet her.*

## RICHARD P. FEYNMAN TO ROBERT L. KAMM, JULY 19, 1985

*Robert L. Kamm arrived at Princeton the same day Feynman had, and he had also worked with him at Los Alamos during the war. "I ate with you at Princeton Graduate College," he recalled. "I was at Mrs. Eisenhart's party when she offered you cream and lemon for your tea." With these and other memories prompted by* Surely You're Joking, Mr. Feynman, *Kamm wrote of his wife, Jane, being badly reprimanded by the chiefs at Los Alamos for leaving her door and safe open. He believed Feynman might owe her an apology.*

Dr. Robert L. Kamm
Birmingham Psychiatric & Medical Associates
Southfield, Michigan

Dear Bob:

Thanks for your letter—it was good to hear from you and I enjoyed your reminiscences. But I'm afraid your mystery is still not solved, because I would never have left a safe open, or a door. I would open the safe, put a

*With Carl on the beach, 1985.*

note in, but always close it up again. I had great respect for all the material and would never have done anything that would leave it available for theft. My pranks were meant to point out the need for greater security.

Sincerely,

Richard P. Feynman


## RICHARD P. FEYNMAN TO DANIEL E. KOSHLAND, JR., SEPTEMBER 3, 1985

*Feynman was asked by Dr. Koshland of* Science *magazine to give his perspective on the new theories on "strings."*

Dr. Daniel E. Koshland, Jr.
*Science*
Washington, D.C.

Dear Dr. Koshland:

Please accept my apology for the delay in replying to your letter of June 17; I have been out of the office since the 1st of June. However, my

*Richard and Gweneth's Silver Wedding Anniversary,*
*September 1985.*

response to your request for an article on "strings" is that I don't believe in them, but then I haven't studied them well enough to know why I don't believe in them. Such prejudice would not make an appropriate article.

Sincerely,

Richard P. Feynman


## RICHARD P. FEYNMAN TO MRS. HARRY GARVER, SEPTEMBER 3, 1985

*Mrs. Harry Garver was another of Feynman's longtime secretaries. I was in high school—eleventh grade—at the time of my father's response to her letter.*

Mrs. Harry Garver
Oroville, California

Dear Bette:

Thank you for writing—it was nice to hear from you. I'm gray-haired but not a grandfather—my kids are slow.

Sincerely,

Richard P. Feynman


## RICHARD P. FEYNMAN TO BERNICE SCHORNSTEIN, SEPTEMBER 5, 1985

*"In any case, there was a party, you and she and Florence walked home arm in arm down the middle of the street, singing at the top of your lungs. My mother thinks you 'probably wouldn't' remember her, tho from photographs of that era I don't see how anyone could forget her." Mrs. Pauli Carnes, daughter of Bernice Schornstein (née Lesser), wrote to remind Feynman of this friend from long ago.*

*Mrs. Schornstein also wrote, asking for an autograph for her cousin's son.*

Mrs. Bernice Schornstein
Scottsdale, Arizona

Dear Bernice:

Of course I didn't remember you very well from the few clues left in your letter—but your daughter supplied a few others and now I remember two beautiful teasers, each successively going upstairs to "get into some-

thing more comfortable" while I was held down by the other. We certainly had a great time, me, you and your cousin Florence.

It seems that the spirit of fun and good humor I admired that day has reappeared in your daughter—who was careful to assure me that I need not be alarmed by receiving a letter from her, as she greatly resembled another man she knew as "daddy." Apparently she is left with the impression that we went much further than we did—times have changed, we should have been born later.

Sure, I'll be glad to sign the book for your cousin's son. What a nice memory to be reminded of!

Sincerely,

Richard P. Feynman

## DR. KLAUS STADLER TO RICHARD P. FEYNMAN, OCTOBER 4, 1985

Dear Professor Feynman,

Today I would like to introduce myself as the editor responsible for the German edition of your book "Surely You're Joking, Mr. Feynman." The Piper Verlag is very glad to have got the German rights of your book. Your colleague in Munich, Professor Harald Fritzsch, was very much involved in the discussion about the book. If you have no objections I will ask Harald Fritzsch to write a preface for the German edition.

Professor Fritzsch also advised us to abridge the book a little bit. He thinks that some parts are not so important for the German reader. In the next days I hope to get a list from him with suggestions for several cuttings. Today I want to ask you to give us permission for minor cuttings. Of course I will send you the list with Professor Fritzsch's suggestions as soon as possible.

Recently we got the information about your new book, *Q.E.D.*, which will be published by Princeton University Press. During my stay at the Frankfurt Book Fair I am going to contact Princeton University Press. I hope to get an option for the German rights.

Do you have any other plans for publications, which would fit into our list of science books? I would be very glad to hear about your plans.

A last question for today: Is there a chance of your coming to Germany, when the German version of *Surely You're Joking, Mr. Feynman* will be published? Couldn't that be combined with a lecture at one of the universities or Max-Planck-Institutes?—We intend to publish the book in fall '86.

I look forward to hearing from you.
With kindest regard,
Sincerely yours,
Klaus Stadler

## RICHARD P. FEYNMAN TO KLAUS STADLER, OCTOBER 15, 1985

Dr. Klaus Stadler
Wissenschaftliches Lektorat
R. Piper Verlag
Munich, West Germany

Dear Dr. Stadler:

Your letter about my book *Surely You're Joking...* suggests that some parts might be cut out as not being so important to a German reader.

This shows a complete misunderstanding of the nature of my book. There is nothing at all in it that would be "important" to a German reader, or to any other reader for that matter. It is not in any way a scientific book, nor a serious one. It is not even an autobiography. It is only a series of short disconnected anecdotes, meant for the general reader which, we hope, the reader will find *amusing.* There should be no pretense of importance. Please. Your advertising should make this clear—otherwise there will be bad reviews by readers who have been disappointed. (Many reviews here have been very good—nearly all the bad ones were by people who expected more and were disappointed.)

I should have been much happier if you gave the book to a department dealing with more general books rather than "Wissenschaftliches Lektorat," and to a translator known for his sense of humor and a healthy disrespect for pompousness and "importance" so that he would be more attuned to the character of the book. Science it is not! Naturally, the book might be improved if some parts were cut out, but for some reason other than they lack importance. Otherwise everything should be cut out.

The new book you mention, QED: *The Strange Theory of Light and Matter,* is an entirely different matter. It is a serious scientific book intended for the (very) intelligent layman or the young person interested in finding out what advanced physics is like. I am very happy to hear you are interested in it.

Sincerely,
Richard P. Feynman

## RICHARD P. FEYNMAN TO EDWARD TENNER, NOVEMBER 14, 1985

*In a March 21, 1985, letter regarding the promotional copy for* QED, *Princeton University Press assured Helen Tuck, Feynman's secretary, that no reference would be made to Feynman's status as Nobel Laureate. "Additionally, we have removed the reference to his 'legendary humor.'" There remained other kinks to be worked out before publication, however.*

> Mr. Edward Tenner
> Princeton University Press
> Princeton, New Jersey

> Dear Mr. Tenner:
> Dr. Mautner showed me the cover for my new book "QED."
> It is truly beautiful and dignified. I am very pleased, as is everyone I show it to. It has class, they say. (To be frank I must add that when they find an advertisement for another book on the inside flyleaf, they are surprised and think that is a bit low class. I don't know, for I don't know what is customary, and I know and like Rudy Peierls. But I hope you don't describe my book on anybody else's cover!)
> I am glad to see it coming out now. I am very curious as to what the reception will be. How understandable it is, really. It is hard for me to tell.
> Thank you.
> Sincerely,
> Richard P. Feynman

*Mr. Tenner wrote back to give credit to Mark Argetsinger, the book's designer. He also assured Feynman that the notice of Peierls's* Bird of Passage *would be removed from future printings, and he would let the production department know that Feynman did not want his own book imposing on the flap copy of others.*

## RICHARD P. FEYNMAN TO JAMES T. CUSHING, OCTOBER 21, 1985

*Professor Cushing sent Feynman a preliminary draft of a paper on Heisenberg's S-Matrix program.*

Professor James T. Cushing
Department of Physics
University of Notre Dame

Dear Professor Cushing:

I read your interesting historical paper on the S-Matrix, but I have nothing to add or comment. I had always thought that the S-Matrix program of Heisenberg's was of relevance for the subsequent work.

By the way, I always find questions like that in your last sentence odd. It seems to me that the answer is: if Heisenberg had not done it, someone else soon would have, as it became useful or necessary. We are not that much smarter than each other. (Except perhaps Einstein's general theory—or was Hilbert already hot on the trail, independently? I don't know history too well. When do you think it might have been invented if there were no Einstein?)

Sincerely,

Richard P. Feynman

## STEPHEN WOLFRAM TO RICHARD P. FEYNMAN, SEPTEMBER 26, 1985

*In the early 1980s, Stephen Wolfram turned his energies from traditional areas of fundamental physics to creating the new field of complexity research. Some physicists and science administrators were skeptical about this new direction.*

Dear Feynman,

First, thanks very much for your letter on the cryptosystem. I managed to break my addiction to studying the thing for a while, but am now getting back to it. I would like to try and understand systematically how far one can get with the kind of approach you were using: in particular, whether it allows the seed to be found in polynomial time. But I must say I am still reasonably confident that the system is at least hard to break. I have a new idea for showing that breaking it would be equivalent to solving an NP-complete problem; I'll let you know if this works out.

I thought I would send the enclosed stuff that I have just written. It is not about science (which is what I would prefer to write about), but rather about the organization of science. I am being treated increasingly badly at IAS, and really have to move. I can't see anywhere that would really be nice to go to, and would support the kinds of things I am now interested in. So

I am thinking of trying to create my own environment, by starting some kind of institute. It would be so much nicer if such a thing already existed, but it doesn't. There are a few plans afoot to create things perhaps like this, but I think they are rather misguided. You probably think that doing something administrative like this is an awful waste of time, and I am not sure that I can disagree, but I feel that I have little choice, and given that I am going to do it, I would like to do it as well as possible. Any comments, suggestions, etc., that you might have I would very greatly appreciate.

Best wishes,

Stephen Wolfram

## RICHARD P. FEYNMAN TO STEPHEN WOLFRAM, OCTOBER 14, 1985

Dear Wolfram:

1. It is not my opinion that the present organizational structure of science inhibits "complexity research"—I do not believe such an institution is necessary.

2. You say you want to create your own environment—but you will not be doing that: you will create (perhaps!) an environment that you might like to work in—but you will not be working *in* this environment—you will be administering it—and the administration environment is not what you seek—is it? You won't enjoy administrating people because you won't succeed in it.

   You don't understand "ordinary people." To you they are "stupid fools"—so you will not tolerate them or treat their foibles with tolerance or patience—but will drive yourself wild (or they will drive you wild) trying to deal with them in an effective way.

   Find a way to do your research with as little contact with non-technical people as possible, with one exception, fall madly in love! That is my advice, my friend.

   Sincerely,

   Richard P. Feynman

*Wolfram did not follow Feynman's advice. Not only did he establish an institute but he also founded the company Wolfram Research, makers of the widely used Mathematica software system. Contrary to Feynman's expectations, Wolfram has been a successful CEO for many years. Within this environment he has managed to pursue ambitious directions in basic science, particularly*

*through his 2002 book* A New Kind of Science. *He has also been happily married since the early 1990s.*

## THOMAS H. NEWMAN TO RICHARD P. FEYNMAN, NOVEMBER 11, 1985

*In autumn 1985, electrical engineering professor R. Fabian Pease at Stanford University and his graduate student Thomas H. Newman informed Feynman that they were ready to claim the first prize offered in his 1959 "There is Plenty of Room at the Bottom" challenge. In the following letter and documents, they provided information about their achievement.*

Professor Feynman,

The photos enclosed are additional TEM pictures of the page of text we reduced 25,000 times in linear scale. By now you probably have reviewed the original pictures. We can supply verification of the scale on the contact prints if you want this. The TEM magnification is calibrated to within 10 percent; this accuracy could be improved by taking a picture of a TEM calibration standard at an identical magnification and then comparing the negatives side by side.

Attached is a description of the procedures we used for sample preparation, exposure and development, and inspection. Please let us know if you would like any additional supporting documentation.

Thomas H. Newman

R. Fabian Pease

Stanford University

Stanford, California

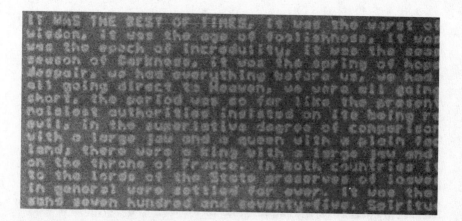

## RICHARD P. FEYNMAN TO THOMAS H. NEWMAN, NOVEMBER 19, 1985

Dr. Thomas H. Newman
Stanford University
Stanford, California

Dear Dr. Newman:

Congratulations to you and your colleagues. You have certainly satisfied my idea of what I wanted to give a prize for. Others have apparently made as small or smaller marks, but no one tried to print an entire page. And on a 512 x 512 dot printer! Each dot is only about 60 atoms on a side. I can't quite manage to imagine the square 1/160 mm on a side onto which all that is printed. It would be 20 times too small on a side to see with the naked eye. Only ten wave lengths of light. The entire *Encyclopaedia Britannica,* perhaps 50,000 to 100,000 pages of your size would be on less than 2 mm on a side—the head of a small plain pin.

Your description of the square silicon nitride windows was a bit incomplete. How big are the windows? Is each window a page, or (less probably?) a letter? Can application to computers be far behind?

As promised long ago, I am enclosing a check for $1,000 for your accomplishment.

Sincerely,

Richard P. Feynman

*So concluded an important chapter in the early history of nanotechnology.*

## RICHARD P. FEYNMAN TO MICHAEL ISAACSON, DECEMBER 20, 1985

Dr. Michael Isaacson
Physics Department
Cornell University
Ithaca, New York

Dear Dr. Isaacson:

It seems I let you down. I told you there was no more prize.

I knew of your work, of course, and it always interested me. Tom Van Sant is my personal friend and he told me of what you did for him. I have

often talked on small things and exhibited a slide of that wonderful eye of yours as the smallest artist's drawing ever made.

When I received an entire page from Stanford I was so delighted I sent them the prize even though I knew people had made things smaller. I forgot that I had explicitly told you there was no prize.

That is a hell of a way to run a railroad! I guess I am not a railroad man.

The first time I was interviewed, after Stanford announced it, I told them others had made things still smaller, mentioned that you and Van Sant made an eye together, but that this was the first full page of text I had seen. They didn't print the complete interview.

Well, Merry Xmas. I am enclosing a Christmas present for you. To encourage you to continue your good work.

Your friend.

Richard P. Feynman

## RICHARD P. FEYNMAN TO JOAN T. NEWMAN, NOVEMBER 10, 1986

*Joan Thomas Newman, the mother of the Thomas Newman who won the tiny writing prize, wrote Feynman to express her appreciation and gratitude: "Thank you for encouraging creativity in your own mind and in the minds of your students and readers." Although she herself had never delved into the domain of physics, she was immensely proud of her son and his accomplishments ("winning your award could not have happened to a better person").*

Dear Mrs. Newman:

Well, what a pleasant surprise to get a letter of appreciation from the parents of a scientist. I am glad to hear of how it looks from the point of view of his proud mother, who really doesn't understand what he is doing. I know. I had a wonderful proud mother who never understood what I was doing either. How could "breaking my head" be fun? And how can Tom, working so hard in the laboratory, be having fun? But her support made my accomplishments possible—and I am sure it is the same in your family.

Sincerely,

Richard P. Feynman

. . . .

## RICHARD P. FEYNMAN TO ARMANDO GARCIA J., DECEMBER 11, 1985

*On September 18, 1985, Mr. Amando Garcia J., a teacher in a high school in La Victoria, Venezuela, sent Feynman a five-page, single-spaced, typewritten letter, seeking help in answering an objection to the law of energy conservation, posed by twins who had been in a class he taught. Mr. Garcia had been revered by the students in his school before he failed to give a satisfactory answer to the twins' objection. "The issue became a debate in class for weeks," Mr. Garcia wrote, "taking us nowhere; my class lost its traditional prestige. . . . Some of my ex-students still drop by once in a while and ask me with irony: 'have you solved the twins' objection yet?'"*

*The twins' objection had to do with a man who lowers a weight (for example, a heavy barbell) from overhead to the ground. The law of conservation says that the weight's large potential energy must be converted into some form of energy inside the man—that is, the weight must do work on the man's muscles as he lowers it. However, everyday experience says this cannot be so. We know from experience that the man's muscles actually have to do work on the weight in lowering it rather than vice versa, so energy cannot be conserved. Mr. Garcia cited another example: we know from experience that when a man climbs up stairs and then descends, his muscles must do work in both the up trip and the down, but energy conservation says they should do work only going up, not going down.*

*Mr. Garcia pleaded with Feynman to explain how to reconcile these everyday experiences with the law of energy conservation.*

Mr. Armando Garcia J.
Aragua, Venezuela

Dear Mr. Garcia:

Thanks for your letter and its questions about energy.

Your trouble comes from dealing with an open system like a man who is sweating, breathing in and out (heating air and changing some $O_2$ for $CO_2$) digesting food, etc., and moving about doing (or having done on him) mechanical work. To find the total energy change after some operation we would have to measure all kinds of things to access the net energy change. We must include the heat to the air, the chemical energy change in the changed air gasses, the change in the amount or form of the partly digested food, the energy difference of the form of the water sweated (from liquid to vapor), the change in energy from the shift in position of weights

moved, etc. This last, the mechanical part is numerically a very small amount compared to the other changes.

So nobody can, without careful check and measurement, demonstrate that the total energy of the system has changed (or changed by amounts not accounted for by flow of energy as heat or chemicals in or out). Attention to just the tiny mechanical contributions plus or minus for a man working is completely insufficient. Observations that it seems to be a similar effort to lower or to raise a weight are insufficiently accurate—they both are efforts with large energy expenditures (to heat, change in body chemistry, etc. etc.) only differing slightly by the small mechanical differences—too slightly to be noticeable.

(For some numbers, we are expending [to keep warm and move about] about 100 watts from our food combining with oxygen.) You can check this because in one day that is 100 x 86400 sec ~ $10^7$ joules ~ = 2300 kilo calories = 2300 food "calories" because one food calorie is 1000 physics calories, and people consume around two thousand food calories a day. Now suppose a man lifts 10 kilograms up one meter. He has done 10 x 1 x 9.8 = 100 joules of work. But this amounts to the average energy to stay alive (100 watts) for only one second.

That is why we do not so directly feel the difference of going up and down stairs (but we all do, do we not, admit that it would be easier to go down many stairs than up).

To get accurate we must measure all the energy in and out over a long enough period, say one full day, breakfast to breakfast, so we can assume the animal is in about the same condition internally before and after. We can do that with animals, like rats, by monitoring all their gasses in and out and food intake. When this is done it is found to check out.

In fact it was by such experiments on animals that the law of conservation was discovered in the first place. It was done by a Doctor Mayer. Much more accurate measurements on simpler physical and chemical systems have continued to confirm it. Today we can see it work for individual atomic collisions, and since more complex systems are the results of hosts of such collisions, it ought to work there too.

Your twins may claim it is wrong by qualitative arguments about how things seem to feel. But if they test their claim with measurements made carefully enough to check for inadvertent outputs or inputs or permanent changes in the objects inside, they will find, I think, what Dr. Mayer first found, and what others have since found, that energy is indeed conserved. The consumption of $O_2$ is higher when lifting than when releasing weights.

Your examples of times when the law was doubted (such as by Rohr) are good. There is no harm in doubt and skepticism, for it is through these that new discoveries are made. The doubts can, and therefore must, be tested and resolved by experiment. It is true that energy is a scalar quantity, like temperature, that has no direction. But measured from some arbitrary level it can be either plus or minus—surely the changes have a sign. In lifting a weight the weight's energy is increased (and the rest of the world's decreased) and in lowering it the signs are reversed.

I judge from your letter that in Venezuela you are teased badly if you are a professor and say you don't know or are not sure. I am glad that I am not so teased because I am sure of nothing, and find myself having to say "I don't know" very often. After all, I was born not knowing and have only had a little time to change that here and there. It is fun to find things you thought you knew, and then to discover you didn't really understand it after all. My students often help me do that (for example, by bringing up apparent difficulties like your twins did). And then they have to help me finally understand it better.

Anyway, I hope this helps some. Good luck to you and your students, teaching each other.

Sincerely,

Richard P. Feynman

## RICHARD P. FEYNMAN TO L. DEMBART, JANUARY 15, 1986

Mr. L. Dembart
Science Writer
Los Angeles Times
Los Angeles, California

Dear Mr. Dembart:

Thank you for mentioning my name in your editorial. If you have no intention of writing a longer article, would you please consider the following letter for the "Letters to the Editor" page?

"You reported in an editorial 'The Wonder of It All' about a proposal to explain some small irregularities in an old (1909) experiment (by Eötvös) as being due to a new "fifth force." You correctly said I didn't believe it—but brevity didn't give you a chance to tell why. Lest your readers get to think that science is decided simply by opinion of authorities, let me expand here.

If the effects seen in the old Eötvös experiment were due to the "fifth force" proposed by Prof. Fischbach and his colleagues, with a range of 600 feet it would have to be so strong that it would have had effects in other experiments already done. For example, measurements of gravity force in deep mines agree with expectations to about 1% (whether this remaining deviation indicates a need for a modification of Newton's Law of gravitation is a tantalizing question). But the "fifth force" proposed in the new paper would mean we should have found a deviation of at least 15%. This calculation is made in their paper by the authors themselves (a more careful analysis gives 30%). Although the authors are aware of this (as confirmed by a telephone conversation) they call this "surprisingly good agreement," while it, in fact, shows they cannot be right.

Such new ideas are always fascinating, because physicists wish to find out how Nature works. Any experiment which deviates from expectations according to known laws commands immediate attention because we may find something new.

But it is unfortunate that a paper containing within itself its own disproof should have gotten so much publicity. Probably it is a result of the authors' "over-enthusiasm."

Sincerely,

Richard P. Feynman

*This letter was subsequently published in the* Los Angeles Times.

## RICHARD P. FEYNMAN TO GWENETH AND MICHELLE FEYNMAN, FEBRUARY 12, 1986

*The following letter was written during Feynman's stint on the presidential commission investigating the* Challenger *Space Shuttle accident.*

Wash. Wed. Feb

2 P.M.

Excuse the paper, I can't find hotel stationary.

Dearest Gweneth and Michelle,

This is the first time I have time to write to you. I miss you, and may figure out later how I can get a day off and visit (when a very boring meeting is scheduled?). This is an adventure as good as any of the others in my

book. You, Gweneth, were quite right—I have a unique qualification—I am completely free, and I there are no levers that can be used to influence me—and I am reasonably straight-forward and honest. There are exceedingly powerful political forces and consequences involved here, but altho people have explained them to me from different points of view, I disregard them all and proceed with apparent naïve and single-minded purpose to one end, first why, physically the shuttle failed, leaving to later the question of why humans made apparently bad decisions when they did.

As you know, Monday at home at 4 P.M. they told me I was on the committee and I should fly in (Tuesday night) for a Wednesday meeting. All day Tuesday I educated myself with the help of Al Hibbs and technical guys he brought around to help me (I knew nothing about the shuttle considering, before, that it was some sort of boondoggle). I was preparing myself technically for the job, and the preparation was excellent—I learned fast.

We had an "informal get together" on Wed. when we were advised by the Chairman on how important press relations were, and how delicate. We had a public meeting Thurs February 5th which used up the whole day, the entire committee being briefed on facts about the Challenger and its flight. I spent the night making a plan of how I—and we should proceed—facts to get—a long list of possible causes—make some calculations of loads and so on and so on, ready to roll.

Fri February 6th, a General Kutyna on the commission tells us how such accident investigations have been made in the past—using a Titan failure for an example. Very good. I was pleased to see that much of my plans of February 5th are very similar, but not as methodical and as complete. I would be happy to work with him, and a few others want to do that too—others suggest they could make better contributions by looking into management questions, or by keeping records and writing the report, and so on. It looks good. Here we go.

But the Chairman (Rogers, not a technical man) says that the General's report cannot serve for us because he had so much more detailed data than we will have (quite patently false—because of human safety questions *very* much more was monitored on our flight!) Further it is very likely that we will be able to figure out just what happened (!?), and the co-chairman says the commission can't really be expected to do any detailed work like that—we shall have to get our technical advice from mutter mutter. I am trying to get a word in edgewise to object and disagree but am always interrupted by a number of accidents, someone just comes in to be introduced around, the chairman returns to a new direction, etc. What is

decided is that we, as an entire committee will go down next Thursday to Kennedy to be briefed by people there on Thurs, Fri. During the discussion earlier there are various pious remarks about how we as individuals or better small groups (called subcommittees) can go anywhere we want to get info. I try to propose I do that (and several physicists tell me they would like to go with me) and I have set my affairs so I can work intensively full time for a while. I can't seem to get an assignment, and the meeting breaks up practically while I am talking—with the Vice Chairman's (Armstrong) remark about our own not doing detailed work. So on leaving I say to the chairman, "then I should go to Boston for a consulting job there for the next five days (Sat, Sun, Mon, Tues, Wed)?" "Yes, go ahead." I DON'T HAVE TO EXPLAIN TO YOU WHY THAT DRIVES ME UP THE WALL!

I leave, very dejected. I then get the idea to call Dr. Graham head of NASA who had been my student and who had asked me to serve. He is horrified, makes a few calls and suggests possible trips to Houston (Johnson Airforce Base where the telemeter data comes in) or Huntsville Alabama (where they make the engines). (I turn down Kennedy because we shall go there later and it is too direct a rebellion from the chairman.) To get the chairman to OK this he calls commissioner (a lawyer, son of Dean Acheson, a good friend of Chairman Rogers). Acheson says he will try, as he thinks it is a good idea. Calls back surprised he can't get OK—chairman doesn't' want me to go, "We want to do things in an orderly manner."

Graham suggests compromise. I will stay in Washington, and even tho it is Saturday he will get his men (high level heads of propulsion, engines, orbiter etc.) to talk to me. That seems to be OK altho I get a call from Rogers trying to bring me to heel—explaining how hard his job is organizing this—how it must be done in an orderly manner etc.—do I really want to go to NASA ("Yes"). I point out so far two meetings and still no talk on how we should proceed, who can best do what etc. (Most of the talk at the meeting was by Rogers, on how he knows Washington, that there are serious questions of orderly press relations—tell any reporter he should go to him, Rogers, for answers etc.). He asks me if he should bother all the commissioners and ask them to come to a meeting Monday which he will convene. "Yes," I say. He drops that subject—OK's that I stay in Washington and suggests, "I hear that you are unhappy with your hotel— let me put you into a good hotel." I don't' want any favors, tell him no everything is OK, that my personal comfort is less important to me than action, etc. He tries again, I refuse again (Reminded of "Serve him tea" at London airport).

*Giving testimony for the Presidential Commission for the Challenger Space Shuttle Accident, 1986.*

So Saturday I got briefed at NASA. In the afternoon we talk in fine detail about joints and O-rings, which are critical, have failed partially before and may be the cause of the Challenger failure. Sunday I go with Graham and his family to see the air and space museum which Carl liked so much—we are in an hour before the official opening and there are no crowds—influence; after all, the acting head of NASA.

All this time evenings I eat with or go over to the house of Frances and Chuck. It is a very welcome unwinding, but I don't tell stories because they are with the press and I don't want to spread or be suspected of spreading leaks. I report to Rogers that I have these close relations with press connections and is it OK to visit with them? He is very nice and says, "Of course." He himself had some AP connections, he remembers Frances, etc. I was pleased by his reaction but now as I write this I have second thoughts. It was too easy—after he explicitly talked about the importance of no leaks etc. at earlier meetings. Am I being set up? (SEE DARLING, *WASHINGTON PARANOIA* IS SETTING IN). If, when he wants to stop or discredit me, he could charge me with leaking something important. I think it is possible that there are things in this that somebody might be trying to keep me from finding out and might try to discredit me if I get too close. I thought I was invulnerable. Others like Kutyna, Ride, etc. have some apparent weakness and perhaps may not say what they wish. Kutyna has Air Force interests to worry about. Ride, her job at Johnson Air Force Center, etc., etc. But, as I learned, I must keep watching in all directions—nobody is invulnerable—they will sneak up behind you. So, reluctantly, I shall not visit Frances and Chuck anymore. Well, I'll ask Fran first if that is too paranoid. Rogers seemed so agreeable and reassuring. It was so easy, yet I am probably a thorn in his side.

Anyway, Monday and Tuesday we have a special closed and open meeting respectively because some internal reports saying the joint seals are or might be dangerous appear in the NY Times. Big deal. I knew all the facts in question—I got them at JPL before I started. Very important emergency concerning press relations! At this rate we will never get down close enough to business to find out what happened. Not really—for now tomorrow at 6:15 we go by special airplane (two planes) to Kennedy Space Center to be "briefed." No doubt we shall wander about being shown everything—gee whiz—but no time to get into technical detail with anybody. Well it won't work. If I am not satisfied by Friday I will stay over Sat and Sun, or if they don't work then Monday and Tuesday. I am determined to do the job of finding out what happened—let the chips fall!

I feel like a bull in a china shop. The best thing is to put the bull out to work the plow. A better metaphor will be an OX in a china shop because the china is the bull, of course.

My guess is that I will be allowed to do this overwhelmed with data and details, with the hope that so buried with all attention on technical details I can be occupied, so they have time to soften up dangerous witnesses etc. But it won't work because (1) I do technical information exchange and understanding much faster than they imagine, and (2) I already smell certain rats that I will not forget because I just love the smell of rats for it is the spoor of exciting adventure.

So much as I would rather be home and doing something else, I am having a wonderful time.

Ralph* called me this morning from Sweden to report amazing progress. Christopher Sykes is involved. He comes back next Wednesday and will tell you all about it. TUVA OR BUST!

Love,

Richard

P.S. Sykes is coming to LA around the 17th for a few days. ~~This letter need not be splashed around too much, if you do leave out the personal~~ —forget it, everything is O.K. Uncensored report from the front.

P.P.S. Save the newspapers. I was at the Pentagon this morning—and they sent you clippings of the NY Times.

## RICHARD P. FEYNMAN TO WILLIAM P. ROGERS, SATURDAY, MAY 24, 1986

Dear Mr. Rogers,

I am very sorry I had to leave at noon Saturday, before we had time to discuss fully the problem that our report might seem overly critical of the NASA shuttle program. I would like to explain my views in more detail.

* Ralph Leighton, a longtime friend, drumming partner, and co-author of *Surely You're Joking, Mr. Feynman!* and *What Do You Care What Other People Think?*. Feynman had recently challenged Ralph to reach the lost country of Tannu Tuva, which he remembered from his stamp collecting days in the 1930s. Their adventure is chronicled in Ralph's book *Tuva or Bust! Richard Feynman's Last Journey*.

*Feynman and good friend Ralph Leighton, 1987.*

It is our responsibility to find the direct and proximate causes of the accident, and to make recommendations on how to avoid such accidents in the future.

Unfortunately we have found, as the proximate cause, very serious and extensive "flaws" in management. Not just a crack but a general disintegration. Our report lists them with our evidence for our view.

This raises serious problems for our nation as to how to continue on with the space program. There are very large questions of budget, what other projects to follow to supplement the shuttle ( i.e., Expendables?) to maintain a scientific and more importantly, military strength, commercial space applications, etc. Our entire position and program in space must be reconsidered by people, Congress and the President. We did not discuss these matters— and make recommendations in this larger theater. We were asked, as I see it, to supply information needed to make such decisions wisely.

It is our duty to supply such information as completely, accurately, and impartially as possible. We have laid out the facts and done it well. The large number of negative observations are a result of the appalling condition the NASA shuttle program has gotten into. It is unfortunate, but true, and we would do a disservice if we tried to be less than frank about it. The President needs to know if he is to make wise decisions.

If it appears we have presented it in an unbalanced way, we should give evidence on the other side. Let us include somewhere a series of specific findings that this or that is very good, or recommendations that they keep up the good work in this and that project specifically. One example is the software verification system. I think, altho I seem to be pollyanna and a little dumb—I thought (apparently incorrectly?) that NASA was very helpful and courteous in giving us all the information (and doing tests) we asked for, especially in our accident panel.

A general statement, without presenting the particular evidence in the report, that NASA is after all actually doing a wonderful job would weaken our careful report, and might even make us look foolish, if it contrasted with the mass of evidence we ourselves present in a careful documented way.

Well, I'll be back Monday and we can discuss it with other commissioners more fully next week. Maybe they may explain to me where my point of view is off base and I will change my mind—now I doubt it of course. I feel *very strongly* about this now.

I am satisfied that my "Reliability" contribution need not go in the main report but will go into the appendix more or less intact rather than being lost in the archives. It is a good compromise we have come to.

See you soon,
Commissioner Feynman, Nobel Prize, Einstein Award, Oersted Medal
and utter ignoramus about politics.

## RICHARD P. FEYNMAN TO DAVID ACHESON, DECEMBER 5, 1986

*Fellow commissioner David Acheson wrote a note to Feynman, sending news
of other commission members. He had also heard of Feynman's illness and sent
earnest wishes for a whole and speedy recovery.*

Mr. David C. Acheson
Washington DC

Dear Dave:
Thank you very much for the note with all the news.
I'm recuperating slowly at home. I also hope I see you again sometime, but
nothing short of a subpoena from Congress will get me to Washington again.
Sincerely,
Richard P. Feynman

## RICHARD P. FEYNMAN TO JOHN W. YOUNG, DECEMBER 8, 1986

*Dr. Young wrote to express appreciation for Feynman's part in the Challenger
accident investigation. "We realize this was an extremely difficult task: technically, physically, and emotionally. Your report is thorough and insightful and
will help lead NASA back to safe flight. You have done a great service for the
Astronaut Office and the Nation."*

Dr. John W. Young
Chief, Astronaut Office
NASA
Lyndon B. Johnson Space Center
Houston, Texas

Dear Dr. Young:
Thank you for your compliment about my work on the Challenger
investigation.

I'm sorry I didn't get time to discuss matters with the astronauts more directly and informally and to meet you. We had a division of labor, and I depended on Dr. Sally Ride to give us your opinions, as she did.

I was particularly impressed by the careful analysis exhibition in the testimony of Mr. Hartsfield, yourself, and the other astronauts during one of our public meetings. It seemed that you were the only people thinking about the future, and the causes of things in a clear way. It soon became apparent that the testimony of higher management was a bit muddleheaded about why they weren't told, why the system broke down, etc. They weren't told because they didn't want to hear any doubts or bad news. They were often involved in justifying themselves to the public and Congress—so the correct thing for lower-level bureaucrats was clear—solve it yourself, or at least don't rock the boat. Like the Wizard of Oz, NASA had a great reputation to maintain, but when a few screens were knocked over, it was seen to be imperfect.

I hear the same thing is happening again—complex questions being hidden in innocuous little "bullets," so even the joint-certifying committee is having trouble getting information.

Is there anything you astronauts can do to clean the Augean stable?

Sincerely,

Richard P. Feynman

## RICHARD P. FEYNMAN TO GENERAL DONALD J. KUTYNA, DECEMBER 8, 1986

*The Cook report made claims that the Presidential Commission on the Challenger Space Shuttle Accident was engaged in a cover-up of the government's involvement in launch decisions, hid many of its findings, and helped NASA personnel perjure themselves.*

Major General Donald J. Kutyna
San Pedro, California

Dear Don:

Thanks for sending me the information about the "Cook report." I suppose it is to be expected that some problem like this would come up. Of course it is true that some criticism is sensible, and there are always paranoids waiting for their chance to criticize every investigation. You are right in that we didn't do a perfect job, but of course if the Chairman had lis-

tened to us it may have been closer to perfect. After all, it has been said that the last perfect person was crucified.

Do you believe "there is evidence that not only did NASA personnel perjure themselves before the Commission, and that in closed sessions the Commission actually advised them to do so and helped to concoct their stories," as reported in that article?

I'm recovering pretty nicely from my surgery, and I hope we can arrange to get together soon.

Sincerely,

Richard P. Feynman

*General Kutyna told me that Mr. Richard C. Cook, former NASA resource analyst and author of this report, had discovered something interesting prior to Challenger's launch: "a NASA action to add money to the budget to fix the O-Rings." NASA was concerned about the impact to flight safety that the O-Ring anomalies might potentially pose, but were not totally open relative to these concerns. However, the General saw absolutely no instances of perjury. Mr. Cook misinterpreted NASA's silence as a Comission cover-up, which it was not. The Comisssion found out about NASA's omission during their investigation, and Feynman's Appendix F of the Comission report details some of the missteps and motives within NASA.*

*General Kutyna emphasized that "the specific direction of the Commission was to find the technical cause of the accident, and not to go on a witch hunt to blame individuals."*

## LEIGH PALMER TO RICHARD P. FEYNMAN, JANUARY 1, 1987

Dear Prof. Feynman,

My son, David, tells me that you are not well, and I am very sorry to hear that. I wanted to tell you about something that happened last semester for which you were principally responsible, and which illustrates a very strange (but avoidable) human failing, the effect of prejudice on learning.

For more than twenty years I have wanted to understand the Hanbury-Brown, Twiss "intensity interference" effect. During the last twenty years I had asked several of my colleagues to explain the effect to me in a way that I might assimilate it into my ill-adapted experimentalist brain. I don't even know how many of those guys I asked actually "understood" the effect in the way I interpret that word. I know that their throwing around phrases

like "Bose-Einstein condensation" did not lead me to enlightenment. During this period I had even taught junior and graduate level optics courses, and in several semesters statistical mechanics, including quantum statistics, which I *think* I understand. I had not been able to say the same, however, about Hanbury-Brown, Twiss.

Then my daughter brought me a copy of *QED* from the U. of Washington bookstore. (The book was not available here in Vancouver.) I read it with delight, but somewhat slowly, as I was teaching at the time, and I had only learned of the book through a review in *Scientific American*. My attitude was that of a physicist reading a very well-done popularization. I do not understand QED, however, so I hoped to gain some insight from the book as well. While I was following the very lucid development, filling in what I feel is a remarkably small number of "details," and understanding what I was reading, I came across a footnote. It seems that Hanbury-Brown, Twiss had been explained to me entirely *en passant*. I suddenly had the rush of understanding, though I did go back and read it again to make sure. The most important part of this was that I discovered that this, at least, was an area in which I was not invincibly ignorant.

The lesson in this is clear. While knowledge is most readily assimilated by the prepared mind, that same mind can be quite refractory to penetration if it is "prepared" to believe that it cannot be taught. I have certainly had to overcome this attitude in my own students many times, but I had thought myself immune. It was very good to learn something from a master teacher after the age of fifty, when I had become convinced that I could not ever hope to understand it. You did it by explaining it to me without first letting on what it was you were about to explain! I will remember (and use) this trick for the rest of my life. It does not matter that you had no intention of doing this; what matters is that the technique is effective in the case of a strong prejudice against learning a particular topic.

My family and I will always remember fondly our all-too-brief interactions with you and with your family here in British Columbia. While our David has not had the chance to be in one of your classes at CalTech, he has learned much which must be attributed to your good influence, indirect though it may have been. I hope the New Year brings to you and your family as much peace and pleasure as possible under your difficult circumstances.

With sincere thanks,

Leigh Palmer

## RICHARD P. FEYNMAN TO LEIGH PALMER, JANUARY 12, 1987

Dr. Leigh Palmer
Burnaby, British Columbia
Canada

Dear Dr. Palmer:

Among the greatest mysteries of learning, and teaching, must be this appearance of "blocks." What causes them? How to get around them?

Perhaps good fortune removed your block for *H.B. & Twiss* but have we learned a trick to use? Perhaps our students will be turned off by a professor who "goes off explaining things, before he tells us what he is going to explain." The trouble is in our 'classes' of several students, all different and thinking in different ways. What works on one loses the others. Ahh, but in the rare case that we are tutoring one on one (the only time I really felt I was really an effective teacher) then perhaps the trick may be useful. Thank you for pointing it out to me. I may find a chance to use it consciously.

I remember my visits to Simon Fraser and Vancouver with the greatest of pleasure. Thank you for helping to make it so.

Sincerely,
Richard P. Feynman

## RICHARD P. FEYNMAN TO H.-J. JODL, APRIL 10, 1987

*Professor Jodl wrote to Feynman asking to translate and reprint "What is Science?"*

Professor Dr. H.-J. Jodl
Fachbereich Physik
Universität Kaiserslautern
West Germany

Dear Professor Jodl:

You have my permission to translate and publish the article in your journal. But the world has changed—and I made a remark about "a girl instructing another one how to knit argyle socks." Could you add a foot-

note, by the author (me), to that paragraph: "How wonderfully the world has changed. Today conversations among women on analytic geometry are commonplace."

Sincerely,

Richard P. Feynman

## NIGEL CALDER TO RICHARD P. FEYNMAN, JULY 2, 1987

Professor Richard Feynman

California Institute of Technology

Dear Dick,

It is with great sadness that I have to tell you that our dear friend and colleague Philip Daly is terminally ill with a brain tumour. He had an operation at the end of May, but it was too far gone for the surgeon to do very much about it.

Phil is now at home, compos mentis, and carrying on with some work on his current projects for BBC-TV on 20th Century Science and the History of Broadcasting. I have visited him twice, to help him with this work.

He knows what is wrong with him, but believes he has until the end of the year, and is talking about doing some filming in the USA in September. His wife Pat, privately and probably more accurately, interprets the surgeon's comments as meaning weeks rather than months.

He would be very glad to hear from you I'm sure, if only by way of a brief note or a greetings card. Too strong an expression of sympathy tends to upset him, not just because of his real predicament but because the tumour is affecting his emotions.

I hope you are keeping well yourself, Dick. It was a pleasure and a privilege to meet you in Cambridge last year, and I greatly enjoyed our chat about QED even though my powers of persuasion were not enough!

With best wishes,

Yours sincerely,

Nigel Calder

. . . .

## RICHARD P. FEYNMAN TO PHILIP DALY, JULY 22, 1987

Mr. Philip Daly
Quarry Corner
Bath, England

Dear Philip:

I heard the surgeons have got to you too, like they got to me. Good luck in your recuperating period—mine was a little long and when I complained, the surgeon said he never had a patient yet who didn't think he was recuperating too slowly. So have patience.

Are you coming to the U.S. soon? If you are near Los Angeles you are welcome to stay with us. Unfortunately we don't plan to go to England this summer. Gweneth sends regards.

Sincerely,

Richard P. Feynman

## VINCENT A. VAN DER HYDE TO RICHARD P. FEYNMAN, JULY 3, 1986

Dear Dr. Feynman:

OK, right from the start this might seem to be a strange letter. But once you see what I'm trying to do maybe it will not sound so strange. First off, I have this 16 year old son, step-son really, that is fairly bright. No genius you understand, but a lot smarter than I am in math and such. Like everybody else he is trying to figure out what life is all about. What he doesn't know yet is that nobody ever figures out what life is all about, and that it doesn't matter. What matters is getting on with living. So, here is this kid, bright, very good in math and chemistry and physics. Flying radio controlled model airplanes, and reading books about wing design that have a lot of equations in them that I sure don't know how to solve.

But at the same time he's trying to grow up and figure himself and his world out a little. A bit over weight, a little shy, not a whole lot of self-confidence. So he makes up for it by coming on a little strong, playing macho sometimes. Trying to figure out what kind of a man he's going to be. Trying to work out how to handle high school. He's going to be a junior come fall, so college is not far away. He'd love to get into some neat school, but with his grades the way they are that could be a problem.

Now, I don't want to be a pushy parent. Whatever he wants to do is fine

*At a game of chess.*

with me. I started out in electrical engineering in 1960 because my father wanted an engineer, and ended up in criminology. So I know what it's like to be pushed around by a parent and have those expectations forced on you. All I want is that he do whatever it is he wants to do to the best of his ability. It's almost a matter of honor in a way. If you can do something well, you have some sort of obligation to yourself to do it the best you can. I'm afraid that's a concept not thought highly of in a lot of circles, now or ever, but how can an intelligent person live with themselves if they aren't doing something they love to the best of their ability?

Anyhow, after talking to his teachers for the past two years a real pattern emerges. It seems that he picks all the science up fast, sees how you do a thing, and then he wants to go on as fast as he can on his own. Some of the teachers really encourage that, which is great. But. . . it turns out that everybody grades on the basis of how you score on the tests, and the tests only cover what they teach to everybody. Martin, that's his name, sees the basic stuff as too easy for him and hence it's beneath him to hand in the routine day to day assignments. He'd rather be doing the neat fun stuff that the rest of the class never gets to do. The trouble is that a lot of the grade comes from doing the routine stuff, not the exotic stuff, so his grades are down. That, of course, is a bummer. His teachers get after him, I harass him more than I should, and he feels bad. Bah and humbug all the way around. In the non-science courses it's even worse, because he knows that a lot of the stuff is bull and indoctrination. You get the picture.

Well now. A few months ago I came across this book. Interesting (different) title and the guy on the cover looks like a standup comic, not a physicist. Both Martin and I read the book. VERY funny. But we notice almost every story has some point to it. This isn't just a book of funny stories; it's a book about how the world works! Clever. We also follow the news about the Challenger tragedy and the Rogers Commission. And here's the same fellow as wrote the book, helping put NASA back on the track to the stars, and not mincing words in doing it. Great.

So I get to thinking. Here's this guy. My kid has read his book, followed the news and all, and the guy is a Nobel Prize winner too. The sort of fellow that kids with a bent toward science look at and go WOW. And I have this "problem." Now, you obviously know a lot about science, and if the book is any indication you know a lot about how people work too. And who knows what it is that would make a smart 16 year old kid stop for a minute and think about what it is that he really wants in his life (at least for

awhile) and what it is going to take to get it. So. . . . Maybe you could write to this kid. Tell him what you think "about life"; what does it mean to "do science," what do you have to do to train yourself to be whatever it is you want to be. I don't know, tell him whatever you want to tell him. Just knowing that somebody "out there" understands and cares a little can make a big difference sometimes. It helps keep the wings straight and the nose up. Thanks.

Sincerely,

Vincent A. Van Der Hyde

PS It is a good book, hope you write another for the "popular press."

## RICHARD P. FEYNMAN TO VINCENT A. VAN DER HYDE, JULY 21, 1986

Mr. V. A. Van Der Hyde

Juneau, Alaska

Dear Mr. Van Der Hyde:

You ask me to write on what I think about life, etc., as if I had some wisdom. Maybe, by accident, I do—of course I don't know—all I know is I have opinions.

As I began to read your letter I said to myself—"here is a very wise man." Of course, it was because you expressed opinions just like my own. Such as, "what he doesn't know yet is that nobody ever figures out what life is all about, and that it doesn't matter." "Whatever he wants to do is fine with me"—provided "he does it to the best of his ability." (You go on to speak of some sort of obligation to yourself etc., but I differ a little—I think it is simply the only way to get true deep happiness—not an obligation— "to do something you love to the best of your ability.")

Actually if you love it enough you can't help it, if anyone will give you a little freedom. Even in my crazy book I didn't emphasize—but it is true— that I worked as hard as I could at drawing, at deciphering Mayan, at drumming, at cracking safes, etc. The real fun of life is this perpetual testing to realize how far out you can go with any potentialities.

For some people (for me, and, I suspect, for your son) when you are young you only want to go as fast as far and as deep as you can in one subject—all the others are neglected as being relatively uninteresting.

But later on when you get older you find nearly everything is really interesting if you go into it deeply enough. Because what you learned as a youth was that some one thing is ever more interesting as you go deeper. Only later do you find it true of other things—ultimately everything too. Let him go, let him get all distorted studying what interests him the most as much as he wants. True, our school system will grade him poorly—but he will make out. Far better than knowing only a little about a lot of things.

It may encourage you to know that the parents of the Nobel prize winner Don Glaser (physicist inventor of the bubble chamber) were advised, when their son was in the third grade, that he should be transferred to a school for retarded children. The parents stood firm and were vindicated in the fourth grade when their son turned out to be a whiz at long division. Don tells me he remembers he didn't bother to answer any of the dumb obvious questions of the earlier grades—but he found long division a little harder, the answers not obvious, and the process fascinating so began to pay attention.

So don't worry—but don't let him get too much out of hand like Don Glaser. What advice can I give him? He won't take it, of course. But the two of you—father and son—should take walks in the evening and talk (without purpose or routes) about this and that. Because his father is a wise man, and the son I think is wise too for they have the same opinions I had when I was a father and when I was a son too. These don't exactly agree, of course, but the deeper wisdom of the older man will grow out of the concentrated energetic attention of the younger. Patience.

To answer your questions in your last paragraph more explicitly.

Q: What do you have to do to train yourself to be whatever it is you want to be?

A: There are many roads all different that have been taken by many different scientists. The road I took is the one your son takes—work as hard and as much as you want to on the things you like to do the best. Try to keep the other grades from going zero if you can. Don't think of what "you want to be," but what you "want to do." Luckily he knows that already, so let him do it. (But keep up some kind of a minimum with other things so that society doesn't stop you from doing anything at all.)

Q: What is it that would make a smart 16-year-old stop for a minute and think..."

A: Nothing, now, I hope. But to fall in love with a wonderful woman, and to talk to her quietly in the night will do wonders.

Stop worrying, Papa. Your kid is wonderful. Yours from another Papa of another wonderful kid.

Sincerely,

Richard P. Feynman

*At the time of this printing, Mr. Van Der Hyde reported that his son had met a wonderful woman in college, was now married with two children, and was in the last year of his Ph.D. program in physical oceanography at the University of Hawaii. When considering the long-term effects of Feynman's letter, Mr. Van Der Hyde decided that it was, of course, impossible to know how significant it was.*

*He continued, "But I know it was important to me, as a parent, and I know that my son has never forgotten how one of the 'greats' took a few minutes just for him."*

# INTERVIEW WITH RICHARD P. FEYNMAN FOR "VIEWPOINT"

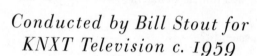

## Conducted by Bill Stout for KNXT Television c. 1959

**Stout:** "All things must change to something new, to something strange." The words of Longfellow, words well suited to this new and strange age of space. We might also go back to Omar Khayyam, who wrote in the *Rubayat,* "And that inverted ball they call the sky, where under crawling cooped we live and die, lift not your hands to it for help, for it as impotently moves as you and I."

Today on "Viewpoint," the relationship of you and me to the mysterious world of the scientist. Our guest, the distinguished young theoretical physicist from the California Institute of Technology, Dr. Richard Feynman.

Dr. Feynman, in the 1920s, the English writer Elizabeth Bibesco said, "It is sometimes the man who opens the door who is the last to enter the room." Do you think this is true of scientists today? Are you really participating as people, you and your colleagues, in this brave new world you're helping to create? As social beings?

**Feynman:** Well, the world we're creating for ourselves is not a social world.

**Stout:** What do you mean?

**Feynman:** The reason that I do science, the motivation for it, is strictly speaking and honestly *not* the usual motivation for helping human

beings. The main motivation is the curiosity and interest to find out about the world we're in. So we're opening a door, if I can use that analogy, and walking into the room of ignorance ourselves and finding out what there is there. Now, it turns out, of course, when you find out something about how the world works, that you can use that knowledge for some application to make a better paint or to make a bomb or to do something like that. And the rest of the world makes use of the results of the scientific work. So in that sense, we're opening the doors for ourselves.

**Stout:** You don't really care, then, about later practical applications?

**Feynman:** Well, as a human being, naturally I care, certainly.

**Stout:** But as a scientist?

**Feynman:** Well, the main motivation of my science is not there directed, the main motivation. There are other motivations. After all, there's only a human being, so you like to live and see that this stuff has applications—rather not have bad applications, but has all kinds of applications. It's also interesting to see the ingenuity that's used in developing some of these things and to see that something that you discovered can help man; that always is a great advantage. Of course, you can't cash in on that advantage, because you can see, at the same time, how what you discovered can be used against or make trouble for people. So you can't feel comfortable that you've discovered something that's useful, because you find at the same time, you've discovered something that's horrible simultaneously. Because any new idea can be used either way.

**Stout:** The bomb, for instance.

**Feynman:** The release of nuclear energy in fission, for instance, can be used to make a bomb, or it can be used to make power.

**Stout:** Is development like that of nuclear fission and the bomb troublesome to scientists who had a part in bringing it into the open?

**Feynman:** First let me answer from the scientific standpoint, and then from the human standpoint. First, regarding the discovery of fission, that is no

trouble at all. That's very exciting, that a nucleus is so big that when you touch it, it splits into two pieces, like a drop of water. That's quite an interesting thing and very amusing. That it's possible, then, to release energy and to get large numbers of neutrons and to make radioactivity to use it for different experiments, that's also very interesting.

On the other hand, as a possibility of a large release of power, now it's often said that science makes the bomb. It isn't. It's engineering which makes the bomb. The reason to make the bomb was a military reason during the war. It's true that the scientists worked on it, but they weren't working on science; they were being engineers during the war. They were taken away from the laboratories in order to do this. They could have been taken away from the laboratories in some other crisis, say, to make power for industry or for some other purpose. Whatever the purpose is that you applied the fission to, that's different.

Now, with regard to our own things as human beings, naturally—I myself, for example—worked on the bomb during the war. Now how do I feel about that? I have a philosophy that it doesn't do any good to go and make regrets about what you did before but to try to remember how you made the decision at the time. At the time, the reason I left the laboratory was because of the great importance of fighting a war, which everybody recognized at the beginning of the war. And this extreme danger that we thought would exist if the Germans obtained the weapon, because of the madman that was governing their country. And if the scientists in Germany could have developed this thing, then we would be helpless, and I think it would be the end of the civilization at that time. I don't know how long the civilization is going to last anyway. So the main reason why I did work on it at the time was because I was afraid that the Germans would do it first, and I felt a responsibility to society to develop this thing to maintain our position in the war.

**Stout:** Do you and other scientists of your sort—not engineers who make the practical applications . . .

**Feynman:** Well, we're hard to distinguish. The same individuals can do both things sometimes.

**Stout:** But in the work you're doing now, you're a scientist. You're not interested in making a better automobile or better paint or a new kind of frozen food product.

**Feynman:** You can easily distinguish it right away. The work that I do, none of it is secret. It is internationally known; it's transmitted by letter, back and forth, and by magazines all over the world. The Russians tell us what they're doing; I tell them what I'm doing, and so forth. It's an international endeavor; so there's one distinction. And the motivation is the curiosity around the world. Excuse me for interrupting.

**Stout:** No, no. You're explaining the point I was going to ask about. When we go beyond this free exchange on the scientific level, we run into the engineering applications, the building of bombs, rockets, and all the rest. And presumably there's a great deal of security that prevents freedom of exchange. Do you work on any secret projects?

**Feynman:** No.

**Stout:** Out of choice?

**Feynman:** Yes, out of choice.

**Stout:** You just don't want to?

**Feynman:** Well, in the first place, I don't want to because I want to do scientific research—that is, to find out more about how the world works. And that is not secret; that work is not secret. There's no secrecy associated with it. The things that are secret are engineering developments which I am not so interested in, except when the pressure of war, or something else like that, makes me work on it.

**Stout:** I wondered if your reason for not being interested in these engineering developments and for not working on secret projects today might be a hangover from the work you did during the war on the bomb.

**Feynman:** Yes, I am definitely anti-working in secret projects.

**Stout:** For what reason?

**Feynman:** Well, there are several reasons. I don't think things should be secret, the people developing this. It seems to me very difficult for citizens to make a decision as to what's going on when you can't say what you're

doing. And the whole idea of democracy, it seems to me, was that the public, where the power is supposed to lie, should be informed. And when there's secrecy, it's not informed.

Now, that's a naïve point of view, because if there weren't secrecy, there'd be the Russians who would find out about it. On the other hand, there's some awfully funny things that are secret. It becomes secret that we know what the Russians are keeping secret from us, for instance, or something like that. It seems to me that things go too far in the secrecy.

**Stout:** The reports, for instance, that we knew in advance about Soviet plans for the first Sputnik, but we didn't tell ourselves we did.

**Feynman:** Yes, for instance.

**Stout:** So that all of us had to be surprised by the Russians, rather than being informed by our own government in advance.

**Feynman:** And that's a good example, because the effect of the knowledge that the Russians were doing, this was very serious and would have made a great difference. It was a shock. And it was an important thing to know, yet we didn't know about it for a long time.

The other thing is that it's hard to work in secrecy; there's a kind of schizophrenia involved in the sense that you have to remember what it is that you know that you're not allowed to say. And what happens is you lean over backwards to avoid saying things until you become inarticulate on certain subjects, because if you start to talk about a subject which you heard something secret, you're afraid to say anything about it for fear that maybe that was one of the things that was secret. So I don't like secrets.

**Stout:** I gather from all of this you see many drawbacks to the security system. What about loyalty oaths and a kind of ideological policing that seems to go with government security, and the increasing spread of government work in the fields of science and engineering? What's your opinion?

**Feynman:** That's a very complicated subject of political nature. I'd rather say I don't know. I have strong opinions, but I don't feel that they're any more valuable than others' opinions.

**Stout:** Except that you have worked in the fields and presumably dealt with these problems.

**Feynman:** That's true, but I don't know very much about it.

**Stout:** You mentioned the strength of democracy in the informing of the people. Do you feel that you and other scientists—getting away from the technologists working on immediate top-secret projects—but are scientists, like yourself, and your colleagues at Caltech, doing all they could, do you think, to inform people of the changes being made? If not the world of today, at least the world of tomorrow.

**Feynman:** No they're not doing all they could, because if they stopped working on the science altogether and spent all their time trying to inform people what they just finished doing, then, of course, they would do more. Don't forget that they have a profession. Furthermore, that they went into this with the reason that they're interested in nature, and not in informing people. A lot of scientists have gone into science because they're not too interested in the relations of human beings. That is, that's not their central interest, so it becomes work to a certain extent to inform. That's not a fair answer, because there are all kinds of different kinds of people, and there are many fellows who want to inform. In fact, we all do, more or less. We inform our students; we teach. Whenever we get an opportunity to give lectures and so forth, we try to. It's very difficult to inform, because there's an enormous amount of information that's been gathered in the last two or three hundred years of science, and people are pretty ignorant of it. It takes a lot of patience to try to explain some of the things because they usually ask you what you are doing now. And what you're doing now is research on the very, very front edge of something that has a tremendous backlog of information and so forth for the last three hundred years of research. And it's very difficult to carry over the whole backlog to explain why that problem is interesting.

**Stout:** Well, we hear a lot, and we hear it all the time, about science and the need for scientists to communicate with people in general. Is it possible, in the kind of science you do—very rarified, theoretical kind—is it possible for you to tell people? Or do you think people really want to know? Do they just want to know what you're doing that may make the paint on next year's automobile a little tougher and a little brighter?

**Feynman:** They have a tendency in that direction. But not only that. There are a large number of people that want to know about the way that the world works. There's three kinds of questions: One is how does it affect me? Another question is, how does it affect you? What are you doing? Because they're interested in me personally. And then there's a question, What are the stars made out of? Now, only the last kind are a real pleasure to talk to. Because the last kind have the same kind of curiosity as motivates you in the sciences. And so you kind of feel a kind of sympathy right away to that kind of question. And those people you can spend a lot of time with and tell them all about the stars, and they grin back and they ask more questions. And that's lots of fun. But when they ask you, "Exactly what are you doing," because they want to know about you, that's not quite so interesting. When they want to ask how is it going to affect me, that's still duller, because actually the work that I do I have no idea how it's going to affect anybody. I can't answer it at all.

**Stout:** A selfish interest in what you're doing.

**Feynman:** If that means something negative, yes.

**Stout:** No, I don't think selfishness is necessarily negative.

**Feynman:** Yes, but that's what it is. It's in relation to society and its effects. What is the effect of the work I'm doing on society, in a certain sense, in a mechanical sense. How is it going to affect their lives in the sense of how is it going to change the size of the automobiles, or whether or not they can cook by pressing a button? But not how is it going to affect their lives in the sense of their intellectual lives—that is, how is it going to affect how they think about the world; their philosophy about where we came from and where we're going and what the earth is, and is it spinning in space or is it on the back of an elephant. You see, these ideas are the results of science. I think that the most interesting result of science and the greatest influence on people had to do with the ideas which are generated as a result of these investigations, and not only on the technical things. The fact that we can transmit these pictures over the air, that's very interesting. But even more interesting is what it is that's carried the waves, or how it is the relation of electricity and light and so on—I find that very exciting. And that's a not-social-responsibility level; that's intellectual.

**Stout:** In our kind of society, can you expect a great many people to have this pure curiosity about the color of the ground?

**Feynman:** I don't know what you can expect, but people have it. I have a lot more faith in people than is usually . . . I talk to a lot of people; I go around a lot, and people aren't technically inclined. And I find they're interested, yes. They're not just interested in themselves. There are a lot of people that are, but there are quite a few people that are interested in the world outside, so to speak.

**Stout:** Aside from the purely scientific kind of interest—the makeup of the stars or the ground or the air—do you find that science has an unsettling effect on religious people? On the religious beliefs of people? Can you be a scientist, knowing as much as you know, and be religious, in the accepted, practicing, conventional sense?

**Feynman:** I can only speak for myself. I personally am unable to. I mean, from what I learned about science, I can't be practicing in the conventionally religious sense. It doesn't fit together. It seems to me that the ideas of conventional religion—like in the Bible and so forth—are very limited. They didn't realize the tremendous extent of the world, or the length of time in which things have been going on. It seems to me impossible, in a certain sense, that so much attention could be paid to man as is advertised in the usual religion, and so little attention paid to the rest of the world. It doesn't seem to me that this fantastically marvelous universe, this tremendous range of time and space and different kinds of animals, and all the different plants, and all these atoms with all their motions and so on, all this complicated thing can merely be a stage so that God can watch human beings struggle for good and evil—which is the view that religion has. The stage is too big for the drama. So I believe it's not the right picture.

**Stout:** Is this a conclusion you reached after you matured in your scientific work?

**Feynman:** I don't know how mature I was, at fifteen or sixteen. But I was brought up religiously; I went through. . .

**Stout:** And you changed your own mind on a reasonable basis?

**Feynman:** Yes, well, I don't know how reasonable, but not emotionally. I changed my own mind.

**Stout:** According to your own reasoning, is what I meant.

**Feynman:** Yes.

**Stout:** On the basis of what you learned and accepted and believed.

**Feynman:** Yes. There's another feature: That a person who learns science gets into a difficulty of another kind, which I could mention, which isn't only with the information. It's not only that when we look at the world, it looks different than it seems to have looked from the religious view. The religious theory of the world—God made it and likes to help people, and so on, when they need it, and so forth—that picture doesn't fit with what you see. It seems not to fit. That's one thing. But here's another thing. When you watch for the development of science, you see that it's necessary always that the great development—because when you say you don't know the answer to something, and you start to investigate all the details; you ask yourself a lot of questions about it. You say you're not sure of anything—we're not sure of anything in the sciences; the thing is we don't know. As we learn more, we get more and more sure, more or less, that is still more likely to be sure, or that such-and-such an idea is more and more likely to be false. But it's very difficult to get any inspiration from the theory of God, if you say to yourself, "Well, I consider it very likely to be true." Yet, this is the kind of view that you're supposed to take, or that you find useful to take on all the world—which is that it's more or less likely to be true. So even if I were to be very strongly prejudiced toward God, you see, at a certain stage, I would take the view, "Well, that's an interesting theory that I think is very likely to be true." Now pray to. It's very hard.

**Stout:** No dogma?

**Feynman:** You can't accept something absolutely. You're not sure. Once you get that feeling, you lose the inspirational value of the religion. Now that's a problem of present day, because certainly there's a great yield in religion to make people do better, to remind them of how they feel—that they feel inadequate; they haven't done the best thing, and they walk out of the church ready to do better.

**Stout:** A force for good.

**Feynman:** A force for good, an inspiration. Now, how can we maintain that inspiration without a metaphysical theory, without necessarily believing certain ideas, such as that Christ rose from the dead, or something like that. Is that necessary in order to believe that the idea that you should help your neighbor, or that you should do unto your neighbor as he does to you; is it necessary to believe that Christ rose from the dead, in order to live like a Christian? According to the conventional religion, it is. But I think that there's a problem for a scientifically trained man, which is he's got no place to go to get inspiration to remember Christ's teachings in his conventional society. He's surrounded by what he considers a certain amount of superstition.

**Stout:** Of course you mentioned earlier that most scientists are not primarily interested in society as a group of human beings living together.

**Feynman:** As scientists. But as human beings. . .

**Stout:** As human beings they are.

**Feynman:** Yes.

**Stout:** So that, even though a scientist on a scientific level may reject the standard concept of religion, he can still be a decent human being who wants to be nice to his neighbor and his wife and his children.

**Feynman:** Oh, sure. There is nothing in Christ's teachings which are unbelievable from the scientific standpoint. Nothing, nothing that I learned about the distance to the stars teaches me that the golden rule is not likely to be true, or that I ought to kill or not kill. It's got nothing to do with it. But the people who develop religion have put something in. In addition to those other propositions, about what moral values are, they put something else in—that I would also believe in certain miracles performed by the same man, Jesus. I don't think those miracles are true. I still can think that the teachings of this great philosopher are worth paying attention to. The science does not make any immorality. It wants the religion to not say anything about the metaphysical structure of the world—that is to say, how it came about, how old it is, whether or not it's possible to have a virgin birth. Why does it have to know that? Why is it necessary?

**Stout:** Of course, with a great many people who are not scientists, isn't it true that the findings of science have shattered the basis of religion? And people are not so selective in deciding what they will and will not believe; they've simply walked away from all of it.

**Feynman:** Because the religions have tied together two things: They, for instance, if they want to teach the Ten Commandments, they're not satisfied to teach the Ten Commandments because it's the experience of mankind or something that these are a good way to proceed. But they teach the Ten Commandments because these things were given to Moses through lightning. Now, when the science comes along, it suggests that it's possible that maybe these things weren't given to Moses through lightning. A person who doesn't think too far says, "Oh, then the whole thing is nuts. And I'm afraid to think that possibility, because maybe then the Ten Commandments have no basis at all." But that's not necessarily so. It's perfectly possible that the moralities could have come from men. It could have been that Moses was an ordinary man and that he wrote these things. And I still could believe and still behave the same way. And what I think has happened is that the religions have put together two different kinds of ideas and welded them so thoroughly—namely, the theory of how the Ten Commandments arose, and the belief that you ought to follow them—that when the science comes along and challenges one end of this—namely, how the Ten Commandments arose—people get nervous that they're challenging the other end of it; namely, that they have. But it's the religion who's tied them together unnecessarily; there's no real connection. And that's the way I feel; that's a personal, philosophical view of the relation of religion and science. I'm very much on the one side. I want you to understand that all the scientists don't think the same way, naturally, because when we get out of our own field, we don't know what we're talking about. And I can be way wrong on this particular thing about religion, but you asked me what I personally thought, and this is what I personally think.

**Stout:** I assume there are some scientists, even in your own field, perhaps colleagues of yours, who are practicing religion.

**Feynman:** Certainly there are.

**Stout:** And they make the two concepts fit together.

**Feynman:** They find some way to do it, and how they do it, I don't know. But they find some way to do it. I haven't figured it out. My view of the world is different.

**Stout:** In our talk today, there were some tones from the writing of Irwin Shaw, who once wrote that, "I got a religion that wants to take heaven out of the clouds and plant it right here in the earth where most of us can get a slice of it." Or perhaps the well-known words of John Henry Newman, who wrote, "Lead kindly light amid the encircling gloom. Lead thou me on. The night is dark and I am far from home. Lead thou me on. Keep thou my feet. I do not ask to see the distance seen; one step is enough for me."

*Interview transcript courtesy of KCBS-TV Los Angeles.*

# "GRAVITY, THE DISRESPECTFUL," BY EARL UBELL, SCIENCE EDITOR

## New York Herald-Tribune
### *February 5, 1961*

The loudspeaker system honked and squealed. One of its horns crashed from its nest in the ceiling to the floor. But neither disturbance could distract the standing-room-only audience of hundreds of physicists in the Terrace Room at the Hotel New Yorker last week.

Intently they followed the insistent mathematical rhythm of four discourses on a subject that is at once a glory of and an agony to the human mind: the general theory of relativity, the theory that describes gravity. At times, they laughed. To most of them, physics is fun.

And since Dr. Richard P. Feynman, of California Institute of Technology, was talking about gravity when the loud-speaker nearly conked one of his colleagues, his listeners roared when he reminded them of the seventeenth-century idea of gravity with his quip:

"Ah—action at a distance!"

Maybe the humor is too special. But Dr. Feynman's attempt to slog through a sea of equations projected on a screen had something of the heroic about it. He was traveling a new course through an old subject.

### Easily Grasped?

The same feeling of the grandeur of physics swept through the audience as it heard Dr. Peter Bergman, of Syracuse University, tell how he and his students also attacked the gravity theory; and as they listened to Dr. John

Wheeler, of Princeton, create a weird geometry of space, and to Dr. Thomas Gold, of Cornell, tell of a universe in which matter—atoms—springs into being as if materialized out of nothingness. All concern gravity and relativity.

Actually, the main ideas behind relativity can be easily grasped by persons not too prejudiced by their common sense. When Dr. Albert Einstein created the theory, he erected its structure on a simple foundation. The idea can be expressed by imagining a man in an elevator.

The elevator begins to move. The man in it will feel himself pushed toward the floor or raised toward the ceiling by the acceleration. The interesting fact is this: the man cannot tell whether the push he feels comes from the movement of the car or from the passage nearby of a body exerting gravity. The force produced by the acceleration and gravity are said to be equivalent.

Starting with this concept, Einstein built up a mathematical description of the universe in which space becomes bent near bodies with mass. It is really a kind of geometry in which straight lines are replaced by suitable curves. Furthermore, the idea of a "force of gravity" disappears. Bodies move toward each other because they tend to move along the curves of the Einsteinian geometry.

### In Pretty Good Shape

Well, the theory is in pretty good shape. It explains a lot about the universe; how a light ray is bent when it passes near a star of heavy mass; how the orbit of the planet Mercury departs from that predicted by Newton's laws, how the color of light changes as it comes out of a dense star, and perhaps, Dr. Gold's created matter.

But now the physicists want to bring the general theory of relativity in tune with the rest of physics. One way would be to write down a single group of equations to describe the universe, from the tiniest atom to the hugest collection of stars.

This was the main effort of Einstein in his later years. He and his students produced at least a dozen such so-called Unified Theories. They all foundered. None fulfilled the hopes held for them.

Now Dr. Wheeler is actively engaged in producing such a theory. He has novel geometries to picture both suns and atom particles. One such geometry consists of a space full of "worm holes." Such holes are ultra-tiny. In the space of this dash—you could lay $10^{60}$ of them end-to-end. And nobody knows if it is a unified theory.

But another approach is taken by Drs. Feynman and Bergmann—through two different paths, to be sure. They try to get the general theory of relativity to "work" together with quantum mechanics, the theory that describes the behavior of atoms, electrons, protons, and most other tiny particles.

### More Non-Common Sense

Quantum mechanics is also a non-common-sense theory of the behavior of atoms. It rests on the idea that atoms transfer energy to one another only in discrete packets called quanta. You can't have a half a quantum. It's all or nothing at all.

This leads to all sorts of strange effects. For example: if you try very hard to pinpoint the position of an electron, you are left in doubt as to how fast it may be moving and in what direction. If you try to sneak up on an electron and catch its speed and direction precisely, then you can't tell where it is.

Drs. Feynman and Bergmann asked themselves: Can the gravitational theory be treated as if it were a quantum theory? Do gravitational bodies transfer "gravitational action" to and fro in quanta of gravity? Is there such a thing as a gravitron—the quantum of gravity?

Working within the general relativity theory, Dr. Bergmann has been trying to "quantize" the theory, to put it in a quantum form. The going has been rough, although progress has been made.

Dr. Feynman attacked it from another view. As an expert in quantum mechanics, he tries to squeeze the general relativity theory through the steps used in quantum mechanics to handle gravitrons as if they existed.

What *will* be the end result of these diverse efforts?

Perhaps a Unified Theory will pop out. Perhaps the physicists will learn how to make gravity and quantum mechanics work together. And perhaps, working on the idea that where there is difficulty there must be treasure, the physicists will find a surprise.

*Appendix III*

# "THE FUTURE OF PHYSICS"
# BY
# RICHARD P. FEYNMAN

*As published in* The Technology Review,
*1961–1962*

As I listened to Professor Cockcroft, Professor Peierls and Professor Yang, I found that I agreed with almost everything they said. I don't agree with Professor Yang's idea that the thing is getting too difficult for us. I still have courage. I think that it always looked difficult at any stage. On the other hand, I agree, as you will see, with something about this pessimism, and I don't think that I can add anything sensible to anything that the other speakers said. So in order to proceed, I have to add something that is not sensible, and if you will excuse me therefore, I am going to try to say something quite different than what they said.

First of all, to make the subject not infinite, I am going to limit myself very much and discuss only the problem of the discovery of the fundamental laws in physics—the very front line. If I were talking about that which is behind the front line, things like solid-state physics, and other applications of physics and so on, I would say very different things. So please appreciate this limitation of the discussion.

I do not think that you can read history without wondering what is the future of your own field, in a wider sense. I do not think you can predict the future of physics alone with the context of the political and social world in which it lies. If you want to predict, as Professor Peierls does, the physics a quarter of a century in the future, you must remember that you are trying to predict the physics of 1984.

The other speakers seem to want to be safe in their predictions, so they predict for 10, perhaps 25, years ahead. They are not so safe because you will catch up with them and see that they were wrong. So, I am going to be really safe by predicting 1,000 years ahead.

What we must do according to the method of prediction used by the other speakers is to look at the physics of 961 and compare it to the present, 1961. We must compare the physics even a century before the age when Omar Khayyam could come out the same door as in he went, to the physics of today as we open one door after the other and see rooms with treasures in them, and in the backs of the rooms five or six doors obviously going into other rooms with even greater treasures. This is a heroic age. This is an exciting time of very vital development in the fundamental physics and the study of the fundamental laws. It is not fair to compare it to 961, but to find another heroic age in the development of physics, the age of, perhaps, Archimedes, of Aristarchus; say, the third century B.C. Add 1,000 years and you find the future of their physics: the physics of the year 750! The future of physics depends on the circumstances of the rest of the world and is not merely a question of extrapolation of the present rate of progress into the future. If I go a thousand years, I have a difficult problem. Is it possible that it's all going to be over?

One of the most likely things, from a political and social point of view, is that we have soon a terrific war and a collapse. What will happen to physics after such a collapse? Will it recover? I would like to suggest that the physics, fundamental physics, may possibly not recover, and to give some arguments as to why not.

In the first place it is very likely that if there were sufficient destruction in the Northern Hemisphere the high-energy machines, which seem to be necessary for further research, would become inoperative. The machines themselves may be destroyed, the electrical power to operate them may be unavailable, and the industrial technology needed to repair or maintain them may no longer exist, at least for a while. Experimental physics techniques are the quintessence of our technological and industrial abilities, and so they must suffer some temporary setback.

Can physics slide back temporarily and then recover? I don't think so. Because, in order to have this heroic age an exciting one, one must have a series of successes. If you look at the grand ages of different civilizations, you see that people have an enormous confidence in success, that they have some new thing that is different, and that they are developing it by themselves. If one were to slide back, you would find for a while, no great suc-

cesses. You would be doing experiments that were done before. You would be working on theories that "the ancients" knew very well. What could result would be a lot of mouthing and philosophizing; a great effort to do the physics in the sense that one should do it to be civilized again, but not really to do it. To write, instead, commentaries, that disease of the intellect, which appears in so many fields. Physics is technically too hard to recover immediately. There would be practical problems at that time that would occupy the attention of intelligent people. The difficulty is that there would be no fun in it. The new discoveries wouldn't come for a while. The other feature is that it would not be useful. No one has yet thought of a use of the results of the experiences we have with the high-energy particles. And finally, it is possible that antagonism is produced by the terrible calamity; there might be a universal antagonism toward physics and physicists as a result of the destruction which people might blame on the scientists who made it possible. Another thing to remember is that the spirit of research may not build itself up again because this spirit is concentrated in the Northern and advanced industrial countries, and this spirit does not exist fully in the other countries.

Well, I said 1,000 years. Maybe in that much time there will be another renaissance. What kind of machinery could there be for a recovery from this thing? (I said I wouldn't talk about anything sensible. I can't.) Some success somewhere must be the cause of a new renaissance. Where will this success lie? Perhaps in other fields. Perhaps in some other field than physics one would find a new age developing a success above "the ancients" and then getting a new confidence and growing. When this grows, it can pass its enthusiasm to physics. Or perhaps there would be a new aspect to physics, some other point of view, or some other completely different thing. That I cannot tell.

Another interesting possibility is that the renaissance may lie in some nation or people discovering a success by using a scientific attitude as a kind of morality, in society, government and business. You know what I mean—when someone says something, looking for what it is that they are saying, not why they are saying it. Propaganda would be a dishonest thing, and no one would pay attention to someone who would say something not for the content of the idea they want to get across, but because they want to show that they are big, or good, or some such reason. It is possible that if any success results from using such a scientific attitude, a country would be encouraged to go on from this success in its society and develop a re-interest in the scientific problems.

Well, now let's take the opposite view. Suppose there is no collapse. How, I don't know, but suppose there is no collapse. Then what? Suppose we can imagine a society somewhat like our own continuing for a thousand years. (Ridiculous!) What would happen to fundamental physics, the fundamental problems, the study of the laws of physics?

One possibility is that a final solution will be obtained. I disagree with Professor Yang that it's self-evident that this is impossible. It has not been found yet, but if you were walking through a building to get from one side to the other, and you hadn't yet reached the door, you could always argue, "Look, we've been walking through this building, we haven't reached the door, therefore there is no door at the other end." So it seems to me that we are walking through a building and we do not know whether it is a long infinite building, or a finite building, so a possibility is that there is a final solution.

What I mean by a final solution is that a set of fundamental laws will be found, such that each new experiment only results in checking laws already known, and it gets relatively more and more boring as we find that time after time nothing new is discovered that disagrees with the fundamental principles already obtained. Of course, attention would then go to the second line about which I am not speaking. But the fundamental problem will have been solved.

I would say that one thing that would happen if such a final solution were found, would be the deterioration of a vigorous philosophy of science. It seems to me that the reason that we are so successful against the encroachment of professional philosophers and fools on the subject of knowledge, and the way of obtaining new knowledge, is that we are not completely successful in physics. We can always say to such people, "That was very clever of you to have explained why the world just *has* to be the way we have found it to be so far. But what is it going to look like to us tomorrow?" Since they are absolutely unable to make any predictions, we see that their philosophy does not have real understanding of the situation. But if the solution is all present, how many people are going to prove it had to be four dimensions, it just had to be this way because of such and such and so forth. And so the vigor of our philosophy, which is a vigor which comes from the fact that we are still struggling, I think that may fail.

What other possibilities are there? Suppose that the building we are walking through is infinite, as Professor Yang feels. Then there will be a continual exciting unfolding. We will rush through this house, one door

after another, one treasure after another. A thousand years! Three unfoldings in sixty years is fifty unfoldings in a thousand years. Is the world going to have fifty exciting revolutions of our basic physical ideas? Is there that much treasure in fundamental physics? If there is, it will become somewhat boring. It will be boring to have to repeat it twenty times, this fact that things change always when you look deeper. I do not believe that it can last 1,000 years of active investigation. Well, if it doesn't stop (I mean if you can't get the final solution), and if I can't believe that it will keep on being excitingly developed for fifty revolutions, what else is there?

There is another possibility, and that is, that it will slow up. The questions will become more difficult. How will it look then? The strong couplings are analyzed, the weak couplings partially analyzed, but there are still weaker couplings that are harder to analyze. To obtain useful experimental information has become extremely difficult because the cross-sections are so tiny. Data comes in slower and slower. The discoveries are made more and more slowly, the questions get harder and harder. More and more people find it a relatively uninteresting subject. So it is left in an incomplete state with a few working very slowly at the edge on the question of what is this third-order tensor field that has a coupling constant 10 to the power of -30 times smaller than gravity?

It is possible, of course, that what we call physics will expand to include other things. I believe, for example, that physics will expand, just as Professor Peierls says, into the studies of astronomical history and cosmology. The laws of physics, as we presently know them, are of this kind. Given the present condition, what is the future? The laws are given by differential equations in time. But there must be another problem: what determines the present condition? That is, what is the whole history of the development of the universe? One way to see that this may someday be a part of physics and will not always be called astronomical history is to note there is at least a possibility that the laws of physics change with time. If the laws of physics change with absolute time, then there will be no way to separate the problems of formulating the laws and of finding the history. I think that it is very likely that cosmological problems will be enmeshed in physics.

Finally, I must remind you that I limited myself to talking about the future in fundamental physics. I'd say that there will be an important return from the front line into the applications and the development of the consequences of the laws. This will be a very exciting thing, and I would say quite different things about its future than I would say about the future of the fundamental laws.

We live in a heroic, a unique and wonderful age of excitement. It's going to be looked at with great jealousy in the ages to come. How would it have been to live in the time when they were discovering the fundamental laws? You can't discover America twice, and we can be jealous of Columbus. You say, yes, but if not America, then there are other planets to explore. That is true. And if not fundamental physics, then there are other questions to investigate.

I would summarize by saying that I believe that fundamental physics has a finite lifetime. It has a while to go. At the present moment it is going with terrific excitement, and I do not want to retire from the field at all. I take advantage of the fact that I live in the right age, but I do not think it will go on for a thousand years.

Now, to finish, I would summarize two points. First, I did not talk about applied physics or other fields, about which I would give a considerably different talk. And second, in these modern times of high-speed change, what I am forecasting for a thousand years will probably occur in a hundred.

Thank you.

# EXTRA!
# *CALIFORNIA TECH*
# EXTRA!

---

## *Associated Students of*
## *the California Institute of Technology*
## *Friday, October 22, 1965*

### DR. RICHARD FEYNMAN NOBEL LAUREATE!

October 21, 1965 (9 A.M.)

Professor Richard Feynman:
Royal Academy of Sciences today awarded you and Tomonaga and Schwinger jointly the 1965 Nobel Prize for physics for your fundamental work in quantum electrodynamics with deep ploughing, consequences for the physics of elementary particles. Prize money each one-third. Our warm congratulations. Letter will follow.

—Erik Rundberg
The Permanent Secretary

Erik Rundberg
Your cablegram has made me very happy!

—Richard P. Feynman

Earlier, at 3:45 A.M.
"Hello, Dr. Richard Feynman? May I congratulate you on your Nobel Prize."

"Look, this is a heck of an hour—"

"But aren't you pleased to hear that you've won the Prize?"

"I could have found out later this morning."

"Well, how do you feel now that you know that you've won it?"

"Look, some other time. . . "

And so Richard P. Feynman, Ph.D., FRS, and Richard Chase Tolman Professor of Theoretical Physics at Caltech, first sleepily learned that he was an awardee of the 1965 Nobel Prize in physics.

Later yesterday morning, as growing realization brought greater excitement, Feynman learned that Schwinger and Tomonaga shared the award with him, and will also be making the December 10 trip to Stockholm. All three have received the Prize as the result of simultaneous, independent theoretical work conducted during 1947–1949 in quantum electro-dynamics.

Though the results of the three were later shown to be equivalent, Feynman introduced the pioneering "Feynman diagram," a powerful tool greatly simplifying quantum-dynamical calculations. As Feynman himself explained:

"It was the purpose of making these simplified methods of calculating more available that I published my paper in 1949, for I still didn't think I had solved any real problems, except to make more efficient calculations. But it does turn out that if the efficiency is increased enough, it itself is practically a discovery. It was a lot faster way of doing the old thing."

This "old thing," as Feynman described it during a press conference held at 10:30 A.M. in the Atheneaum, was the solution of Dirac's equations, formulated in 1929. Previous attempts to get more accuracy through second-order approximations led to infinite solutions. What the three Nobel Prize winners did, in the words of Feynman, was "to get rid of the infinities in the calculations. The infinities are still there, but now they can be skirted around. . . . We have designed a method for sweeping them under the rug."

Later in the morning Feynman went through another press conference. In his words:

A group came who couldn't get to the press conference because they were late. This guy comes into my office, and says to me: "I'll tell you what I'm going to ask you, so you're ready when the cameras start. One of the questions is: 'What applications does this paper have in the computer industry?'"

I said, "The answer to that will be 'none.'"

"Well, then, does it have application?"

"It hasn't got any—"

"Oh, you're kidding, sir."

"No." I knew that this interview was going to be a mess.

"Well, I'm going to ask you also to comment on the statement that your work was to convert experimental data on strange particles into hard mathematical fact."

"No, I'm not going to comment on that."

Finally, "All right. What time did you hear about the Award?"

"OK, now turn on the cameras!"

In the afternoon, spirited undergraduates raised a "Win big, RF" banner on the dome of Throop. And naturally, Feynman was the center of attention at a packed Physics Department seminar tea held in Bridge at 4:15, where he was formally inducted into the Nobel elite by Dr. Carl D. Anderson. "I feel," confided Feynman, "that the Nobel Committee was very wise in its Prize selection." Three hip-hip-hoorays followed.

Feynman told how a telephone caller from New York had asked him to comment on the New York school system. "It was all right when I was going to it 30 years ago," he answered. And Feynman has already decided how to spend his one-third of the $55,000 Prize money. "I'll use it to pay my income tax for the next three years, so that my income is tax-free."

The *California Tech* visited Feynman at his home in the evening for a special interview. In describing some of his more recent work, Feynman told how his quantum theory of the gravitational field is "so far along, but not perfect." As of late, Feynman has turned his attention to the rules of strong nuclear interaction.

Later Feynman told about phoning Tomonaga:

"Congratulations."

"Same to you," replied Tomonaga.

"How does it feel to be a Nobel Prize winner?"

"I guess you know."

"Can you explain to me in laymen's terms exactly what it was you did to win the prize?"

"I am very sleepy."

*Appendix V*

# "NEW TEXTBOOKS FOR THE 'NEW' MATHEMATICS" BY RICHARD P. FEYNMAN

*As published in* Engineering and Science,
*Volume 28, No. 6*
*March 1965*

As a member of the California State Curriculum Commission last year, I spent considerable time on the selection of mathematical textbooks for use in a modified arithmetic course for grades 1 to 8 in California's elementary schools.

I have carefully read all of the books submitted by their publishers for possible adoption in California (18 feet of shelf space, 500 pounds of books!). Here I should like to describe and criticize these books in a general way, particularly with regard to the mathematical content—what it is we are trying to teach. I shall omit important matters, such as whether the books are written so that it is easy for the teacher to teach well from them, or the student to read them. Many of the books finally selected by the State for adoption do still contain some of the faults described below. This is because one could only select from what was submitted by the publishers, and few really good books were submitted. Also, budget limitations prevented the adoption of most of the supplementary books that the commission recommended in order to try to compensate for the faults of those basic books that were selected.

Why do we wish to modify the teaching of mathematics in the schools? It is only if we see this clearly that we can judge whether or not the new books satisfy the need. Most people—grocery clerks, for example—use a

great deal of simple arithmetic in their daily life. In addition, there are those who use mathematics of a higher form—engineers and scientists, statisticians, all types of economists, and business organizations with complex inventory systems and tax problems. Then there are those who go directly into applied mathematics. And finally there are the relatively few pure mathematicians.

When we plan for early training, then, we must pay attention not only to the everyday needs of almost everyone, but also to this large and rapidly expanding class of users of more advanced mathematics. It must be the kind of training that encourages the type of thinking that such people will later find most useful.

Many of the books go into considerable detail on subjects that are only of interest to pure mathematicians. Furthermore, the attitude toward many subjects is that of a pure mathematician. But we must not plan only to prepare pure mathematicians. In the first place, there are very few pure mathematicians and, in the second place, pure mathematicians have a point of view about the subject which is quite different from that of the users of mathematics. A pure mathematician is very impractical; he is not interested—in fact, he is purposely disinterested—in the meaning of the mathematical symbols and letters and ideas; he is only interested in logical interconnection of the axioms, while the user of mathematics has to understand the connection of mathematics to the real world. Therefore we must pay more attention to the connection between mathematics and the things to which they apply than a pure mathematician would be likely to do. I hear a term called "new mathematics" used a great deal in connection with this program. That it's a new program of mathematics books is, of course, true, but whether it is wise to use "new," in the sense of very modern, mathematics is questionable. Mathematics which is used in engineering and science—in the design, for example, of radar antenna systems, in determining the position and orbits of the satellites, in inventory control, in the design of electrical machinery, in chemical research, in the most esoteric forms of theoretical physics—is all really old mathematics, developed to a large extent before 1920.

A good deal of the mathematics which is used in the most advanced work of theoretical physics, for example, was not developed by mathematicians alone, but to a large extent by theoretical physicists themselves. Likewise, other people who use mathematics develop new ways to use it, and new forms of it. The pure mathematicians have in recent years (say, after 1920) turned to a large extent away from such applications and are

instead deeply concerned with the basic definitions of number and line, and the interconnection of one branch of mathematics and another in a logical fashion. Great advances in this field have been made since 1920, but have had relatively little effect on applied, or useful, mathematics.

### What we're after

I would consider our efforts to find new books and modify the teachings of arithmetic as an attempt to try to make it more interesting and easier for students to learn those attitudes of mind and that spirit of analysis which is required for efficient understanding and use of mathematics in engineering, science, and other fields.

The main change that is required is to remove the rigidity of thought found in the older arithmetic books. We must leave freedom for the mind to wander about in trying to solve problems. It is of no real advantage to introduce new subjects to be taught in the old way. To use mathematics successfully one must have a certain attitude of mind—to know that there are many ways to look at any problem and at any subject.

You need an answer for a certain problem: the question is how to get it. The successful user of mathematics is practically an inventor of new ways of obtaining answers in given situations. Even if the ways are well known, it is usually much easier for him to invent his own way—a new way or an old way—than it is to try to find an answer by looking it up. The question he asks himself is not, "What is the right way to do this problem?" It is only necessary that he get the right answer.

This is much like a detective guessing and fitting his answer to the clues of a crime. In terms of the clues, he takes a guess as to the culprit and then sees whether that individual would be likely to fit with the crime. When he has finally suggested the right culprit, he sees that everything fits with his suggestion.

### Any way that works

What is the best method to obtain the solution to a problem? The answer is, any way that works. So, what we want in arithmetic textbooks is not to teach a particular way of doing every problem, but, rather, to teach what the original problem is, and to leave a much greater freedom in obtaining the answer—but, of course, no freedom as to what the right answer should be. That is to say, there may be several ways of adding 17 and 15 (or, rather, of obtaining the solution to the sum of 17 and 15) but there is only one correct answer.

What we have been doing in the past is teaching just one fixed way to do arithmetic problems, instead of teaching flexibility of mind—the various possible ways of writing down a problem, the possible ways of thinking about it, and the possible ways of getting at the problem.

This attitude of mind of a user of mathematics is, it turns out, also really the attitude of mind of a truly creative pure mathematician. It does not appear in his final proofs, which are simply demonstrations or complete logical arguments which prove that a certain conclusion is correct. These are the things that he publishes, but they in no way reflect the way that he works in order to obtain a guess as to what it is he is going to prove before he proves it. To do this he requires the same type of flexible mind that a user of mathematics needs.

In order to find an example of this, since I am not a pure mathematician, I reached up on the shelf and pulled down a book written by a pure mathematician. It happened to be *The Real Number System in an Algebraic Setting*, by J. B. Roberts, and right away I found a quotation I could use:

"The scheme in mathematical thinking is to divine and demonstrate. There are no set patterns of procedure. We try this and that. We guess. We try to generalize the result in order to make the proof easier. We try special cases to see if any insight can be gained in this way. Finally—who knows how?—a proof is obtained."

So you see that mathematical thinking, both in pure mathematics and in applied mathematics, is a free, intuitive business, and we wish to maintain that spirit in the introduction of children to arithmetic from the very earliest time. It is believed that this will not only better train the people who are going to use mathematics, but it may make the subject more interesting for other people and make it easier for them to learn.

In order to take this discussion of the character of mathematics away from the abstract, and to give a definite illustration, I'll choose material from the first and second years of school as an example; simply the problem of adding.

We suppose that children learn to count and that after a while they are very adept at counting. And now we wish to teach them addition. May we remark immediately that a child who can count well, say to 50 or 100, can immediately solve a problem such as $17 + 15 = 32$. For example, if there are 17 boys in the class and 15 girls, how many children are there in the class? The problem doesn't have to be given in the form of an abstract addition, it is simply a matter of counting the boys, counting the girls, and counting the class. We have, as a summary of the result: 17 boys plus 15 girls equals 32 children.

This method could be used to add any pair of whole numbers but, of course, it's a rather slow and cumbersome method for large numbers, or if a very large number of problems comes up. Similar methods exist, such as having a set of counters or fingers and counting the things off with them. Another way is to count the numbers in the head. For example, after a while it is possible for a child to be able to add 3 to 6 by thinking to himself, 7, 8, 9. A more practical method is to learn by rote some of the simpler combinations, such as 3 + 6, so that if they come up often it is not necessary to do the counting.

Counting large numbers—of pennies, for example—can be simplified by counting in groups instead of counting all the pennies. You can make little piles of five and count the piles of five; or better, you can make the piles of five into ten and then count the piles of ten first, and the number left over. Adding numbers could then be done more easily by adding the groups and the leftovers together.

Other ways of getting at addition facts (or doing additions, I should say) is to have a line on which the numbers are marked, or a thing like a calendar on which a whole series of numbers are written, one after the other. Then, if you want to add 3 to 19, you start at 19 and count off three more spots along the line and come to 22. Incidentally, if these numbers are written as dots equally spaced along a line—called the number line—this becomes very useful later for an understanding of fractions and also of measurements; for inch-rulers and other things like thermometers are nothing but a number line written along the edge of the ruler. Therefore, putting the numbers on a line is useful not only for learning addition in the first place, but also for understanding other types of numbers.

(Another special trick to remember, at a very elementary level, is that it is possible to determine which is the greater of two numbers without actually counting the numbers. If we have two rather large groups of things, it is easy to find out which group is larger by matching the things in pairs and seeing which group has objects left over. This is the way the number of molecules in different gases was first compared, by the way.)

### Addition—the old way

In older books, addition is handled in a very definite way, without any variety of tricks or techniques. First we learn the simpler sums by drawing pictures of ducks—5 ducks and 3 ducks, swimming, makes 8 ducks, and so forth—which is a perfectly satisfactory method. Then these numbers are memorized, which is again satisfactory. Finally, if the numbers are bigger

than 10, a completely different technique is used. It is first explained how to write the numbers that are larger than 10 and then rules are given for the addition of two-column numbers, without carrying at first. It is not until the third grade that carrying can be done for the first time.

The dissatisfaction with the old text is not that any of the methods used to teach addition are unsatisfactory; they are all good. The trouble is, there are so few methods allowed that only a rigid and formal knowledge of arithmetic can result.

For example, a problem such as 29 + 3 is not a legitimate problem for two years, for it is not given in the first and second grades, and the child is presumably unable to do that problem because, of course, for it you must carry. On the other hand, if you really understood what addition was, you could obtain the result of 29 + 3 not very long after you have learned to count. Very early—in other words, in the first grade—you could do it by simply thinking 30, 31, 32.

It is true that this method is slow, but if no other method is available, then this is the method which ought to be used. It should be permitted to be used. It should be one of the possible things that a child might do when he has to add by hook or by crook in a difficult problem. As he gets older he may increase his efficiency at doing the problems by using other methods, but it should be possible at the very earliest age to do addition problems with any reasonable numbers. There is really nothing different about adding 3 to 6 and adding 3 to 29; it is just that the technical, and generally more efficient, way we finally use when we get older is somewhat different.

### A limited approach

In understanding the meaning of addition of two numbers—the meaning of sum, and how to get at it—there is no difference in the two problems. So the objection to the standard text is that only one method is given for making additions—namely, when the numbers are small, memorize them; when the numbers are larger, formally add them in a vertical column, two numbers together, and not carrying until after the second year. This is entirely too limiting for two years of study. If a child will not learn, or is unable to learn, the formal rules, it should still be possible for him to obtain the result of some simple problem by counting or making a number line or by other technical methods.

In order to develop the kind of mental attitude which is required later, we should also try to give as wide a mathematical experience as possible. The sum should not appear always in the same form. There is no reason

why every sum should be written 17 with 15 underneath and a line drawn to obtain 32. A problem such as 17 +    = 32, to fill in the blank, is a somewhat different variety but exactly the same type of question with numbers. So let the first-grade child cook up a way to obtain the answer to this problem. This is exactly the type of problem he will have to solve later if he becomes an engineer. I don't mean he will have to learn how to subtract. What I mean is that he has to deal with a new form of an old type of situation. The problem is to fill in a blank by any method whatsoever. However, when the blank is finally filled in, it must be correct.

We would not usually be interested in engineering or in physics, in how a man obtained the result that 15 will go into the blank as long as he finally shows that 15 does work by simply adding the 15 and 17 and seeing if it comes out to 32.

(The only time we would be interested to know how he obtained the 15 is if this is the first time that such a problem has ever been done, and no one has ever known a way to do it previously—or if it seems likely that this type of problem will appear again and again in the future because of a new technical development, and that we would like to have a more efficient method. Then it would be worthwhile to discuss the methods of obtaining the 15.)

So this problem, 17 + a blank = 32 is an analogue of the general problem of applied mathematics, to find a way to fill in a blank number by any method whatsoever. It is a problem that could be given very early in the first year, leaving a freedom for the children to try to obtain the solution by any method they want, but of course not permitting wrong answers. The thing has to be checked out at the end.

### Developing freedom

Here is another example of developing freedom, of a somewhat more complex form. Two times an unknown number + 3 is 9. What is the unknown number? This is, of course, algebra, and there are very definite rules for solving such a problem—subtracting 3 from both sides and dividing by 2. But the number of algebraic equations that can be solved by definite rules is very small.

Another way is to try various numbers for the blank until one is found which fits. This way should be available to children at a very early age. In other words, problems should be put in many different forms. Children should be allowed to guess and to get at the answers in any way that they wish, in terms of those particular facts which they happen to memorize. Of

course, it is necessary as time goes on for them to memorize the ordinary addition facts, the ordinary methods of making additions, multiplications, divisions, and so on, in addition to being allowed a freedom about the solution and the form of various problems that are given to them.

Later, in more advanced work in engineering, when we have more complex algebraic equations, the only available method is, in fact, to try numbers. This is fundamentally a method that is of great power and will only have to be learned later by the student or the engineer. The old teaching, that of every problem there is a definite fixed method, is only true for the simplest problems. For the more complex problems which actually arise there is *no* definite method, and one of the best ways to solve complex algebraic equations is by trial and error.

Another exercise which involves a greater degree of freedom is guessing a rule. This type of problem appears in more complex forms later, but a simple example, and a typical engineering or scientific problem, is the following: in a series of numbers 1, 4, 7, 10, 13, what is the pattern of rule by which they are being generated? The answer could be given in several ways. One is by adding 3 each time. Another is that the nth number is $3 \times n + 1$.

The key, then, is to give a wide variety of mathematical experience and not to have everything in a limited and stringent fixed form. This is not an argument about teaching methods. The point is not that it will then be easier to teach the regular arithmetic (although, for all I know, that may be so). The point is, it will be teaching a new subject in a sense—an attitude of mind toward numbers and toward mathematical questions which is precisely that attitude of mind which is so successful later in technical applications of mathematics.

It will not do simply to teach new subjects in the old way. For example, it has been recommended that numbers written in a different base than 10 be discussed in the early grades. This could serve to illustrate the freedom in mathematics to generalize, and help toward a deeper understanding of the reason behind the carrying rule in arithmetical operations. For this, a mention and an explanation with a few examples might delight some students. But if the matter is not understood by some of the slower members of the class, it is senseless to drill it in with interminable exercises, changing from one base to another. For such students, for whom a short exposure doesn't "take," more practice in the usual rules of base 10 computation is surely more sensible than drilling to perfection calculation in base 5 and 12.

## Words and definitions

When we come to consider the words and definitions which children ought to learn, we should be careful not to teach "just" words. It is possible to give an illusion of knowledge by teaching the technical words which someone uses in a field (which sound unusual to ordinary ears) without at the same time teaching any ideas or facts using these words. Many of the math books that are suggested now are full of such nonsense—of carefully and precisely defined special words that are used by pure mathematicians in their most subtle and difficult analyses, and are used by nobody else.

Secondly, the words which are used should be as close as possible to those in our everyday language; or, as a minimum requirement, they should be the very same words used, at least, by the users of mathematics in the sciences and in engineering.

Consider the subject of geometry. It is necessary in geometry to learn many new words related to the mathematics. For example, one must learn what a triangle is, a square, a circle, a straight line, an angle, and a curved line. But one should not be satisfied solely to learn the words. At least somewhere, one should learn facts about the subjects to which the words refer, such as the area of the various figures; the relations of one figure to another; how to measure angles, possibly the fact that the sum of the angles of a triangle is 180 degrees; possibly the theorem of Pythagoras; or maybe some of the rules that make triangles congruent; or other geometrical facts. *Which* facts may be decided by those having more experience with curriculum, for I am not intending here to make any specific suggestions of what should be included and what not. I only mean to say that the subject of geometry, if it is taught at all, should include a reasonable knowledge of the geometrical figures over and above what the conventional names are.

Some of the books go a long way with the definition of a closed curve, open curve, closed regions, and open regions, and so on—and yet they teach no more geometry than the fact that a straight line drawn in a plane divides the plane into two pieces. At the end of some of these geometry books, look over to find, at the end of a long discourse, or a long effort at learning, just what knowledge of geometry has been acquired. I think that often the total number of facts that are learned is very small, while the total number of new words is very great. This is unsatisfactory. Furthermore, there is a tendency in some of the books to use most peculiar words—the words that are used in the most technical jargon of the pure mathematician. I see no reason for this.

It will be very easy for students to learn the new words when, and if, they become pure mathematicians and discourse with other mathematicians on the fundamentals of geometry. It is very easy indeed to learn how to use such words in a new way when one is older. A great deal of the objection that parents have to the so-called new mathematics may well be merely that it sounds rather silly to them when they hear their children trying to explain to them that a straight line is a "curve." Such arguments in the home are absolutely unnecessary.

### Precise language

In regard to this question of words, there is also in the new mathematics books a great deal of talk about the value of precise language—such things as that one must be very careful to distinguish a number from a numeral and, in general, a symbol from the object that it represents. The real problem in speech is not *precise* language. The problem is *clear* language. The desire is to have the idea clearly communicated to the other person. It is only necessary to be precise when there is some doubt as to the meaning of a phrase, and then the precision should be put in the place where the doubt exists. It is really quite impossible to say anything with absolute precision, unless that thing is so abstracted from the real world as to not represent any real thing.

Pure mathematics is just an abstraction from the real world, and pure mathematics does have a special precise language for dealing with its own special and technical subjects. But this precise language is not precise in any sense if you deal with the real objects of the world, and it is overly pedantic and quite confusing to use it unless there are some special subtleties which have to be carefully distinguished.

### A fine distinction

For example, one of the books pedantically insists on pointing out that a picture of a ball and a ball are not the same thing. I doubt that any child would make an error in this particular direction. It is therefore unnecessary to be precise in the language and to say in each case, "Color the picture of the ball red," whereas the ordinary book would say, "Color the ball red."

As a matter of fact, it is impossible to be precise; the increase in precision to "color the picture of the ball" begins to produce doubts, whereas, before that, there was no difficulty. The picture of a ball includes a circle and includes a background. Should we color the entire square in which the ball image appears or just the part inside the circle of the ball? Coloring the ball

red is clear. Coloring the picture of the ball red has become somewhat more confused.

Although this sounds like a trivial example, this disease of increased precision rises in many of the textbooks to such a pitch that there are almost incomprehensibly complex sentences to say the very simplest thing. In a first-grade book (a primer, in fact) I found a sentence of the type: "Find out if the set of the lollypops is equal in number to the set of girls"—whereas what is meant is "Find out if there are just enough lollypops for the girls."

The parent will be frightened by this language. It says no more, and it says what it says in no more precise fashion, than does the question: "Find out if there are just enough lollypops for the girls"—a perfectly understandable phrase to every child and every parent. There is no need for this nonsense of extra-special language, simply because that type of language is used by pure mathematicians. One does not learn a subject by using the words that people who know the subject use in discussing it. One must learn how to handle the ideas and then, when the subtleties arise which require special language, that special language can be used and developed easily. In the meantime, clarity is the desire.

I believe that all of the exercises in all of the books, from the first to the eighth year, ought to be understandable to any ordinary adult—that is, the question that one is trying to find out should be clear to every person. It may be that every adult is not able to solve all of the problems; perhaps they have forgotten their arithmetic, and they cannot readily obtain 2/3 of 1/4 of 1 1/3, but they at least should understand that the product is what one is trying to obtain.

By putting the special language into the books, one appears to be learning a different subject, and the parent (including highly trained engineers) is unable to help the child or to understand what the thing is all about. Yet such a lack of understanding is completely unnecessary and *no gain whatsoever* can be claimed for using unusual words when usual words are available, generally understood, and equally clear (usually, in fact, far clearer).

### New definitions—and no facts

I believe that every subject which is presented in the textbook should be presented in such a way that the purpose of the presentation is made evident. The reason why the subject is there should be clear; the utility of the subject and its relevance to the world must be made clear to the pupil.

I would take, as an example, the subject of sets. In almost all of the textbooks which discuss sets, the material about sets is never used—nor is any

explanation given as to why the concept is of any particular interest or utility. The only thing that is said is that "the concept of sets is very familiar." This is, in fact, true. The idea of sets is so familiar that I do not understand the need for the patient discussion of the subject over and over by several of the textbooks if they have no use for the sets at the end at all.

It is an example of the use of words, new definitions of new words, but in this particular case a most extreme example, because *no facts whatever* are given at the end in almost all of the books. A zookeeper, instructing his assistant to take the sick lizards out of the cage, could say, "Take that set of animals which is the intersection of the set of lizards with the set of sick animals out of the cage." This language is correct, precise, set theoretical language, but it says no more than "Take the sick lizards out of the cage." The concept of things which have common properties by being a member of two groups (such as the Chinese Communists, or of a larger number of groups such as East German refugee children) does involve intersections of sets but one does not use that language. No lack of precision results from this. And, besides, people who use mathematics in science, engineering, and so on, never use the long sentences of our imaginary zookeeper.

If we would like to, we can and do say, "The answer is a whole number less than 9 and bigger than 6," but we do not have to say, "The answer is a member of the set which is the intersection of the set of those numbers which is larger than 6 and the set of numbers which are smaller than 9."

It will perhaps surprise most people who have studied these textbooks to discover that the symbol » or « representing union and intersection of sets and the special use of the brackets  and so forth, all the elaborate notation of sets that is given in these books, almost never appear in any writings in theoretical physics, in engineering, in business arithmetic, computer design, or other places where mathematics is being used. I see no need or reason for this all to be explained or to be taught in school. It is not a useful way to express one's self. It is not a cogent and simple way. It is claimed to be precise, but precise for what purpose?

### Making the "new" mathematics more worthwhile

In the "new" mathematics, then, first *there must be freedom of thought;* second, *we do not want to teach just words;* and third, *subjects should not be introduced without explaining the purpose or reason,* or without giving any way in which the material could be really used to discover something interesting. I don't think it is worthwhile teaching such material.

# "TWO MEN IN SEARCH OF THE QUARK" BY LEE EDSON

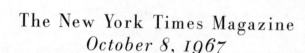

## The New York Times Magazine
### *October 8, 1967*

In the last couple of years an intensive hunt has been going on all over the world for an elusive quarry known as the quark. This is no Alice-in-Wonderland adventure; far from it. The hunters are some of the world's leading physicists. The hunting grounds: almost anywhere from the high atmosphere to the bottom of the sea to the inside of the latest atom smashers. One enthusiastic researcher at the University of Michigan even has been grinding up oysters on the theory that an oyster eats almost anything—so why not a quark?

Despite this painstaking search, the quark so far has remained as hard to track down as Lewis Carroll's Snark. There is an excellent reason. According to modern theoretical physics, the quark, if it exists at all, is the simplest particle in the universe, out of which almost everything else is made.

Capturing this incredibly wraithlike substance may not help us produce superbombs—the implications of gaining new elementary facts about nature depend on how man uses them, whether for war or for peace, and are not ascertainable for years, anyway. But to physicists, the excitement in confronting the quark is something more sublime than the discovery of a new application to everyday life. It is the immediate realization that through it we may uncover the missing linkage to our understanding of the structure of matter throughout the universe.

· · · ·

The men largely responsible for sending scientists on this wild quark chase are two California Institute of Technology physicists named Murray Gell-Mann and Richard Feynman. Each has won a fistful of high honors. In 1965, Feynman shared the Nobel Prize for his achievements in explaining some of the abtruse mechanisms in the subatomic world. Many physicists think Gell-Man is next. One California scientist calls the two men "the hottest properties in theoretical physics today."

What makes them radiate so fiercely on the high-energy physics circuit is a fine blend of showmanship and brilliance. Take the quark hunt, for instance. "Dick and I were batting around some aspects of theoretical physics," Gell-Mann recalls. "We started to get excited about a new theory and threw out words for our ideas. There have been some crazy ones in physics lately. The theory depended on a triplet of particles, with the right characteristics, and we needed a word for it. I started to say 'squeak,' 'squark,' and it came out 'quark.' We loved the word as soon as it was uttered. Much to my surprise, I found the line 'Three quarks for Muster Mark' in James Joyce's *Finnegans Wake*. Nothing could have fitted better."

Of course, both men could have rejected the word for something sane and august and ending with "on" like "electron" or "neutron," or even used "ace, deuce, trey" as later suggested by physicist George Zweig of Cal Tech, who came to a similar conclusion about the nature of the fundamental particle, but Gell-Mann admits to a certain puckishness in accepting and publicizing the quark. "Maybe it will help C. P. Snow to bridge his two cultures," he says with a smile.

Gell-Mann and Feynman—who claim they work together separately—are nearly unique in modern physics. In a field noted for quiet introversion, they generate a kind of charisma which draws students and faculty members to Cal Tech to an extent that had not been seen on the campus since the days of Robert Oppenheimer. The lectures of both men play to S.R.O. audiences. When Feynman won the Nobel Prize, one blasphemous sophomore admirer celebrated by slipping his picture into a plaque of the Last Supper in place of Jesus.

Although they have comparable intellectual impact, Feynman and Gell-Mann create their sparks in entirely different ways. Feynman, a lean, intense, dark-haired man of 49 who is beginning to fear that he may soon be regarded as a premature elder statesman, is a natural showman, full of exuberance for his subject. His lectures are couched in pithy, often rough-cut phrases—"I always try to say things differently," he says—and he uses

hand gestures and intonations the way Billy Rose used beautiful women on the stage, spectacularly but with grace.

Occasionally he sets the stage. In one popular lecture on color vision, for instance, Feynman flooded the entire stage with a rainbow of light. "Why use a tiny prism on a small table, which people have to strain to see?" he says. "Nature is too interesting to stick in a corner." But generally he needs no such props. The subject matter to him is so glorious that it is only necessary to see it honestly, as it really is, for everyone to respond to his fervor. "I always come to Feynman's lectures," says an old-time faculty member, "because I am sure there will be at least one good surprise."

Gell-Mann is less flamboyant but equally compelling on the platform. Eleven years Feynman's junior, he is round-faced and bespectacled, and looks somewhat like a jovial neighborhood storekeeper. In the classroom, his lectures are clear, smooth and amusing, and draw upon an extraordinary erudition. But he really excels among small groups. If Feynman can be compared to a star who basks in the warmth of large audiences, Gell-Mann seems to enjoy the give-and-take intimacy of a seminar of bright, articulate graduate students. Like Oppenheimer, he prefers a small but devoted following—and, indeed, he has declared often that a teacher lecturing in front of a class represents a primitive form of education.

Despite this personality difference, the two men work well together—which means, according to a friend, that they argue continuously and loudly but, after a few moments, come to an understanding and move forward rapidly. This rapport was almost interrupted a few years ago when Gell-Mann for several reasons thought seriously of leaving Feynman and switching to Harvard. He decided not to, according to gossip among physicists, when he found that Harvard was willing to meet all his demands except one—changing its name to Gell-Mann Institute.

Since that time, the two men have mellowed, but an undercurrent of rivalry occasionally crops up. "Dick is always calling up to see whether Murray is working," says Mrs. Margaret Gell-Mann, an attractive blond from Birmingham, England. "If I say he's in the garden, Dick is happy for the rest of the day. But if I tell him Murray is doing physics, then Dick gets nervous and immediately wants to come over."

When Dick married an English girl a few years ago (his third wife), he made a point of calling Murray and saying: "I looked around to see what you and I didn't have in common. I saw that you had an English wife and a brown dog, so I went out and got both."

One quality both men do have in common is an extraordinary capacity to make physics lucid and highly romantic. In part, the romance is made easy by the nature of the world of physicists. In the pecking order of this society, the theoretical physicist is the glamour boy. After him comes the experimentalist, the fellow in the laboratory, and then the engineers and applied physicists who make such things as sonar, rockets and hydrogen bombs.

Right now, the theoretical physicists have captured the front and center of scientific interest because they seem to be on the threshold of answering a long-asked, almost childlike question: What are things really made of? Have we at last come down to the last foundation stone from which we can build anything: a table, a human being, or universe? Or must we go on looking at smaller and smaller pieces, and going deeper and deeper into a bottomless pit?

To answer these questions, one has to remember that from the very start of civilization philosophers have wanted to find a simple idea that would unite everything we experience in the world around us. So there has always been a search for the building block, like the cell or gene in biology. The fifth-century B.C. Greek Democritus receives the honor of declaring that the simplest thing out of which everything else is made is an atom (atamos means "uncut"), and this idea sufficed for 2,000 years.

In the 19th century scientists came to realize that the atom was not the ultimate particle after all. Inside the atom was a nucleus, with electrons orbiting around it like the planets around the sun. Then, in the 20$^{th}$ century, scientists began to concentrate on the nucleus, and saw that this was no simple item either. It contained smaller things, such as neutrons and protons, which must be held together by a very strong force, perhaps the strongest in the universe. So the question for years was: What was this "strong interaction," this glue that held the nucleus together?

In 1935, a Japanese physicist named Hideki Yukawa provided the first educated guess. He theorized that there was another particle in the nucleus, which he called a meson. It would act as a carrier of force between the proton and the neutron, so that energy would be exchanged among the particles pretty much as a football is passed between players in a game. Two years later Dr. Carl Anderson of Cal Tech discovered a particle that he thought might be Yukawa's carrier. Physicists exulted. Things seemed neat again— until it was found that the new particle failed to obey predictions expected from the laws of physics, and indeed created such a mess in attempts to relate theory and experience that Columbia's I. I. Rabi threw up his hands

and said: "Who needs it?" It took five years for physicists to realize that the new particle was not the Yukawa meson, but an unrelated particle known as a muon. After the war, as bigger and more powerful atom smashers appeared, a number of mesons were discovered, including the one suggested by Yukawa. This constituted a major advance in understanding the strong interaction—the interaction which is known to us in everyday life only through the power of the A-bomb.

By then the atom smashers were revealing all kinds of new and peculiar particles, whose mass was created out of the energy of motion. Most of them died out in incredibly brief periods—on the order of a billionth of a second—but their trails could be photographed, and their presence raised new questions. How did they disintegrate? Were they complex structures in themselves?

Feynman was one of the Young Turks of physics who in the nineteen-fifties addressed themselves to these thorny questions. (He had already earned a formidable reputation in theoretical physics because of his efforts in another area—applying quantum mechanics to electromagnetic radiations. For this work, 16 years later, he was to share the Nobel Prize with Dr. Julian Schwinger of Harvard and Dr. S. Tomonaga of Japan.)


Feynman's interest centered around a phenomenon that had intrigued scientists for years—the emission of fast-moving electrons from radioactive substances. This process, which goes under the name of "beta decay," had introduced physicists to a new force in the nucleus, a force quite different from the "strong interaction" that holds it together. Feynman was fired by the challenge of this insight and the subsequent revolutionary developments.

For one thing, physicists in the nineteen-fifties were discovering that the new force—or "weak interaction," as it came to be called—was far more widespread and manifested itself in many more reactions than beta decay. Indeed it was on a par with the strong interaction, and with the two other universal forces known to science; electromagnetism, which keeps electrons perpetually spinning around a nucleus, and gravity, the weakest force of all. The weak interaction, which is now known to be involved in the decay of many strange particles, is 100,000 times weaker than the electromagnetic force. However, it is exceedingly more powerful than gravity, which is so incredibly feeble compared with the strong interaction that it takes a fraction with 42 zeros to describe its relative strength. This kind of comparison

delights Feynman. "Isn't nature wonderful," he says, raising his hands ecstatically, "to make something with 42 zeros!"

Another important development in physics in the nineteen-fifties was the overthrow of one of the fundamental laws in nature. It had long been thought that nature operates with a number of conservation laws, ranging from the familiar indestructibility of energy and matter to the conservation of lesser-known properties of the atom, such as those that explain the stability of the proton and the reason why some particles are created in atom smashers only in multiplets. These laws were regarded as unchanging and universal. One of the most important of them was the one known as "the law of the conservation of parity." It said that if an object had a mirror image, that image would obey the laws of physics just like the real object. To satisfy this law, particles in the subatomic world could exit in one of two ways. The particle and its mirror image could be completely identical—like the word MOM, which reads the same in the real world and the looking-glass world. Or there could be two particles, a "left-handed" one and a "right-handed" one—one the mirror image of the other, like the words MAY and YAM.

Since the strong interactions obey this rule of parity, it was assumed hat the weak interactions should also. But a striking thing was discovered in atom-smashing experiments involving a weak interaction. A particle was found that did not have an image fitting the pattern of MOM.

Were there actually two particles, with reversed symmetry, like MAY and YAM? But that suggestion did not jibe with further experiments, which continued to point to the existence of just one particle. In an inspired moment, Feynman and another physicist, Martin Block, offered the view that the law of parity might have failed for this particular weak interaction.

It was a prophetic suggestion. In an epic-making paper, two Chinese-born scientists, C. N. Yang and T. D. Lee, then of the institute for Advanced Study at Princeton and Columbia University, suggested that perhaps *all* weak interactions violate the law of parity. They proposed experiments, which were carried out and proved them right. Their intuition won them the Nobel Prize in 1957. More important, they turned nuclear physics topsy-turvy.

Freed from the confines of parity, Feynman and Gell-Mann (along with E.C.G. Sudarshan and R. E. Marshak) tackled the problem of finding a way to describe the law of the weak interactions. In 1957, they developed a theory that shows how this force depends on various properties of the particles,

such as their directions of spin, thus providing what is now generally regarded as a major contribution to our understanding of the nucleus.

Feynman says that the discovery of this new law was the most exciting thing in his life, far more exciting than his earlier work that led to his Nobel prize. "I won the prize for shoving a great problem under the carpet," he says, "but in this case there was a moment when I knew how nature worked. It had elegance and beauty. The goddamn thing was gleaming." So shiny was the new law that when several eminent physicists conducted experiments that seemed to vitiate it, Feynman insisted the experiments must be wrong. And so indeed they proved to be.

Gell-Mann experienced the thrill of another major and beautiful discovery a few years later. By then the number of particles emerging from the nucleus was increasing fantastically. Almost 100 had been counted by 1962. They had been classified into two main groups known as leptons (weakly interacting particles), or weaklies, as Feynman calls them, and hadrons, or stronglies. Examples of the first group are positive and negative electrons, muons and neutrinos; the second group includes neutrons, protons and pions. (Just to confuse matters, each of these particles has an antiparticle carrying an opposite charge which annihilates the particle on contact.)

A subdivision of the hadrons is a group of particles which are known as "strange particles" because, instead of dying out as they should, they live to a relatively ripe old age. To round out this picture, there is also the photon, a particle that carries the electromagnetic force—and, supposedly, there is a graviton for gravity, though this particle is yet to be found.

To bring some order into this nuclear grab bag, Gell-Mann introduced two new concepts. The first, which had developed as early as 1952, was a quality which he called "strangeness." As with the quark, it had a literary counterpart, this time in Sir Francis Bacon's line: "There is no excellent beauty that hath not some strangeness in the proportion."

Each particle could be assigned a degree of strangeness, depending on the number of steps in its disintegration, and thus it could be distinguished from its neighbors—just as a neutron can be distinguished from a proton by its different electric charge. (Gell-Mann did not know until later that a Japanese scientist named Nishijima working independently in Tokyo came to the same conclusion at about the same time.) With the differences

known, Gell-Mann set about seeing in what ways the particles were similar and whether they could all be slipped into a neat, organizational chart, more or less as Mendeleev had done with his periodic table of the chemical elements in the 19th century.

Gell-Mann recalls that he and Feynman tried one pattern after another without luck. Finally, one scheme seemed to work. The particles seemed to fit into families of eight or ten members, with similar characteristics of strangeness, electric charge and other properties, such as mass and spin. Even while Gell-Mann was doing this, Yuval Ne'eman, an Israeli military attaché in London, studying for his doctorate in physics while trying to buy guns for his country, came upon the same scheme.

However, one family of particles in the table had only nine members instead of ten. It seemed incomplete. Gell-Mann thereupon predicted that, if a new member were found, it would have certain properties as indicated by its relatives in the same family of particles. Experimental physicists took up the hunt.

A couple of years later, a team of 33 scientists at Brookhaven National Laboratory hit pay dirt. They had bombarded nuclei and peered at more than 100,000 photographs of the interactions. In one of those pictures they saw the track of the missing particle, which was called omega-minus. It had a life expectancy of a 10-billionth of a second, and it had the basic properties predicted by Gell-Mann. The existence of the omega-minus was shortly verified by scientists at the University of Maryland and thereafter independently at CERN, the great nuclear establishment in Switzerland.

Gell-Mann, with his flair for literary analogy, called his chart the Eightfold Way, as in the Buddhist dictum: "This is the noble truth that leads to cessation of pain. This is the noble eightfold way—right views, right intentions, right speech, right action…" More prosaic physicists call it the SU–3 theory because it is a symmetrical structure based on a triplet of fundamental particles.

The Eightfold Way hit physics like a bombshell. Gell-Mann occasionally marvels at the apparent simplicity of nature that is revealed in the scheme. "Why should an esthetic criterion be so successful so often?" he asks. "Is it just that it satisfies physicists?" Then, echoing Feynman, he declares: "I think there is only one answer—nature is inherently beautiful."

In recent months, a number of physicists have tried to go beyond the Eightfold Way and explain why nature operates with so many particles in this

rather neat and poetic style. At Harvard, Dr. Julian Schwinger claims to have developed a simple mathematical theory to explain it all. Others, following Gell-Mann, have constructed a "quark model" in which neutrons and protons behave as if they were made of quarks—and, more than that, as if they were made of three quarks, each of which may come in three forms. The reason why a number of scientists are seriously hunting real quarks (though Gell-Mann himself is not sure that they exist) is that theoretically they cannot decay into something else and thus a stable quark must be somewhere around, and may indeed have lasted since the birth of the universe.

Not all physicists agree, of course. At the University of California in Berkeley, Dr. Geoffrey Chew has taken a radical approach to the entire problem. He has developed a theory, known whimsically as the "bootstrap theory," which says that there really is no rock-bottom fundamental particle like the quark or anything else; indeed, that the strong particles are made of one another, pulling themselves into existence, so to speak, by their own bootstraps. "At first glance, this theory and our quark model may seem contradictory," Gell-Mann says, "but they may actually be quite compatible—and both may even be right—especially if the quark should turn out to be, as is likely, a useful mathematical figment rather than a concrete building block of matter."

Although Gell-Mann and Feynman were both born in New York City of middle-class families, they came to theoretical physics by different routes. Feynman traces almost everything in his make-up to the influence of his father.

"When I was a boy," he says, "Dad and I took long walks in the woods and he showed me things I would never have noticed by myself. He told me about the world and how it looked many years ago. He would say, 'See this leaf? It has a brown line; part of it is thin and part thick. Why?' And when I tried to answer, my father would make me look at the leaf and see whether I was right and then he would point out that the line was made by an insect that devotes its entire life to that project. 'And for what purpose? So that it can leave eggs which turn into new insects.'

"My father taught me continuity and harmony in the world. He didn't know anything exactly, whether the insect had eight legs or a hundred legs, but he understood everything. And I was interested because there was always this kick at the end—a revelation of how wonderful nature really is."

Dick's love of science flowered rapidly, and occasionally it got an unexpected boost. In Far Rockaway High School, which he found pretty dull, a

teacher disciplined him for talking too much in class and not paying attention. He was sent to the back of the room and given a book. "After you
read this," the teacher said, "you can open your mouth again."

"So I learned calculus," says Feynman. He went on to M.I.T., where he
graduated with honors in 1939, and did his graduate work at Princeton. In
World War II, he ended up as a group leader at Los Alamos, working on
theoretical aspects of the material in the A-bomb. He was present at the
first test explosion at Alamogordo in July, 1945.

As befits one who challenges the fundamental laws of nature, Feynman is
an inherent iconoclast. He loves to play the bongo drums, and he is probably the first physics professor in history—certainly the first Nobel
Prizewinner—to have a picture of himself on the drums included in the
introduction to a three-volume college textbook of physics.

Gell-Mann's extracurricular tastes run in more erudite directions. With
his wife, who studied archeology at Cambridge, he often enjoys digging for
artifacts in such places as Greece and Palestine. He is also a devotee of lin-
guistics; he has studied many languages, including some odd dialects of
Africa and the Middle East. "I like diversity," he says, "and I like the natural
history behind diversity. Why are there so many different tongues, so many
different birds, even so many different human neuroses? It's interesting to
find the pattern behind them."

Gell-Mann, the son of a language teacher and a prodigy who entered
Yale at 15, says he came into physics almost inadvertently. "I had to fill out
an application form," he recalls, "where I had to list my future occupation. I
started to put down archeologist but my father said I'd never make a living
in archeology, and suggested engineering instead. I couldn't stand engineering, so I put down the closest thing, physics."

He went on to get his doctorate at M.I.T., and then to Princeton, where
he worked with Oppenheimer. The Gell-Mann and Feynman merger
occurred in 1954, during a visit by Gell-Mann to Pasadena. Some words
were passed between them, along with howls of laughter over jokes they
both found funny, and the next day, Gell-Man recalls, he was being interviewed by the dean of Cal Tech. He joined the staff as an associate professor
in 1955 and became a full professor the following year.

In the last few years, Gell-Mann has been a missionary for the development
of the world's largest atom smasher, the proposed 200-billion-electron-volt
machine scheduled to be built at Weston, Ill. Along with other top physi-

cists, he believes the $200-million machine is vital to maintain United States supremacy in particle physics.

"I think particle physics is where atomic physics was in the early years of the century," Gell-Mann says. "We're getting an outline of an underlying structure, but there is still no complete theory of either strong or weak interactions which enables us to understand what is really happening at the bottom of everything.

"Recently an experiment performed by Princeton physicists has shown the violation of anther law of symmetry that was thought to be valid (like parity 10 years ago). Some theorists went so far as to speculate that a fifth natural force was involved, but that doesn't seem to be true. How is the violation occurring? Nobody seems to know at the moment, but I think we are on the verge of an important discovery."

Feynman says much the same, but in terms of a metaphor—playing chess with a Martian. "If you don't know the rules," he says, "and you see only parts of the board, how do you know how to play? If you know all the rules, can you tell what's in the Martian's mind when he moves the pieces in a certain way?

"The biggest mystery of physics is where the laws are known, but we don't know exactly what's going on. We don't know the strategy in the middle game. We know castling, or how the different pieces move, and we know a little bit about the end game, but nothing in the middle.

"We get reports from the experimentalists, the watchers of the chess game, and we try to analyze the information. We may even suggest a new experiment. But we're still waiting and hoping for the big strategy. Then maybe we'll really understand how wonderful is nature."

# ACKNOWLEDGMENTS

There are many people to thank for their assistance on this project. Ralph Leighton was a constant help and a valuable source of information as I attempted to put this all together. Thanks also go to Christopher Sykes for his advice, and to James Gleick for providing copies of letters I did not even know existed. I am greatly indebted to these gentlemen for their efforts. They set the bar high with their own books about Richard Feynman, and I felt I was standing on the shoulders of giants. I came to truly appreciate their hard work as I set about this task.

My profuse thanks also go to Dr. Kip S. Thorne and Dr. Steven C. Frautschi for their enormous contributions to this book. Dr. Thorne was generous with his time and very accommodating to paraphrase the scientific questions, as well making the technical letters more understandable. It is now clear to me why he is so highly regarded as a teacher. Dr. Frautschi was equally outstanding for writing the science parts of the chapter introductions. He was able to piece things together based on small amounts of information, and I am greatly impressed with his skill. Both of these Caltech professors were longtime colleagues and friends of my father. I am forever grateful to them for the scholarship they brought to this book.

Helen Tuck, my father's secretary for many years, shared many colorful stories about what it was like to work for my father. I was very pleased with Linda Bustos and the Caltech Public Relations Department for their highly efficient handling of the photos I requested from them. Professor Jenijoy La Belle provided interesting background information on her experience as the first tenured female professor at Caltech. I remember saying hello to

Dr. Wolfram while on late-night visits to Caltech with my father, and my father's impression of Dr. Wolfram was that he was a tireless and hard worker. Dr. Wolfram kindly gave me permission to use the letter he wrote to my father, and he wrote the remarks giving it context. Edward Fredkin was also a big help. My father was best man in Fredkin's wedding, and later Edward named his son Richard after my dad. The two had much respect and affection for each other.

Amanda Cook at Basic Books first embarked on this journey many years ago, and even though she moved on to another publishing house before I started on this project, I greatly benefited from the perspective she, along with Ingrid Finstuen and Maria Rapoport, provided. Vanessa Hartmann was an invaluable help in locating letter writers, or their heirs, for permission to publish their letters to my father. In cases where the correspondents could not be found, names were changed (and in potentially embarrassing situations, names were changed "to protect the innocent"). Megan Hustad was an incredibly knowledgeable, enthusiastic, hardworking, and inspiring editor. She was the glue that held this project together. I am proud to have had the privilege of working with her on this book. And I am always grateful for the involvement of our agent, Melanie Jackson, whose opinion I have respected immensely over the years.

General Donald Kutyna impressed me with his highly detailed recollections of the Challenger Commission, and he was enormously helpful. I also appreciate Michael Gottlieb's help with both temporarily storing the massive quantity of papers during a crucial period and locating my father's eyewitness account of the world's first atomic bomb blast. Mark Salzman took the time to read my work and gave me valuable input. Cameron Richmond was a big help in locating several letter writers who might have otherwise remained out of reach, and Anita Bunn kindly helped me edit hundreds of photos.

I thank my brother, Carl Feynman, for his confidence in me to do this project. My aunt and uncle, Jacqueline and Eric Shaw, were enthusiastic supporters who helped me piece together information. My aunt, Joan Feynman, provided a lot of the early letters, for which I am extremely grateful. Another aunt, Frances Lewine, also contributed, and my cousin, Charles Hirshberg, was a fantastic source of wisdom and hilarity. I would like to thank Jules and Rosalie Greenbaum, Arline's brother and sister-in-law, for their gracious permission to publish her letters and photographs. My husband's family was a great support to me throughout the editing process. My brother-in-law, Pablo Miralles, was an intelligent and sensible

reader whom I often pressed into service when I needed another opinion. Brenda Miralles, my sister-in-law, was a babysitter extraordinaire during the last hectic days. My mother-in-law and father-in-law, Maria and Adolfo Miralles, were wonderfully generous and giving of their time.

Finally, I would like to thank my children, Ava and Marco, for their patience and understanding, and my husband, Diego Miralles, whose judgment I could always rely on, and who kept me company at the computer during the wee hours, for his continuous assistance and support.

I am grateful to the following people for granting permission to publish letters written to my father: Henry Abarbanel, Molly Anderson, Martin Bernal, Hans A. Bethe, Laurie M. Brown, Adrian M. Bronk, Steven Cahn, Nigel Calder, Robert Carneiro, Helen Choat, Lawrence Cranberg, Sir Francis Crick, Robert Coutts, Beulah E. Cox, Martin B. Einhorn, Debra Feynman, Tomas E. Firle, Betsy Holland Gehman, Michael H. Hart, Ben R. Hasty, Richard C. Henry, Marka Oliver Hibbs, Heidi Houston, John Alexander Howard, Jon A. Johnsen, Vera Kistiakowsky, Julia Kornfield, Portia Parratt Kowolowski, Tina Levitan, Joan Thomas Newman, Thomas H. Newman, Clifford S. Mead, David Mermin, Mark Minguillon, Ken Olum, Leigh Palmer, Frank Potter, Ernest D. Riggsby, Tom Ritzinger, Irwin Shapiro, Jeff Stokes, Lewis H. Strauss, Paul Teller and Wendy Teller, Ilene Ungerleider, Vincent Van der Hyde, Jonathan Vos Post, Spencer Weart, Edwin J. Wesely, John A. Wheeler, Jack Williamson, Jane S. Wilson, J. G. Wolff, and Stephen Wolfram.

# PHOTO CREDITS

# INDEX